ASSERTING NATIVE RESILIENCE
Pacific Rim Indigenous Nations Face the Climate Crisis

Edited by Zoltán Grossman and Alan Parker

Foreword by Billy Frank, Jr.

FIRST PEOPLES
New Directions in Indigenous Studies

Oregon State University Press
Corvallis

Front cover photo by Zoltán Grossman. Sunset over the Salish Sea, after the arrival of the Tribal Canoe Journey at the Lummi Nation on July 30, 2007.

Back cover photo by Zoltán Grossman. Tribal Canoe Journey arrival on Seattle's Lake Washington in 2006, hosted by the Muckleshoot Tribe.

The paper in this book meets the guidelines for permanence and durability of the Committee on Production Guidelines for Book Longevity of the Council on Library Resources and the minimum requirements of the American National Standard for Permanence of Paper for Printed Library Materials Z39.48-1984.

Library of Congress Cataloging-in-Publication Data

Asserting native resilience : Pacific rim indigenous nations face the climate crisis / edited by Zoltán Grossman and Alan Parker.
    p. cm.
  Includes bibliographical references and index.
  ISBN 978-0-87071-663-8 (pbk. : alk. paper) -- ISBN 978-0-87071-664-5 (e-book)
  1. Indigenous peoples--Ecology--Pacific Area. 2. Human beings--Effect of climate on--Pacific Area. 3. Traditional ecological knowledge--Pacific Area. 4. Climatic changes--Pacific Area. 5. Pacific Area--Environmental conditions. I. Grossman, Zoltán. II. Parker, Alan.
  GF798.A77 2012
  363.738'74091823--dc23
                        2011052244

© 2012 Zoltán Grossman and Alan Parker
All rights reserved. First published in 2012 by Oregon State University Press
Second printing 2013
Printed in the United States of America

Oregon State University Press
121 The Valley Library
Corvallis OR 97331-4501
541-737-3166 • fax 541-737-3170
http://osupress.oregonstate.edu

Resilience:

1. The power or ability to return to the original form, position, etc., after being bent, compressed, or stretched; elasticity.

2. Ability to recover readily from illness, depression, adversity, or the like; buoyancy.

TABLE OF CONTENTS

GRAPHICS CREDITS

Photographs of authors were provided by the authors. Photographs not credited above are by Zoltán Grossman.

Renée Miller Klosterman Power

Renée passed from this world on Wednesday, August 26, 2009. She left us far too soon, but what a history and legacy she manifested.

Renée was a summa cum laude graduate of the Edward R. Murrow School of Communications at Washington State University with a communications major and a sociology minor. She obtained her masters degree in public administration–tribal governance from The Evergreen State College. She turned much of her energy and creativity to supporting Native American issues through the use of her video production skills.

Renée was one of the first female TV news photographers, and she worked for KING-TV in Seattle. She began her career with the Washington State Traffic Safety Commission, where she developed an innovative statewide leadership program for youth. She was recognized as a national and state leader for her work in facilitating the development of the National State Youth Coordinators network. Renée also worked for Washington State Division of Information Systems as a multimedia production manager. She was involved in video production for twenty-five years—mainly in projects focusing on social issues. She was nominated for an Emmy Award for her production of the video *Journey to the Healing Circle*, which explored fetal alcohol syndrome.

The last two years of her life, Renée resided in Tahoe City, California, with her husband Jared. She wrote for a small newspaper and produced video for the public access station. Renée loved horses, and among her accomplishments was an ascent of Mount Rainier. She loved hiking in the mountains and became an expert in stand-up paddleboarding. She was a bright light for all who came into contact with her, and that light will continue to shine in all who knew and loved her.

Renée fought and won her first battle with breast cancer in 2004, but when it returned in 2009 she lost a courageous fight. She was a true warrior in every sense of the term. She was a true friend to so many colleagues, and her presence is missed by all who came into contact with her. She leaves behind her loving husband, Jared Power; her daughter, Kourtnei Nibler; her mother, Barbara Miller; her sister, Tina Anderson; her brother, Eric Klosterman; and many aunts, uncles, cousins, nieces, and nephews.

Renée and friend Sunny.

FOREWORD: LOOKING AHEAD

Billy Frank, Jr.

Chairman, Northwest Indian Fisheries Commission (NWIFC)

Billy Frank, Jr. (Nisqually)

As a lifetime resident of the Pacific Northwest and the descendent of people who have lived here for thousands of years, I can tell you many things about the land, the water, and the life that has been sustained here for a long, long time.

I can tell you that our people have seen many changes over the past few hundred years, brought on by the growth of a population foreign to us, the removal of our native trees, the pollution of our pure waters, and so many other changes in the land. I can tell you, from the experience of my own life and the lives of my ancestors, that those who deny the existence of climate change are wrong. Those who think we are not going through major change due to man's impact on the land, the waters, and the air are wrong.

These things exist, and I don't need to rely on "contemporary science" to prove it; tribal members have the benefit of knowing and understanding traditional science, or traditional knowledge, something we possessed long before contact with Europeans.

It may seem ironic that today's tribes have nonetheless invested heavily in contemporary science as well—so much so that, percentage-wise, our investment in habitat restoration and other aspects of natural resource management far outweighs that of non-tribal governments. Ironic, that is, until one remembers that virtually all of the techniques used today have their roots in Indian Country. And ironic, that is, until you remember that we do have long-term memory, and thus we understand the necessity of having long-term vision and the fact that without natural resources such as clean water, nothing else survives.

Now that mainstream society's pollution of the atmosphere and the oceans is leading to climate change impacts unlike anything that exists in our memory or the stories and lessons left to us by our ancestors, it should come as little surprise that the Indigenous peoples of the world gathered long before the non-tribal governments of the world did and developed a consensual agreement that actually called for meaningful change, unlike the non-Indigenous nation-states of the world. Our principles and proposed solutions have followed the same principles of stewardship we have always followed, and have placed our spiritual connection with Mother Earth at the top of our priority list where it belongs. This is where this priority belongs for everyone, Indian and not, because we are all born with the responsibility to take care of our planet and the life that exists here. We are all dependent on the health of our ecosystem, whoever we are and whatever we do.

Once people understand this, we will all be able to join hands in dealing with the environmental challenges that face us. Where we must adapt, we will be able to do so. Where we can help Nature to prevent human tragedies that will otherwise occur—by curtailing environmental damage, conserving resources, and restoring habitat for fish and wildlife—we will do so, together. First, we know that we must be aware of the existence of climate change. It cannot be ignored or hidden away. We must adapt to those changes that are upon us and do what we can to prevent the impacts we can—for the benefit of all existing life and for the generations to come.

Let me say it again. Climate change does exist. It is making changes now and it will continue to make changes, unlike anything we have ever experienced before—and it will affect us, the Indigenous people, first and hardest. It will affect those of us who live on the rivers, whose lives are most directly dependent on fish and wildlife, first and foremost. But it will ultimately

The Medicine Creek Treaty was signed in a treaty council on the Nisqually River Delta in 1854. A single tree on the treaty grounds stood for more than a century and a half, symbolizing the continuing rights of the people. The snag is shown in 1914 (top), 2002 (middle), and after a hurricane-force windstorm toppled it in 2006 (bottom). The Nisqually have taken the remaining wood for a tribal carving.

affect us all, and we need to be prepared for it. You owe it to yourself to understand as much as you can about it. For example, its impacts will not always be the warmer weather that has caused the ancient glacier on our sacred Mount Rainier to melt away as it has. Sometimes it will increase the severity of polar winds. It's the height of arrogance to say with certainty what will happen when you mess with Mother Nature. Nonetheless, we need to open our eyes to the potential impacts of climate change and do what we can to deal with them and to prevent them when we can. We need to work together to do these things—for the sake of our children and the children still to come. No matter who you are, or what you do, your help is needed and your future is at stake.

In a time when the glaciers are melting, storms are brewing, and parts of villages are falling into the sea in Alaska, the baby steps are not enough. Major cooperative efforts are needed. If there is one thing that people whose hearts have been removed from the land and become fixated on money have demonstrated with consistency through the years, it is that they turn to long-term wisdom and principles such as genuine stewardship only when they are brought to the brink of obvious total destruction. At this point, major industries are beginning to capitalize on the assets their limited vision can see in clean energy, but they're still exerting major pressure on world leaders to keep the door wide open for them to continue business as usual so they can drain the last dollar out of the Earth. As the price of kowtowing to such pressures and ignoring the need for definitive action becomes more and more clear, let's hope the political leaders of the world can gain some of the insight of the Indigenous leaders.

Otherwise, the citizens of the various countries of the world will be forced to live with the consequences of their greed and their disrespect for future generations and our mother, the Earth.

Now let's imagine a different scenario. Suppose people everywhere open their eyes and suddenly understand that their future and the future of their children and their children to be depend on making good decisions—wise decisions with long-term vision—seventh generation–type decisions—based on respect for the Earth that sustains them. Suppose they decide, once and for all, that what they think matters and that the leaders they elect to office work for them—not for fancy lobbyists employed by big business. Suppose they say, "Enough is enough," loud and clear. By letter. By phone. By fax. In person. In the newspapers. On TV and radio. In the streets (peacefully, of course). Suppose they decide that they are not going to take no for an answer—that they decide they will be heard, and they will be listened to, and there will be action. No more nonsense. There will be clean energy. There will be clean water. There will be clean air. What do you suppose would happen?

The truth is that now is the time for definitive action. Now is the time for people to realize that their future, and the future of all the various forms of life on this planet, is at stake. And it's time for action.

A chum salmon tries to get across Skokomish Valley Road Nov. 11, 2001 to re-enter the Skokomish River north of Shelton, Mason County, and continue upstream. It and others, upper right, seemed to wait for the wake from passing vehicles to dash across the road from flooded fields.

INTRODUCTION

Alan Parker

Director of the Northwest Indian Applied Research Institute (NIARI); Faculty in the Master of Public Administration program, The Evergreen State College (Olympia, Washington)

Zoltán Grossman

Faculty in Geography and Native American & World Indigenous Peoples Studies, The Evergreen State College

Indigenous nations are on the frontline of the climate crisis, around the continent and the world. Native peoples are the first to experience climate change, and the peoples who feel it the deepest, with economies and cultures that are the most vulnerable to climate-related catastrophes.

Native nations of the Arctic and Subarctic are already feeling catastrophic effects of warmer temperatures in the melting of sea ice, permafrost, and glaciers, and an increase in fires, insects, flooding, and drought patterns. South Pacific Indigenous peoples are finding their islands inundated by rising sea levels, erosion from intense storms, and saltwater intrusion into freshwater supplies. Pacific Northwest tribes have already been deeply affected by flooding, reduced glaciers and snowpack, seasonal shifts in winds and storms, the northward shifting of species on the land and in the ocean, and many other factors that affect tribes' ability to continue practicing treaty rights.

These changes have drastically affected Indigenous peoples' hunting and fishing lifeways, economic infrastructure, water and housing availability, forest and agricultural resources, and even human health. Native harvesters in direct contact with natural resources are describing today the same drastic shifts in the environment that Western scientists had predicted would occur in the future. The real-time observations based in the traditional ecological knowledge of Native Science is alerting the world to the changes to come, often years before the studies of Western Science are able to confirm them.

This scale of change will present severe challenges to all tribal cultures, resources, and well-being. Yet tribal peoples also possess cultural tools that have historically enabled Indigenous nations to mitigate or adapt to sudden environmental changes, the political sovereignty to implement new models of responding to the new changes, and a sense of community that leads people to band together to meet challenges.

The focus of this anthology is on the Indigenous nations of the Pacific Rim. The region can be defined politically (as the countries that border the Pacific Ocean) or geographically (as the Pacific Basin, whose many watersheds empty their rivers and streams into the ocean). Indigenous nations along the Pacific Rim place great value on the health of the water, the watersheds, and the species (such as salmon and whales) that migrate through the waters. People living on the Pacific Coast are also commonly affected by climate change, including the shifting cycles of ocean currents, and their effects on storms, droughts, and food supplies.

Climate change is already here

The latest global scientific evidence compels us to recognize that climate change is accelerating at a much faster rate than predicted even a year ago. Like the Katrina disaster in 2005 (itself worsened by warmer Gulf of Mexico temperatures), we can see climate change coming. We can see it is going to devastate us if we are not prepared, so we have to go out and meet it. The people of the world, and especially Native communities, no longer have five to ten years to begin planning. *We must begin today!*

The consensus of leading climate scientists is that it is no longer a question whether there is human-caused climate change. The Intergovernmental Panel on Climate Change (IPCC) 2007 report put the range of climate-induced temperature increases expected in this century at 2°–5°C (some models estimate even higher). Mitigation measures to prevent disastrous effects must be taken. However, scientists believe that we're now locked into unavoidable climate change, and there are considerable lag times in the impacts of greenhouse gases, many taking twenty to two hundred years to reach the upper atmosphere to degrade.

These changes may not be as slow as once predicted. Scientists now have strong evidence of abrupt climate change, with sudden and dangerous shifts that can lead to catastrophic loss of human lives and property. The primary impacts on Indigenous peoples are prolonged droughts or excessive rainfall and other weather shifts brought on by warming, diminishing and disappearing sources of fresh water, changes in habitat for wildlife that impact cultural sustainability for Native communities, and impacts on food sources.

Climate change is a potential culture killer

Native rights are primarily place-based rights, based on the longtime occupation of Indigenous territories. Climate change shifts and disrupts plant and animal habitats, and in doing so forces cultures to adapt to these conditions, or die. Species adapt to rising temperatures by shifting their ranges farther north or to higher elevations. Many species driven entirely out of their habitats and feeding areas may face extinction. Other species are migrating into new areas, and competing with or displacing native and culturally important species.

The salmon habitat that tribes and First Nations have fought to protect is already being affected by less rainfall and snowpack in the mountains, melting glaciers, warmer water temperatures, and shifts in ocean currents. Gravel spawning beds for salmon are being scoured out by spring floods caused by sudden snowmelts. Climate change may also directly threaten species, such as in the "dead zone" on the U.S. Pacific Northwest coast, where some fish and crabs are being starved of oxygen by wild swings in ocean upwellings of phytoplankton, which in turn may be tied to shifting seasonal wind patterns. Treaty-guaranteed rights to hunt, fish, and gather may be rendered meaningless by these changes, or may require adaptation by transferring harvesting rights to new species.

The loss of culturally important species on which traditional knowledge depends will make it more difficult for elders to practice and pass their knowledge to the next generation. Some climate stresses will fall directly on the elders, who are particularly vulnerable to heat waves, food stress, and water stress. It is precisely these traditional elders who possess the greatest knowledge of how to survive with local natural resources, and who can recognize subtle shifts in nature. The 1993 hantavirus outbreak in the southwest U.S., for example, was a mystery to scientists until Navajo elders noted that increased rainfall had caused an explosion in the population of mice, who feed on piñon nuts. (The increased rainfall had been caused by El Niño ocean temperature oscillations, which are expected to intensify as climate change further warms the Eastern Pacific.)

In our region, where the ocean meets the coast and the mountains, we are especially vulnerable to extreme variations in weather. The concern is not primarily that there is more wind and rain, but that it comes with increased intensity in shorter bursts. Wind systems coming from the south (rather than the west) are bringing less rain and snow in the mountains. Every winter we've been seeing larger "megastorms" with flooding, mudslides, blizzards, and hurricane-force winds. Though scientists debate the role of climate change in individual storms, the patterns of storms across the country can be intensified by climate instability.

Climate change impacts are expected to impose hardships on Indigenous nations. They will directly impact Native economies through loss of economically important plant and animal species, and through increased costs of defending against climate change impacts. Loss of traditional economic activities, economic revenue, economic opportunities, and the practice of traditional culture are expected to increase the social and cultural pressures on Native peoples. Increased outmigration of tribal youth to seek economic opportunities, expected under severe climate impacts (or their daily lives limited to the indoors), could lead to further erosions of tribal economies and culture.

Devastation of tribal economies and health

Climate change not only changes the average temperature, but also changes temperature extremes, which in turn are expected to increase the severity and frequency of storms, floods, droughts, and other natural hazards in different regions—leading to increased defense and mitigation costs for tribal nations. The most common impact is expected to be a 3–4 foot (or more) rise in sea level; the levels have risen only about 1 foot in the past century. Coastal estuaries, wetlands, and marshes, already degraded by coastal development, will be highly affected.

At least three Native communities in Alaska are being forced to relocate as melting sea ice and permafrost and increased storms have led to coastal erosion. In Washington, the Quileute and Hoh tribes have secured land from the adjacent Olympic National Park (with congressional legislation) to relocate to higher ground because of the risk of tsunamis and storm surges, which will only be made worse by higher sea levels. To keep this in perspective, a moderate 14-inch rise in ocean levels would inundate 40 percent of Puget Sound mudflats, wiping out a significant habitat for shellfish and waterfowl.

Climate change can affect water availability through changes in rainfall, and lowering of water infiltration into aquifers and reservoirs. Decreased rainfall in mountains lowers snowpack. Shorter winter seasons lead to earlier and more rapid spring melt, resulting in flooding. The vegetation changes and loss of ground cover decrease water-holding capacity and infiltration, and increase erosion. Atmospheric carbon interacts with water chemistry to acidify fresh and marine waters. Increased runoff can cause irreversible changes in river and stream dynamics, increase loads of sediments, heavy metals, and sediment-borne diseases, and destroy aquatic habitats. These changes would raise costs for tribal nations in storing and cleaning water supplies, and increase water rights conflicts and litigation costs related to defending reserved water rights.

Fires are expected to increase due to increased droughts, lightning strikes, and forests impacted by new levels and kinds of infestations. Forest vulnerability from insect infestations can already be seen in the growing spruce beetle and mountain pine beetle crisis. The spruce bark beetle alone has infected more than 10 million acres in southern Alaska and British Columbia. Fires destroy or modify habitat and culturally important species, and increase timber-management and firefighting costs.

Climate change is also expected to increase the frequency and severity of human, wildlife, plant, livestock, and crop diseases and pests. Climate instability is known to change patterns of fire, and to shift plant populations (such as those used for basketmaking materials) northward and up mountain slopes. Shorter and more mild winters favor disease and pest build-up; more ticks, mosquitoes, rodents, and other carriers of human diseases are migrating north. Heat and environmental stress further increase vulnerability to diseases and pests. These will cause increased tribal costs in health services and veterinary services, and in losses to tribal natural resource enterprises.

Indigenous advantages

On one hand, Indigenous peoples are on the frontline of climate change—the first to feel its effects, with subsistence economies and cultures that are the most vulnerable to climate catastrophes. On the other hand, Indigenous people can also be viewed as the most historically adaptable and resilient, because of traditional ecological knowledge, political sovereignty, and community ties. Unlike much of the non-Native population, Indigenous peoples still have community, and leadership that is responsive to community.

Traditional ecological knowledge, derived from Indigenous relationships to the natural world, can serve as an early-warning system for the rest of humanity. Native harvesters have recognized the glaring and undeniable changes on the land and water, before Western scientists even began to study them. Harvest-

ers are starting to share information with each other about harmful or desirable species shifting northward, and thinking about alternate supplies of the resources, as well the effects of these shifts on traditional culture. Some tribes are turning back to traditional agricultural systems, which provide healthier foods that are also more resilient to drought or floods. Indigenous nations can begin to cooperate and trade with each other in foods, to become less reliant on grocery chains in times of shortage.

Political sovereignty allows Indigenous nations to develop methods of responding to climate change that can become models for other Native and non-Native communities. Tribes and First Nations cannot wait for the state, provincial, or federal governments to protect or save us. Treaty rights are the strongest tool to protect salmon habitat and to guarantee access to freshwater. In Washington state, for example, a federal court cited treaty rights in its 2007 ruling that culverts that harm salmon migration be repaired or replaced. Some U.S. tribes are starting to convert from fossil fuels to cleaner and locally controlled renewable energy—such as Tulalip's Qualco Energy biogas plant—and sell it to cities looking for "green energy." Other tribes (such as Nisqually and Swinomish) are cooperating with local governments in land use, water use, and emergency planning, to survive the growing threat of sea-level rise and intensifying winter storms.

A sense of community enables us to care for and support each other in trying times. Native family ties are strong, especially when elders can pass their knowledge on to the youth. Tribal youth can help educate their communities about the threats of climate change and about proactive and positive ways to plan for the future. Indigenous nations need to use innovation and ingenuity in developing new solutions to the climate crisis. The communities that survive will be those that planned ahead and worked with their neighbors for their mutual benefit. *Resilient tribal cultures have survived colonialism, epidemics, assimilation, urbanization, and industrialization. Using traditional strengths, Indigenous peoples can face this new assault of climate change head-on.*

The climate crisis imposes a special duty on tribal leaders and members (particularly elders and youth) to come together and share information on what climate change means specifically to tribal communities, what behavior patterns will emerge in the general population and governance institutions around us, and what preparations Indigenous nations can begin to make in light of this sharing of data and insights. Preparations should include both individual tribal planning and planning for joint efforts. Effective, responsive tribal planning could serve as a model for neighboring non-Native communities. As President Obama told tribal chairs meeting at the White House in November 2009, "We have a lot to learn from your nations in order to create the kind of sustainability in our environment that we so desperately need."

Indigenous nations are in a uniquely vulnerable position in regards to the climate crisis. The limited Native land base provides few opportunities to relocate or expand to cope with a changing climate. Treaty rights and reserved rights are fixed to specific parcels of land and waterways, so that it is unclear what tribal rights to resources might shift away from their reserved lands. Even if tribal rights can be expanded to include species and other resources that migrate off reserved lands, this will impose extreme hardships and problems of access for Native elders, members, and enterprises. Tribal governments could elect to integrate climate adaptation measures (such as water conservation, crop rotation, housing shifts, etc.) into tribal planning, and cooperate with other Indigenous peoples in the process. Harvesters may have to be trained by other practitioners on how to interact with new species coming into their area. This is one area where Indigenous climate change concerns intersect with tribal intergovernmental cooperation.

Indigenous nations are also often vulnerable to the loss of government funding. Economic downturns associated with the climate crisis could impact funds available for Native programs. In the U.S., climate impacts on non-reservation trust lands could also negatively affect rents and receipts from those lands. Tribal nations could take measures to increase mitigation and adaptation actions on their lands. Mitigation must be restorative, and look to historical baselines, not current baselines, for the environmental and hydrological processes that maintain healthy watersheds and communities. U.S. and British Commonwealth law on reserved rights and treaty rights place Indigenous nations in a unique position to pressure governments across the Pacific Rim to take actions to protect those rights.

U.S. tribes can also use diplomatic rights associated with government-to-government relationships. Indigenous non-governmental organizations (NGOs) have historically made recommendations at each conference of the UN Framework Convention on Climate Change (UNFCCC). Indigenous nations should study this record, and determine where they may be strengthened by their formal involvement. U.S. tribes could use their sovereign standing to promote federal actions to prevent, mitigate, and adapt to climate change,

and pressure the federal government to fulfill its trust responsibility by reducing carbon emissions.

Native nations can also bring legal and political pressure to bear upon their settler states, based on potential liabilities for impacts to their trust resources by climate change. They can demand that agencies change laws and policy to recognize tribal rights to shifting species and resources. They can demand a co-management role in any government climate planning, mitigation, or adaptation measures that affect tribal resources. They can promote the development of climate mitigation and adaptation trust funds (climate trusts) for Indigenous climate change defense. Indigenous governments can also request the development of national government policy statements and policy guidelines for agencies regarding climate impacts on Native resources.

Indigenous and national governments can work jointly on assessments, monitoring, prevention, and mitigation of impacts of climate change to Indigenous resources through the establishment of permanent institutions or agencies. The actions of Indigenous nations should create integrated, holistic solutions that address health, housing, transportation, labor, economy, production, population growth, consumption, environment, development, and the full range of climatological, hydrological, environmental, and ecological relationships.

These solutions must address problems at multiple environmental and societal scales, and devise action appropriate to each scale. They must be flexible to respond to changes in the environment and in scientific and local knowledge, and should be designed to monitor and respond to the effectiveness of their objectives. The only ways to guarantee this effectiveness are to respect the rights and privileges of the entire range of stakeholders, to base solutions on ecological and cultural sustainability, including input regarding traditional ecological knowledge. Solutions must include mechanisms to ensure the sustained financial and administrative support for their implementation.

The Climate Change and Pacific Rim Indigenous Nations Project

Since 2006, the Climate Change and Pacific Rim Indigenous Nations Project has been coordinated by the Northwest Indian Applied Research Institute (NIARI) at The Evergreen State College in Olympia, Washington. Its purpose is to document the existing effects of climate change on Indigenous peoples and homelands in Pacific Rim countries, describe examples of Indigenous nation responses to their circumstances at the local, national, and international levels, and recommend future paths for Indigenous nation governments to consider.

When the project began, NIARI published an eighty-one-page report to tribal leadership on climate change and Pacific Rim Indigenous nations. The following year, the Institute produced the first of two editions of a sixteen-page Community Organizing Booklet for Pacific Northwest tribal members, which translated the technical themes of the report into regular English, to make the complex issues of climate change accessible to tribal members, in order to start local discussions on community responses. Through these publications, NIARI encouraged both tribal government and members to start these discussions on the community level, to start planning for the coming changes ahead.

Indigenous nations have a long history of conflict with and oppression by the various settler states that are successors to earlier colonial governments. The U.S., Canada, Australia, and New Zealand were the only four countries to initially vote against adoption of the United Nations Declaration on the Rights of Indigenous Peoples in 2007, though they are in the process of reversing their stance, with certain preconditions. Few of the Pacific Rim countries have expressed a willingness to hear from their Indigenous communities regarding those impacts of climate change that present unique or distinctive problems and challenges to a subsistence way of life or to preservation of fish and wildlife species essential to their diet and the practice of traditions and ceremonies.

Our project was initiated in the context of Indigenous nations joining together to influence the global discourse on climate change and adaptation strategies, to respond to the inevitable impacts of the climate crisis.

The effort includes Maori in Aotearoa (New Zealand), Native Alaskans, First Nations in Western Canada, and U.S. tribal nations in the Pacific Basin states. A central purpose of our study is to ask how these distinctive Indigenous concerns can be represented, and how Indigenous peoples around the Pacific Rim can cooperate with each other in the same way that the settler states collaborate with each other.

Our inquiry includes an examination of possible mechanisms for Indigenous nation intervention in global forums addressing environmental and species protection. We have concluded that continued reliance only upon NGOs for representation and advocacy of Indigenous nation concerns before such international and global forums is not adequate, in relation to the gravity of the issues and the right of Indigenous nations to speak directly on behalf of their constituents. *Tribal and First Nation governments need to step forward to become directly involved in this process.*

Indigenous nations can also collaborate with each other to develop a united representation of their concerns (such as the climate crisis), independent of settler-state governments. In recent years, alliances have been developing that transcend the U.S.-Canada boundary and reinforce pre-colonial relationships among Native peoples who straddle the boundary, such as the Coast Salish Gathering, the Great Lakes Water Accord, and the Yukon River Inter-Tribal Watershed Council. Climate change has become one of the emphases of the unfolding process to develop an Indigenous Nations Treaty—an independent compact among sovereign Native governments around the Pacific Rim.

The United League of Indigenous Nations Treaty being signed by representatives of Native nations from the U.S., Alaska, Canada, Australia, and Aotearoa/New Zealand, at the Lummi Nation, Washington, on August 3, 2007.

The Indigenous Nations Treaty

Indigenous people have through time immemorial traveled great distances, sometimes across continents, sometimes across oceans, to explore and meet with each other. Native peoples' gatherings have been used as an opportunity to engage in trade, to share information at many different levels, and to develop different kinds of alliances. If a critical mass of Indigenous nations decides to make a treaty committing themselves to an alliance on common goals, they will increase their influence and political leverage in proportion to their numbers.

An Indigenous Nations Treaty was first proposed in 2002 during meetings of the National Congress of American Indians (NCAI) in the U.S. and the Assembly of First Nations (AFN) in Canada. Indigenous government representatives discussed the history of treaty making with colonial governments and concluded that a common history of treaty relations with the British Crown further supported the idea of a treaty among Indigenous nations. A strong motivation for developing unity between Indigenous nations and a common strategy on common issues was to act independent of the settler states.

The United League of Indigenous Nations was formed during historic treaty negotiations between Indigenous nations from the U.S., Canada, Australia, and Aotearoa (New Zealand). On August 1, 2007, eleven indigenous nations—including representatives from U.S. tribes, First Nations of Canada, Maori Tribal Confedereations of Aotearoa, and Australian Aboriginal nations—convened on the lands of their host, the Lummi Nation, near Bellingham, Washington. Their goal was to create an international indigenous treaty to bind themselves into an alliance that addresses four broad areas of interest: (1) the impacts of the climate crisis on Indigenous peoples of the Pacific Rim, and Indigenous nations' responses to climate change; (2) the protection of cultural properties, including sacred items and traditional knowledge rights; (3) developing economic and trade relations among Indigenous nations; and (4) international border-crossing rights.

Aroha Te Pareake Mead, faculty member in commerce and administration at Victoria University in Wellington, New Zealand, said in 2007,

> For indigenous nations to enter into international treaties with each other is quite consistent with how we've always conducted our affairs. We have traditions of trading with other nations and in engaging in peacekeeping and other forms of foreign policy. States are stepping up their resistance to the sovereignty of indigenous nations, and the United Nations isn't delivering

enough for indigenous peoples. We need to look to each other in order to pave an appropriate development pathway for our future generations. The answers lie within us.

By 2011, the treaty had been ratified by delegates representing more than eighty-four Indigenous nations, and now serves as a mandate for the implementation of political, cultural, and trade rights and protects the exercise of sovereignty by Indigenous peoples. Treaty objectives are specific elements to be achieved through taking concrete, practical steps and actions designed to accomplish the broader goals of treaty participants.

The treaty specifies "Protecting our Indigenous lands, air and waters from environmental destruction through exercising our rights of political representation as Indigenous nations before all national and international bodies that have been charged, through international treaties, agreements and conventions, with environmental protection responsibilities … " The goal of politically unifying the representation of Indigenous nation concerns before international bodies regarding the impacts of climate change would include many of the specific steps recommended for Indigenous nation leaders in the Recommendations chapter in Part IV of this book.

The voices represented in this anthology come from many diverse perspectives and experiences, use different forms of knowledge, and therefore express their concerns using very different languages. Some use the traditional indigenous insights of Native Science, and others use the data-based findings of Western Science. Some are Indigenous leaders or members, and others are allies and neighbors who have worked closely with Native communities. Some are examining the local scale of a single Indigenous homeland or community—studies that can shape our understanding of larger ecosystems and possible response models—and others discuss the national or international scales that in turn shape local realities. But whether we are looking at the forest or the trees, the climate crisis affects us all, and it is increasingly necessary to use every possible angle to research and respond to the challenges. All the contributors to this anthology have something in common: a deep love of the people and lands affected by climate change, and an appreciation of the resilience of human beings in the face of great adversity.

Indigenous nations throughout the Pacific Rim are in a very precarious position in relation to the impacts of climate change. Survival as Indigenous peoples over the years of contact with European explorers and subsequent colonization, urbanization, and industrialization has depended upon an ability to remain connected to the land. These connections have served as a wellspring

Indigenous nation delegates to the United League of Indigenous Nations meeting held at the Lummi Nation in August 2007.

of spiritual energy and have linked Native peoples to the ancestors. These links provide a body of knowledge that defines who Native peoples are in the cosmos and how to structure daily lives in order to survive.

If future generations of Indigenous people are to continue the traditional practices that make culture a source of spiritual nourishment, these vital connections must be maintained. Moreover, the ecological knowledge and resilience possessed by Indigenous people is essential to the ability of all peoples to understand the behavior of the Earth's ecosystem as they attempt to adapt to the climate crisis.

Acknowledgements

We wish to acknowledge all the contributors to this anthology for their hard work and dedication. We wish to recognize the assistance for our project by Tulalip Tribes Natural Resources director Terry Williams and Preston Hardison, whose knowledge and expertise acquired during years of service representing Indigenous concerns at the international level (with the Convention on Biological Diversity) were pivotal to this project. We wish to thank Ted Whitesell of the Masters in Environmental Studies program at Evergreen, and Brett Stephenson, then director of environmental studies at Te Whare Wananga o Awanuiarangi, a leading Maori university in Whakatane, Aotearoa/New Zealand. We also wish to acknowledge the support from Dr. Graham Smith (CEO at Awanuiarangi), NIARI Assistant Directors Jennifer Scott, Virginia Ith, and Aleticia Tijerina, Program Assistant Bonnie Graft, Research Associate Debra McNutt, and Evergreen student interns Shonri Begay and Courtney Hayden (who both contributed to this introduction). We wish to acknowledge the interest and support from Oregon State University Press: Managing Editor Jo Alexander; Assistant Director Tom Booth; Acquisitions Editor Mary Braun, who encouraged and nurtured the development of this anthology; Editorial and Production Assistant Judith Radovsky; Editorial and Marketing Associate Micki Reaman; and freelance editor Catherine Cocks, freelance indexer Sue Marchman, and freelance proofreader Julie Talbot. Thank you to Swinomish Chairman Brian Cladoosby and Debra Lekanof of the Coast Salish Gathering for offering a forum for the project. Finally, we acknowledge our families, for their lifelong love and support.

The Longhouse Education and Cultural Center at The Evergreen State College in Olympia, Washington.

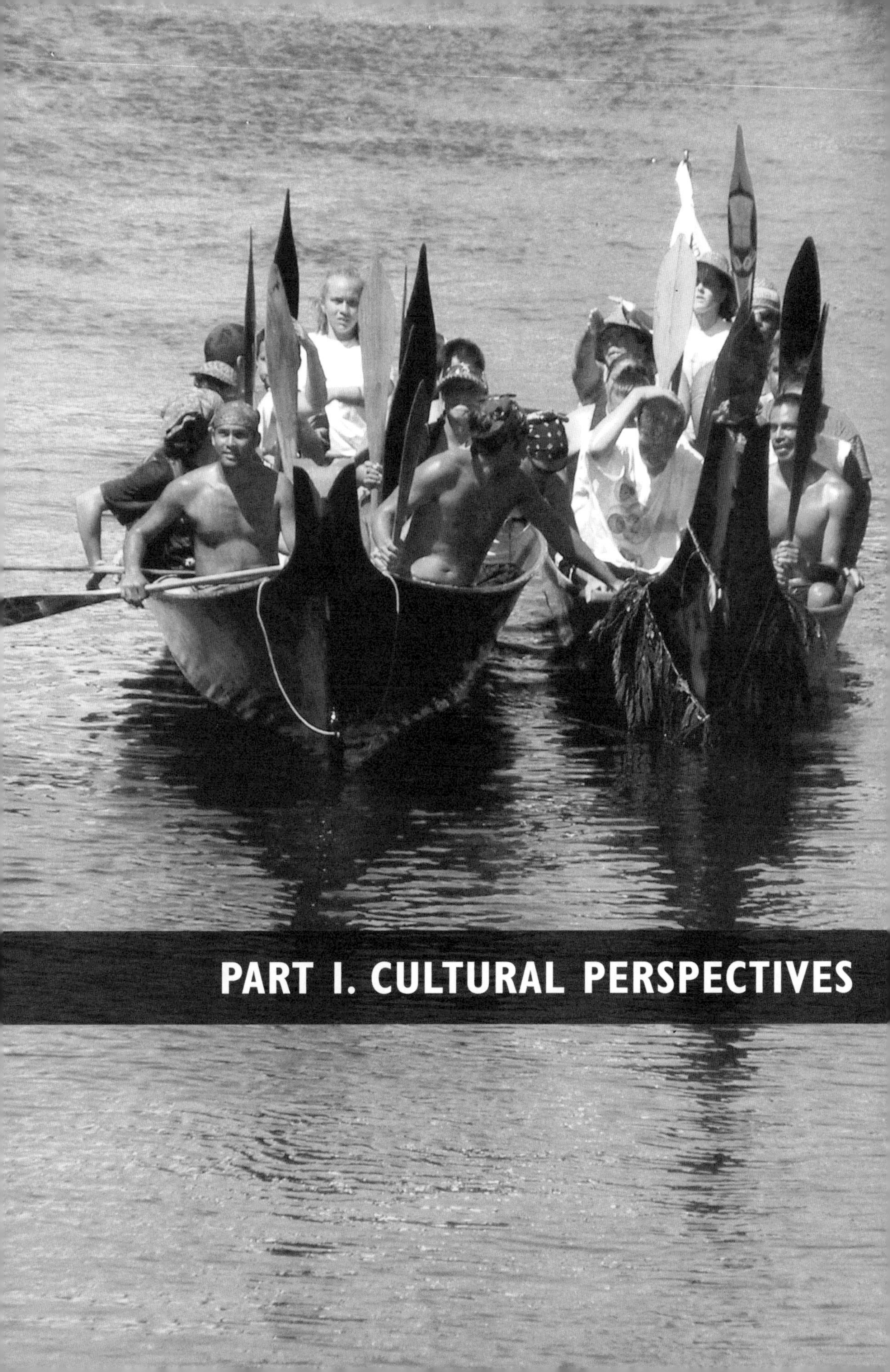

PART I. CULTURAL PERSPECTIVES

CULTURAL PERSPECTIVES

Indigenous peoples, the First Nations of our planet, are the first victims of climate change, but are also the first to grasp and explain the profound meanings of a changing climate. Indigenous voices draw on many millennia of experience with the land and waters and the plants and animals that share Native homelands. Indigenous harvesters are often the first to recognize both the subtle and drastic changes taking place in the circle of life, or what Western science describes as "the environment." The dominant industrialized society ignores these Indigenous voices at its own peril and can now begin to listen before it's too late.

Traditional ecological knowledge, which the Santa Clara Pueblo educator Gregory Cajete calls "Native science," is a set of varied and time-tested practices rooted in each local landscape that provides a guide to the natural laws of interdependence. In this era of rapid climate change, these laws are being thrown into disarray. Native science can identify these massive shifts and provide an early warning long before the academic research studies of Western science can be funded, reviewed, and released. Ideally, Native science and Western science can (and must) work in tandem to reinforce their respective strengths and respond together to the deepening crisis that faces Native and non-Native communities alike.

Indigenous peoples from around the world have prophecies that describe the times that we are in. Many of these prophecies describe changes in the climate resulting from the actions of human beings who neglect the Earth. By conveying this traditional knowledge, the ancestors and elders warned the present generation about the crisis ahead and gave advice about how to survive it. The Sisseton–Wahpeton Dakota musician, actor, and activist Floyd Red Crow Westerman said in an interview before he passed in 2007:

> Time evolves, and comes to a place where it renews again. There is first a purification time, then there is renewal time. We are getting very close to this time now. We were told that we would see America come and go. In a sense, America is dying from within, because we forgot the instructions about how to live on Earth.
>
> Everything is coming to a time in prophecy when man's inability to live on Earth in a spiritual way will come to a crossroad…. It's our belief that if you're not spiritually connected to the Earth and understand the spiritual reality of how to live on Earth, it's likely you will not make it….
>
> Everything is spiritual. Everything has a spirit. Everything was brought to you by the Creator…. We are all from the Earth, and when the Earth, the water, the atmosphere is corrupted, then it will create its own reaction….To me, it is not a negative thing to know that there will be great changes…. It's time. Nothing stays the same.
>
> You should learn how to plant something; that's the first connection. You should treat all things as spirit, and realize that we are one family. It's never something like the end. It's like life—there is no end to life. (Floyd Red Crow Westerman, http://www.youtube.com/watch?v=g7cylfQtkDg)

Ojibwe grandmother Josephine Mandamin (Beed-awsige) began the 2011 Mother Earth Water Walk from Bushoowah-ahlee Point on The Evergreen State College campus in Olympia, Washington. She was one of four elder women walkers who converged two months later at Wisconsin's Bad River Reservation on Lake Superior.

LAND GRAB ON A GLOBAL SCALE

by Dennis Martinez

Founder and co-chairman of the Indigenous Peoples' Restoration Network of the Society for Ecological Restoration International.

Among the English-speaking settler societies—U.S., Canada, Australia, New Zealand—an irrational but powerful myth still prevails. It drove "manifest destiny" and is still alive and well, if usually unconscious.

Divinely inspired colonists wrested lands occupied by Native peoples and bestowed the mixed blessings of civilization on them. The rationalization for dispossession then—and now—was that these "primitive" peoples were not making productive use of their lands. What the colonists did not know, and still do not, is that they took over lands that were largely shaped and maintained by Indigenous peoples through extensive and intensive land care practices that enabled them to not only survive but also thrive.

Enter the 21st century. The work of indigenous dispossession is about to be completed. The last great global land grab and Indigenous asset stripping is happening as I write. (I borrowed these phrases from Rebecca Adamson of First Peoples Worldwide and Andy White of Rights and Resources Initiative at a meeting of the World Bank that I participated in.)

We have a big problem. Some unintended outcomes of well-intentioned climate mitigation measures are below the media radar screen. Land values are dramatically increasing because of demand by northern multinational corporations for land to produce biofuels, plantation monocultures for carbon trading offsets, and transfat substitutes such as palm oil in the developing south.

Indigenous peoples presently occupy 22 percent of the Earth's land surface, are stewards of 80 percent of remaining biodiversity, and comprise 90 percent of cultural diversity. As demand increases the value of indigenous lands—already poorly protected—the rate of loss of indigenous assets and livelihood options becomes more rapid. Adding to these losses are losses of homelands set aside by big environmental NGOs and third-world government elites for conservation reserves and parks through forced evictions. Also disappearing is global genetic diversity maintained by indigenous peoples, which is essential for maintaining the capacity of plants and animals to adapt to climate change.

Disappearing with land and resources is an incalculable wealth of stewardship experience and knowledge. But climate change is here. While the developed north (west) is scrambling for solutions, indigenous peoples are receiving the brunt of the effects of climate change caused by the north. Ignored in the global debate are indigenous cultures that have survived intact for millennia while "great" civilizations have repeatedly collapsed. Indigenous peoples are neither noble nor ignoble.

Some have made environmental mistakes in the past and did not survive. The cultures that survived have done so in proportion as they have learned to adapt. They are just people like everyone else, but people with great practical know-how.

The current economic asymmetry is the result of the myth that wealth will eventually filter down to the poor through so-called free trade and speculative global markets. But as the wealth of a small number of privileged individuals has increased, world poverty has increased fivefold.

The Rio Convention on Biological Diversity (1992), Article 8 (j), and Agenda 21 affirmed that indigenous cultures protect biodiversity and should be compensated for their sustainable practices and products. But the U.S.-dominated Uruguay round of GATT in the same year effectively shut out indigenous peoples from any protection or compensation.

In the meantime the world is losing its best strategy for mitigating climate change—viable indigenous cultures who are the stewards of genetic diversity through traditional land practices. We will also lose the continuing contributions of native knowledge to medicine, sustainable agriculture, health products, lubricants, common foods, wildlife and fisheries management, and more.

The tobacco industry is now liable for costs to states for paying smokers' health bills. Why not hold the developed nations accountable for the damage to ecosystems and indigenous ecosystem peoples who are suffering from climate change that they didn't cause? Where is the accountability? Why not support existing national and international laws and treaties that are simply ignored?

We do not want victimhood. We want parity and compensation through recognition of our substantial contributions to your wealth. It is not an "ethnic" issue.

Indigenous peoples are the miner's canary. It is about the survival of all humans and it is about the loss of the collective heritage of our species. It is all of our lands and all of our assets that are being stolen by economic criminals. They benefit and we pay.

First published April 2, 2008 by the Seattle Post-Intelligencer.

http://seattlepi.nwsource.com/opinion/ 357294_eco-cultural02.html.

©1996-2008 Seattle Post-Intelligencer. Reproduced by permission of Hearst Corporation.

Participants at the Indigenous Peoples Global Summit on Climate Change, held in Anchorage in 2009.

THE ANCHORAGE DECLARATION

Indigenous Peoples Global Summit on Climate Change, Anchorage, Alaska, April 24, 2009

On April 20–24, 2009, Indigenous representatives from the Arctic, North America, Asia, Pacific, Latin America, Africa, [the] Caribbean, and Russia met in Anchorage, Alaska, for the Indigenous Peoples' Global Summit on Climate Change. We thank the Ahtna and the Dena'ina Athabascan Peoples in whose lands we gathered.

We express our solidarity as Indigenous Peoples living in areas that are the most vulnerable to the impacts and root causes of climate change. We reaffirm the unbreakable and sacred connection between land, air, water, oceans, forests, sea ice, plants, animals, and our human communities as the material and spiritual basis for our existence.

We are deeply alarmed by the accelerating climate devastation brought about by unsustainable development. We are experiencing profound and disproportionate adverse impacts on our cultures, human and environmental health, human rights, well-being, traditional livelihoods, food systems and food sovereignty, local infrastructure, economic viability, and our very survival as Indigenous Peoples.

Mother Earth is no longer in a period of climate change, but in climate crisis. We therefore insist on an immediate end to the destruction and desecration of the elements of life. Through our knowledge, spirituality, sciences, practices, experiences, and relationships with our traditional lands, territories, waters, air, forests, oceans, sea ice, other natural resources, and all life, Indigenous Peoples have a vital role in defending and

lies in the wisdom of our elders, the restoration of the sacred position of women, the youth of today, and in the generations of tomorrow.

We uphold that the inherent and fundamental human rights and status of Indigenous Peoples, affirmed in the United Nations Declaration on the Rights of Indigenous Peoples (UNDRIP), must be fully recognized and respected in all decision-making processes and activities related to climate change. This includes our rights to our lands, territories, environment, and natural resources as contained in Articles 25–30 of the UNDRIP. When specific programs and projects affect our lands, territories, environment and natural resources, the right of Self Determination of Indigenous Peoples must be recognized and respected, emphasizing our right to Free, Prior, and Informed Consent, including the right to say "no." The United Nations Framework Convention on Climate Change (UNFCCC) agreements and principles must reflect the spirit and the minimum standards contained in UNDRIP.

Calls for Action

1. In order to achieve the fundamental objective of the United Nations Framework Convention on Climate Change (UNFCCC), we call upon the fifteenth meeting of the Conference of the Parties to the UNFCCC to support a binding emissions reduction target for developed countries (Annex 1) of at least 45% below 1990 levels by 2020 and at least 95% by 2050. In recognizing the root causes of climate change, participants call upon States

to work towards decreasing dependency on fossil fuels. We further call for a just transition to decentralized renewable energy economies, sources, and systems owned and controlled by our local communities to achieve energy security and sovereignty.

In addition, the Summit participants agreed to present two options for action; some supported option A and some option B. These are as follows:

    A. We call for the phase out of fossil fuel development and a moratorium on new fossil fuel developments on or near Indigenous lands and territories.

    B. We call for a process that works towards the eventual phase out of fossil fuels, without infringing on the right to development of Indigenous nations.

2. We call upon the Parties to the UNFCCC to recognize the importance of our Traditional Knowledge and practices shared by Indigenous Peoples in developing strategies to address climate change. To address climate change we also call on the UNFCCC to recognize the historical and ecological debt of the Annex 1 countries in contributing to greenhouse gas emissions. We call on these countries to pay this historical debt.

3. We call on the Intergovernmental Panel on Climate Change (IPCC), the Millennium Ecosystem Assessment, and other relevant institutions to support Indigenous Peoples in carrying out Indigenous Peoples' climate change assessments.

4. We call upon the UNFCCC's decision-making bodies to establish formal structures and mechanisms for and with the full and effective participation of Indigenous Peoples. Specifically we recommend that the UNFCCC:

    A. Organize regular Technical Briefings by Indigenous Peoples on Traditional Knowledge and climate change;

    B. Recognize and engage the International Indigenous Peoples' Forum on Climate Change and its regional focal points in an advisory role;

    C. Immediately establish an Indigenous focal point in the secretariat of the UNFCCC;

    D. Appoint Indigenous Peoples' representatives in UNFCCC funding mechanisms in consultation with Indigenous Peoples;

    E. Take the necessary measures to ensure the full and effective participation of Indigenous and local communities in formulating, implementing, and monitoring activities, mitigation, and adaptation relating to impacts of climate change.

5. All initiatives under Reducing Emissions from Deforestation and Degradation (REDD) must secure the recognition and implementation of the human rights of Indigenous Peoples, including security of land tenure, ownership, recognition of land title according to traditional ways, uses, and customary laws and the multiple benefits of forests for climate, ecosystems, and Peoples before taking any action.

6. We challenge States to abandon false solutions to climate change that negatively impact Indigenous Peoples' rights, lands, air, oceans, forests, territories, and waters. These include nuclear energy, large-scale dams, geo-engineering techniques, "clean coal," agro-fuels, plantations, and market-based mechanisms such as carbon trading, the Clean Development Mechanism, and forest offsets. The human rights of Indigenous Peoples to protect our forests and forest livelihoods must be recognized, respected, and ensured.

7. We call for adequate and direct funding in developed and developing States and for a fund to be created to enable Indigenous Peoples' full and effective participation in all climate processes, including adaptation, mitigation, monitoring, and transfer of appropriate technologies in order to foster our empowerment, capacity-building, and education. We strongly urge relevant United Nations bodies to facilitate and fund the participation, education, and capacity building of Indigenous youth and women to ensure engagement in all international and national processes related to climate change.

8. We call on financial institutions to provide risk insurance for Indigenous Peoples to allow them to recover from extreme weather events.

9. We call upon all United Nations agencies to address climate change impacts in their strategies and action plans, in particular their impacts on Indigenous Peoples, including the World Health Organization (WHO), United Nations Educational, Scientific, and Cultural Organization (UNESCO), and United Nations Permanent Forum on Indigenous Issues (UNPFII). In particular, we call upon all the United Nations Food and Agriculture Organization (FAO) and other relevant United Nations bodies to establish an Indigenous Peoples' working group to address the impacts of climate change on

food security and food sovereignty for Indigenous Peoples.

10. We call on United Nations Environment Programme (UNEP) to conduct a fast-track assessment of short-term drivers of climate change, specifically black carbon, with a view to initiating negotiation of an international agreement to reduce emission of black carbon.

11. We call on States to recognize, respect, and implement the fundamental human rights of Indigenous Peoples, including the collective rights to traditional ownership, use, access, occupancy, and title to traditional lands, air, forests, waters, oceans, sea ice, and sacred sites as well as to ensure that the rights affirmed in Treaties are upheld and recognized in land use planning and climate change mitigation strategies. In particular, States must ensure that Indigenous Peoples have the right to mobility and are not forcibly removed or settled away from their traditional lands and territories, and that the rights of Peoples in voluntary isolation are upheld. In the case of climate change migrants, appropriate programs and measures must address their rights, status, conditions, and vulnerabilities.

12. We call upon states to return and restore lands, territories, waters, forests, oceans, sea ice, and sacred sites that have been taken from Indigenous Peoples, limiting our access to our traditional ways of living, thereby causing us to misuse and expose our lands to activities and conditions that contribute to climate change.

13. In order to provide the resources necessary for our collective survival in response to the climate crisis, we declare our communities, waters, air, forests, oceans, sea ice, traditional lands, and territories to be "*Food Sovereignty Areas*," defined and directed by Indigenous Peoples according to customary laws, free from extractive industries, deforestation, and chemical-based industrial food production systems (i.e., contaminants, agro-fuels, genetically modified organisms).

14. We encourage our communities to exchange information while ensuring the protection and recognition of and respect for the intellectual property rights of Indigenous Peoples at the local, national, and international levels pertaining to our Traditional Knowledge, innovations, and practices. These include knowledge and use of land, water and sea ice, traditional agriculture, forest management, ancestral seeds, pastoralism, food plants, animals, and medicines and are essential in developing cli-

mate change adaptation and mitigation strategies, restoring our food sovereignty and food independence, and strengthening our Indigenous families and nations.

We offer to share with humanity our Traditional Knowledge, innovations, and practices relevant to climate change, provided our fundamental rights as intergenerational guardians of this knowledge are fully recognized and respected. We reiterate the urgent need for collective action.

Agreed by consensus of the participants in the Indigenous Peoples' Global Summit on Climate Change, Anchorage Alaska, April 24, 2009

INTERNATIONAL INDIGENOUS PEOPLES FORUM ON CLIMATE CHANGE (IIPFCC) POLICY PAPER ON CLIMATE CHANGE

Discussed and finalized at the IIPFCC meeting in Bangkok, Thailand, September 26–27, 2009.

Mother Earth is no longer in a period of climate change, but in climate crisis.... Indigenous Peoples have a vital role in defending and healing Mother Earth. We uphold that the inherent rights of Indigenous Peoples...must be fully respected in all decision-making processes and activities related to climate change.

—*Anchorage Declaration (Indigenous Peoples Global Summit on Climate Change, 2009)*

Climate Change Calls for Historic Transformations

1. Climate change, in the light of the current global financial, economic, environmental, and food crises, represents an unprecedented challenge and opportunity for humanity to transform global economic, political, social, [and] cultural relations to live in balance with Mother Earth. Reaching climate equilibrium and justice is inseparable from acknowledging the historical responsibilities of developed countries while promoting social equity between and within nations, maintaining ecological integrity, addressing the climate and ecological

debt, and pursuing an effective transition away from fossil fuel dependency towards a green economy. It requires honouring international commitments to poverty eradication, sustainable development, biodiversity, and human rights. The full and effective participation of indigenous peoples, local communities, and vulnerable groups is key to achieve a just and equitable outcome of the climate negotiations.

2. Climate science, indigenous and traditional knowledge, international solidarity, equity and human rights, widespread social mobilisation, and strong political leadership are all building blocks towards desirable outcomes in Copenhagen [at the UN Conference of the Parties 2009] and beyond.

3. Climate change governance must transcend state-governments' negotiations, to recognize the rights of Indigenous Peoples, which include the full and effective participation in all negotiations by Indigenous Peoples' traditional governments, institutions, and organizations. It must also embrace diverse contributions and intercultural collaboration, recognizing distinct and valuable contributions from children and youth, women, indigenous peoples and local communities. All voices need to be included in climate governance and decision making: we are all learners and teachers together in addressing human-induced climate change.

Indigenous Peoples Are Rights-holders

4. We hold inalienable collective rights over our lands, territories, and resources. Policies and actions that are being negotiated now directly affect our traditional lands, territories, oceans, waters, ice, flora, fauna, and forests, thereby also affecting the survival and livelihoods of over 370 million Indigenous Peoples from all regions of the globe. However, our concerns and views have not been seriously addressed in the climate negotiation processes, least of all those from indigenous women and youth. We reiterate the States' and whole UN system's obligations to uphold regional and international human rights commitments and standards, especially the UN Declaration on the Rights of Indigenous Peoples (UNDRIP). The provisions of the UNDRIP articulate rights which must be respected and safeguarded in all climate decision making and actions. We are therefore holders of collective rights, including sovereign and inherent rights to land and treaty rights, covenants, and agreements. Protecting these rights also strengthens the capacity and resilience of indigenous peoples and local communities to respond to climate change.

5. Respect for the human rights of indigenous peoples and local communities, valuing our traditional knowledge and innovations, and supporting our local mitigation and adaptation strategies are critical and invaluable requirements towards adequate holistic solutions to climate change. As such, our local strategies and priorities must be reflected in National Adaptation and Mitigation Actions (NAMAs) and National Adaptation Plans and strategies of Action (NAPAs), in the development and implementation of which we must participate fully and effectively. The distinct roles and responsibilities of indigenous women and youth will need to be considered, underlining the importance of their inclusion in decision-making and planning processes.

6. Our rights to self-determination and free, prior, and informed consent (FPIC) are the minimum standards to safeguard our rights and interests through the different stages of the project lifecycle, including policy framing, planning and design, implementation, restoration, rehabilitation, benefit-sharing, and conflict resolution.

7. Our governing bodies have the right to enact such laws and regulations as appropriate and adopt mitigation and adaptation plans within their jurisdictional authority as they deem necessary to protect and advance the social, economic, political, and cultural welfare of their communities in matters pertaining to climate change. Each indigenous people's governing body has the prerogative to determine and apply the best available science, including native sciences and conventional sciences, according to their cultural requirements consistent with the right to determine and develop priorities and strategies for the development or use of their lands or territories and other resources.

Indigenous Peoples' Contributions to Ecosystem-based Mitigation and Adaptation

8. We have intrinsic contributions towards addressing the climate crisis and renewing the relationships between humans and nature. For generations, we have managed ecosystems nurturing [their] integrity and complexity in sustainable and culturally diverse ways. Our customary resource management systems have proven to be ecologically sustainable, low carbon economies. These include mobile pastoralism in drylands and rangelands,

rotational swidden agriculture and ecological agriculture in tropical forest regions, [and] the conservation, management, and restoration of other natural ecosystems such as mangroves, savannahs, wetlands, the Arctic environment, and small island ecosystems. Traditional knowledge, innovations, and adaptation practices embody local adaptative management to the changing environment and complement scientific research, observations and monitoring.

9. The climate crisis threatens our very survival, particularly forest-dependent, ice-dependent peoples, peoples in voluntary isolation, and the indigenous peoples of small island states and local communities. Addressing such vulnerabilities requires recognition, respect, and strengthening of the traditional knowledge of indigenous peoples, and strengthening the resilience of ecosystems and Indigenous Peoples and local communities' capacities to adapt to climate change. Ecosystem-based adaptation based on holistic indigenous peoples' systems and rights can deliver significant social, cultural, spiritual, and economic values to Indigenous Peoples and local communities as well as to the biodiversity of indigenous lands and territories. This should be considered with the full participation of indigenous peoples in the planning, design, implementation, monitoring, and evaluation of these measures. The empowerment of Indigenous peoples and local communities is critical to successful adaptation strategies to climate change.

10. Our cosmovision, ways of life and traditional practices have been in existence since time immemorial. Sumak Kawsay [Good Living], Penker Pujustin, and other indigenous visions and values propose a way of life that is respectful, responsible, balanced, and harmonious with nature and offers equity and solidarity as the guiding principles of global well-being. Indigenous worldviews embody an organized, sustainable, and dynamic economic system, as well as political, socio-cultural, and environmental rights. This vindicates a social dimension of democracy that goes beyond formal democracy, where economy becomes a subordinate activity to the development of peoples in the name of humanity, solidarity, and respect for Mother Earth.

Securing Indigenous Peoples' Territories

11. The global economic transition to sustainable, low carbon development will require revitalization of diverse local economies, including support for Indigenous peoples' self-determined development. Economic planning combined with adaptive management to climate change will need to apply an ecosystem-based approach and must fully respect the rights and interests of indigenous peoples and local communities. Securing our rights to our ancestral lands, forests, waters, and resources provides the basis for sustainable local social, cultural, spiritual, and economic development and some insurance against our vulnerability to the impacts of climate change. This is also beneficial towards improving ecosystem governance, ecosystem resilience, and the delivery of ecosystem services.

12. Many forests are within the traditional lands and territories of Indigenous peoples and Indigenous peoples around the world live in and depend upon forests for their survival and to enjoy their fundamental rights to forests and land tenure. They are of cultural, social, economic, and spiritual significance for us and provide benefits for humankind. Accordingly, the rights of Indigenous peoples, including our land and resource rights, must be recognized and respected at all levels (local, national, and international) before we can consider REDD [Reducing Emissions from Deforestation and Degradation] initiatives and projects. The recognition of our rights must be in accordance with international human rights law and standards including the UN Declaration on the Rights of Indigenous Peoples (UNDRIP) and International Labour Organisation (ILO) Convention 169, among other human rights instruments. If there is no full recognition and full protection for Indigenous peoples' rights, including the rights to resources, lands, and territories, and there is no recognition and respect of our rights of free, prior, and informed consent of the affected indigenous peoples, we will oppose REDD and [enlarged] REDD+ and carbon offsetting projects, including CDM [Clean Development Mechanism carbon trading system] projects. All decision-making processes on REDD and REDD+, Clean Development Mechanism, Land Use and Land Use Change and Forests (LULUCF), Agriculture, Forestry and Other Land Use (AFOLU), as well as other ecosystem-based mitigation and adaptation measures and projects, must be conditional to the free, prior, informed consent of Indigenous peoples.

13. Our laws, regulations, and plans shall be recognized as authoritative and determinative as to the risks, values, and benefits associated with measures

to adapt to, or mitigate for, climate change effects within the territorial jurisdiction of tribal governing bodies.

The IIPFCC affirms our global unity and solidarity to realize the enjoyment of our collective rights and the recognition of our vision, indigenous knowledge, and our contributions in solving the climate change crisis.

THE MYSTIC LAKE DECLARATION

From the Native Peoples Native Homelands Climate Change Workshop II: Indigenous Perspectives and Solutions

At Mystic Lake on the Homelands of the Shakopee Mdewakanton Sioux Community, Prior Lake, Minnesota, November 21, 2009

As community members, youth and elders, spiritual and traditional leaders, Native organizations and supporters of our Indigenous Nations, we have gathered on November 18–21, 2009, at Mystic Lake in the traditional homelands of the Shakopee Mdewakanton Dakota Oyate. This Second Native Peoples Native Homelands Climate Workshop builds upon the Albuquerque Declaration [see the following document] and work done at the 1998 Native Peoples Native Homelands Climate Change Workshop held in Albuquerque, New Mexico. We choose to work together to fulfill our sacred duties, listening to the teachings of our elders and the voices of our youth, to act wisely to carry out our responsibilities to enhance the health and respect the sacredness of Mother Earth, and to demand Climate Justice now.

We acknowledge that to deal effectively with global climate change and global warming issues, all sovereigns must work together to adapt and take action on real solutions that will ensure our collective existence. We hereby declare, affirm, and assert our inalienable rights as well as responsibilities as members of sovereign Native Nations. In doing so, we expect to be active participants with full representation in United States and international legally binding treaty agreements regarding climate, energy, biodiversity, food sovereignty, water, and sustainable development policies affecting our peoples and our respective Homelands on Turtle Island (North America) and [the] Pacific Islands.

We are of the Earth. The Earth is the source of life to be protected, not merely a resource to be exploited. Our ancestors' remains lie within her. Water is her

lifeblood. We are dependent upon her for our shelter and our sustenance. Our lifeways are the original "green economies." We have our place and our responsibilities within Creation's sacred order. We feel the sustaining joy as things occur in harmony. We feel the pain of disharmony when we witness the dishonor of the natural order of Creation and the degradation of Mother Earth and her companion Moon.

We need to stop the disturbance of the sacred sites on Mother Earth that she may heal and restore the balance in Creation. We ask the world community to join with the Indigenous Peoples to pray on summer solstice for the healing of all the sacred sites on Mother Earth.

The well-being of the natural environment predicts the physical, mental, emotional, and spiritual longevity of our Peoples and the Circle of Life. Mother Earth's health and that of our Indigenous Peoples are intrinsically intertwined. Unless our homelands are in a state of good health, our Peoples will not be truly healthy. This inseparable relationship must be respected for the sake of our future generations. In this Declaration, we invite humanity to join with us to improve our collective human behavior so that we may develop a more sustainable world: A world where the inextricable relationship of biological and environmental diversity and cultural diversity is affirmed and protected.

We have the power and responsibility to change. We can preserve, protect, and fulfill our sacred duties to live with respect in this wonderful Creation. However, we can also forget our responsibilities, disrespect Creation, cause disharmony, and imperil our future and the future of others.

At Mystic Lake, we reviewed the reports of indigenous science, traditional knowledge, and cultural scholarship in cooperation with non-native scientists and scholars. We shared our fears, concerns, and insights. If current trends continue, native trees will no longer find habitable locations in our forests, fish will no longer find their streams livable, and humanity will find their homelands flooded or drought stricken due to the changing weather. Our Native Nations have already suffered disproportionately the negative compounding effects of global warming and a changing climate.

The U.S. and other industrialized countries have an addiction to the high consumption of energy. Mother Earth and her natural resources cannot sustain the consumption and production needs of this modern industrialized society and its dominant economic paradigm, which places value on the rapid economic growth, the quest for corporate and individual accumulation of wealth, and a race to exploit natural resources. The

Participants at the Indigenous Peoples Global Summit on Climate Change, held in Anchorage in 2009.

non-regenerative production system creates too much waste and toxic pollution. We recognize the need for the United States and other industrialized countries to focus on new economies, governed by the absolute limits and boundaries of ecological sustainability, the carrying capacities of the Mother Earth, a more equitable sharing of global and local resources, encouragement and support of self-sustaining communities, and respect and support for the rights of Mother Earth and her companion Moon.

In recognizing the root causes of climate change, participants call upon the industrialized countries and the world to work towards decreasing dependency on fossil fuels. We call for a moratorium on all new exploration for oil, gas, coal, and uranium as a first step towards the full phase-out of fossil fuels, without nuclear power, with a just transition to sustainable jobs, energy, and environment. We take this position and make this recommendation based on our concern over the disproportionate social, cultural, spiritual, environmental, and climate impacts on Indigenous Peoples, who are the first and the worst affected by the disruption of intact habitats, and the least responsible for such impacts.

Indigenous peoples must call for the most stringent and binding emission reduction targets. Carbon emissions for developed countries must be reduced by no less than 40%, preferably 49% below 1990 levels by 2020 and 95% by 2050. We call for national and global actions to stabilize CO_2 concentrations below 350 parts per million (ppm) and limiting temperature increases to below 1.5°C.

We challenge climate mitigation solutions to abandon false solutions to climate change that negatively impact Indigenous Peoples' rights, lands, air, oceans, forests, territories, and waters. These include nuclear energy, large-scale dams, geo-engineering techniques, clean coal technologies, carbon capture and sequestration, biofuels, tree plantations, and international market-based mechanisms such as carbon trading and offsets, the Clean Development Mechanisms and Flexible Mechanisms [explained above] under the Kyoto Protocol, and forest offsets. The only real offsets are those renewable energy developments that actually displace fossil fuel–generated energy. We recommend the United States sign on to the Kyoto Protocol and to the UN Declaration of the Rights of Indigenous Peoples.

We are concerned with how international carbon markets set up a framework for dealing with greenhouse gases that secure the property rights of heavy Northern fossil fuel users over the world's carbon-absorbing capacity while creating new opportunities for corporate profit through trade. The system starts by translating existing pollution into a tradable commodity, the rights to which are allocated in accordance with a limit set by States or intergovernmental agencies. In establishing property rights over the world's carbon dump, the largest number of rights is granted (mostly for free) to those who have been most responsible for pollution in the first place. At UN COP15 [the 15ᵗʰ Session of the Conference of the Parties of the UNFCCC], the conservation of forests is being brought into a property right issue concerning trees and carbon. With some indigenous communities it is difficult and sometimes impossible to reconcile with traditional spiritual beliefs the participation in climate mitigation that commodifies the sacredness of air (carbon), trees, and life. Climate change mitigation and sustainable forest management must be based on different mindsets with full respect for nature and not solely on market-based mechanisms.

We recognize the link between climate change and food security that affects Indigenous traditional food systems. We declare our Native Nations and our communities, waters, air, forests, oceans, sea ice, traditional lands, and territories to be "Food Sovereignty Areas," defined and directed by Indigenous Peoples according to our customary laws, free from extractive industries, unsustainable energy development [and] deforestation and free from using food crops and agricultural lands for large scale biofuels.

We encourage our communities to exchange information related to the sustainable and regenerative use of land, water, sea ice, traditional agriculture, forest management, ancestral seeds, food plants, animals, and medicines that are essential in developing climate change adaptation and mitigation strategies and will restore our food sovereignty [and] food independence and strengthen our Indigenous families and Native Nations.

We reject the assertion of intellectual property rights over the genetic resources and traditional knowledge of Indigenous peoples which results in the alienation and commodification of those things that are sacred and essential to our lives and cultures. We reject industrial modes of food production that promote the use of chemical substances [and] genetically engineered seeds and organisms. Therefore, we affirm our right to possess, control, protect, and pass on the indigenous seeds, medicinal plants, [and] traditional knowledge originating from our lands and territories for the benefit of our future generations.

We can make changes in our lives and actions as individuals and as Nations that will lessen our contribution to the problems. In order for reality to shift, in order for solutions to major problems to be found and realized, we must transition away from the patterns of an industrialized mindset, thought, and behavior that created those problems. It is time to exercise desperately needed Indigenous ingenuity—Indigenuity—inspired by our ancient intergenerational knowledges and wisdom given to us by our natural relatives.

We recognize and support the position of the International Indigenous Peoples Forum on Climate Change (IIPFCC), operating as the Indigenous Caucus within the United Nations Framework Convention on Climate Change (UNFCCC), that is requesting language within the overarching principles of the outcomes of the Copenhagen UNFCCC 15ᵗʰ Session of the Conference of the Parties (COP15) (and beyond Copenhagen) that would ensure respect for the knowledge and rights of indigenous peoples, including their rights to lands, territories, forests, and resources to ensure their full and effective participation, including free, prior, and informed consent. It is crucial that the United Nations Declaration on the Rights of Indigenous Peoples (UNDRIP) is entered into all appropriate negotiating texts, for it is recognized as the minimum international standard for the protection of rights, survival, protection, and well-being of Indigenous Peoples, particularly with regard to health, subsistence, sustainable housing and infrastructure, and clean energy development.

As Native Nations and Indigenous Peoples living within the occupied territories of the United States, we acknowledge with concern the refusal of the United States to support negotiating text that would recognize applicable universal human rights instruments and agreements, including the UNDRIP, and further safeguard principles that would ensure their [Indigenous Peoples'] full and effective participation including free, prior, and informed consent (FPIC). We will do everything humanly possible by exercising our sovereign government-to-government relationship with the U.S. to seek justice on this issue.

Our Indian languages are encoded with accumulated ecological knowledge and wisdom that extends back through oral history to the beginning of time. Our ancestors created land and water relationship systems premised upon the understanding that all life forms are relatives—not resources. We understand that we as human beings have a sacred and ceremonial responsibility to care for and maintain, through our original instructions, the health and well-being of all life within our traditional territories and Native Homelands.

We will encourage our leadership and assume our role in supporting a just transition into a green economy, freeing ourselves from dependence on a carbon-based fossil fuel economy. This transition will be based upon development of an indigenous agricultural economy comprised of traditional food systems, sustainable buildings and infrastructure, clean energy and energy efficiency, and natural resource management systems based upon indigenous science and tra-

Permafrost melting results in a sinkhole.

ditional knowledge. We are committed to development of economic systems that enable life-enhancement as a core component. We thus dedicate ourselves to the restoration of true wealth for all Peoples. In keeping with our traditional knowledge, this wealth is based not on monetary riches but rather on healthy relationships, relationships with each other, and relationships with all of the other natural elements and beings of creation.

In order to provide leadership in the development of green economies of life-enhancement, we must end the chronic underfunding of our Native educational institutions and ensure adequate funding sources are maintained. We recognize the important role of our Native K–12 schools and tribal colleges and universities that serve as education and training centers that can influence and nurture a much needed *Indigenuity* towards understanding climate change, nurturing clean renewable energy technologies, seeking solutions, and building sustainable communities.

The world needs to understand that the Earth is a living female organism—our Mother and our Grandmother. We are kin. As such, she needs to be loved and protected. We need to give back what we take from her in respectful mutuality. We need to walk gently. These Original Instructions are the natural spiritual laws, which are supreme. Science can urgently work with traditional knowledge keepers to restore the health and well-being of our Mother and Grandmother Earth.

As we conclude this meeting, we, the participating spiritual and traditional leaders, members, and supporters of our Indigenous Nations, declare our intention to continue to fulfill our sacred responsibilities, to redouble our efforts to enable sustainable life-enhancing economies, to walk gently on our Mother Earth, and to demand that we be a part of the decision making and negotiations that impact our inherent and treaty-defined rights. Achievement of this vision for the future, guided by our traditional knowledge and teachings, will benefit all Peoples on the Earth.

Approved by Acclamation and Individual Sign-Ons.

KEY NORTH AMERICAN INDIGENOUS CONCERNS

Native Peoples/Native Homelands Workshop I, Albuquerque, New Mexico, 1998

Arctic Region

➤ Those who live in the Arctic are experiencing shorter winters that disrupt the life cycles of plants and animals that they depend on.

➤ The Yupik people see the winter ice pack receding sooner every year, limiting walruses' [ability] to breed and feed themselves.

➤ Rising water levels from the melting glaciers forced several communities on the Arctic coast and islands to abandon their homes and traditional lands.

➤ Many Arctic communities already have their lands and natural resources polluted by oil spills and oil development that has seriously disrupted the environment and their health.

Eastern Woodlands Cultural Area

➤ Severe ice storms show how dependent we have become on transmitted energy sources.

➤ Damage to traditional foods from shifting climate, natural disasters, and current practices, though those with seeds and traditional knowledge can survive.

➤ Extreme weather events could release more industrial pollution, impacting our diet, polluting our water, plants, fish, and animals.

➤ Culturally significant sugar maples and birch trees will be gone from our territories.

➤ Black ash trees and sweet grass are already disappearing, plants cannot adapt fast enough to changing environments.

➤ Greater imbalance in insect and animal communities: high level of black flies and mosquitoes, predatory fish, and fewer hummingbirds.

➤ Severe water pollution will result from flooding, which will mobilize chemicals applied to the land,

and from droughts, which will concentrate materials already present in our waters.

► More destructive storms will impact communities.

Great Lakes Region

► Climate change provides an ecological risk that disrupts traditional foods of wild rice, berries, and maple syrup for the Indigenous communities that live in northern Minnesota and other Great Lake areas, such as the Anishinaabe.

► Early and rapid winter snowmelt led to flooding of various rivers and lakes causing damage and havoc.

► Dramatic fluctuations in water levels and warmer temperatures of lake waters has affected fish populations and insect populations such as fish kills from increasing dead zones in lakes and severe infestation of disease-spreading insects such as mosquitoes.

Great Plains Region

► Increased extreme weather events such as blizzards and droughts are threatening Great Plains tribal economies where livestock and land extensive agriculture are the primary sources of income.

► Water resources are becoming scarce and depleted before they can be replenished.

► In the past 10 years droughts, blizzards, and flooding have caused six national disaster declarations in the Dakotas.

► Summer heat and severe weather has increased health risks of children and elders.

Southwest Region

► Drought has affected the water table levels and limited water sources that depend on the little rain the region gets to replenish them, causing plants and livestock to die.

► Droughts have caused beetles to suck the saps of trees such as the piñon tree for water and led to tree deaths, some of which are medicinal plants.

► Much of the Navajo (Diné) and Hopi peoples have suffered their lands being desecrated and poisoned by fossil fuel mining companies.

Pacific Coast and Rocky Mountain Regions

► Increased winds not just periodic any more but tend to be constant.

► Violent weather changes where storms wipe out intertidal shellfish and hurt economy, Indian people, and their resources.

► Declining salmon runs.

► Deformed fish.

► Significant decrease in life spans of Indians due to unavailability of traditional foods.

► Transportation and shipping costs affected.

► No more birds, frogs along the river.

► Air pollution due to burning forests.

► Minimum river/stream flows for fish.

► Too much rain saturates soil, causing erosion.

► Erosion due to rising sea levels.

► Contamination of fresh water by saline water.

Regional Recommendations

*Workshops at the Native Peoples/
Native Homelands Workshop I,
Albuquerque, New Mexico, 1998*

Eastern Woodlands Cultural Area

► Hold corporations accountable.

► Incentivize reduction of pollutant emission.

► Advocate respect for treaties as international law.

► Prepare children for the future.

► Do demonstration projects in reforestation, fish farming, renewable energy technologies, and other areas.

► Encourage communities to produce their own energy and food to become more sustainable.

► Control the flow of Tribal money, local economies.

► Create a Spiritual Network that goes beyond organized religion.

► Create an information clearinghouse.

Great Lakes Region

► Work toward sustainability.

► Encourage Congress to support clean and renewable energy initiatives and other "no regrets" strategies.

► Reduce greenhouse gas emissions.

► Assess the impact of current forestry practices.

► Reduce non-point source pollution.

► Promote biodiversity.

► Reestablish Tribal jurisdiction over threatened areas.

- ➤ Use and support indigenous sovereignty to co-manage different regions.
- ➤ Protect spiritual rites.
- ➤ Create a story that lays out all the steps in unsustainable industrial activities such as industrial forestry, i.e., how such practices cause local environmental damage yet are tied to the global economic picture.
- ➤ Establish policies for environmental conservation, especially in food production areas.
- ➤ Preserve historic and cultural practices for food production.
- ➤ Examine how western technology and free market policies have impacted food production and sacred sites.
- ➤ Respect, give thanks, give offerings, and conduct ceremonies.
- ➤ Recognize the importance of sacred sites.
- ➤ Ensure Tribal access and control over sacred sites.
- ➤ Take control over resources.
- ➤ Follow existing rules and laws, including natural laws.

Great Plains Region

- ➤ Save the prairie ecosystem.
- ➤ Develop native land use practices.
- ➤ Plan and develop Tribal food production projects.
- ➤ Use spiritual intelligence.
- ➤ Improve science curriculum.
- ➤ Use available technology and funding to research land.
- ➤ Comparatively analyze differences between beef and buffalo economies.
- ➤ Request funding for sustainable agriculture development and energy resources.
- ➤ Develop intertribal markets and trade agreements to support sustainable development in food production and energy use.
- ➤ Propose "green" energy technology to Tribal councils.
- ➤ Develop Tribal energy efficiency codes and weatherization programs.
- ➤ Address nutritional issues and the Native diet.
- ➤ Protect medicine plants and transplant to safe land area.
- ➤ Use elders to help teachers develop environment and ecology curriculum.
- ➤ Plant and protect trees.

Southwest Region

- ➤ Change the way we all live.
- ➤ Continue adapting.
- ➤ Conclude and adjudicate legal issues.
- ➤ Get regulations in place: e.g., water use contracts.
- ➤ Solidify your finite land/space resources.
- ➤ Quantify and qualify resource management using Indian terms.
- ➤ Improve or eliminate detrimental obstacles.
- ➤ Consider both positive and negative consequences of resource management.
- ➤ Base planning on sound information and good decisions.

Pacific Coast and Rocky Mountain Regions

- ➤ Increase carbon sinks by reducing forest harvesting and increasing planting.
- ➤ Avoid actions that reduce stream integrity and change runoff patterns.
- ➤ Restore stream ecosystems.
- ➤ Use native or natural controls to replace chemicals in agriculture.
- ➤ Give nature a rest from agriculture.
- ➤ Conserve natural habitats.
- ➤ Fight for sensible and sustainable development.
- ➤ Maintain and appreciate wildlife.
- ➤ Develop buffers and plant riparian zones.
- ➤ Promote graduate programs in Native colleges and educational institutions.
- ➤ Develop stronger Native American offices in federal science agencies.
- ➤ Accommodate flood zones and wetlands.

ALASKA: TESTIMONY FROM THE FRONT LINES
Mike Williams

Vice Chairman, Alaska Inter-Tribal Council

Editors' note: This chapter combines testimony by Mike Williams (Yupiaq) before a U.S. House of Representatives Select Committee on Energy Independence and Global Warming hearing on Energy and Global Warm-

Mike Williams

ing Solutions for Vulnerable Communities (October 18, 2007) with his testimony to a U.S. Senate Indian Affairs Committee briefing on Indian Tribes and Climate Change Legislation for Alaska (September 7, 2009).

My name is Mike Williams. I am a Yupiaq from Akiak, Alaska, located on the Kuskokwim River. Currently, I am vice chairman of the Akiak Native Community, a federally recognized tribe, and I also serve as vice chairman of the Alaska Inter-Tribal Council, which represents 229 tribes in Alaska. In addition, I am vice president for the National Congress of American Indians, Alaska Region, and a board member for the National Tribal Environmental Council.

Climate change is having substantial and adverse impacts on Alaska Natives. Our temperatures are increasing, our ice is melting, our animals are becoming diseased and dislocated, our oceans are acidifying, our sea levels are rising, and our villages are sinking. These impacts are affecting our daily existences and every other facet of life, such as traditional and customary hunting practices, travel, and consequently a viable future for our homelands. Our elders, in particular, are deeply concerned about what they are witnessing. In Alaska, unpredictable weather and ice conditions make travel and time-honored subsistence practices hazardous, endangering our lives.

Coastal Impacts

Sea level rise, greater storms, storm surges, flooding, and erosion—all impacted by climate change—are endangering my people, the Yupiaqs. Alaska Native villages are literally being swept away into the sea because of coastal erosion. According to the U.S. Corps of Engineers, at least three tribes must be moved in the next ten to fifteen years, Shishmaref, Kivalina, and Newtok, while according to a Government Accountability Office (GAO) Report, over 180 communities are at risk.

Two reports prepared for congressional requesters by the GAO indicate that 86 percent of Alaska Native villages are threatened by erosion and flooding due to warming temperatures, and thirty-one villages qualify for permanent relocation. Yet Alaska Native villages cannot access some federal program assistance due to prohibitive funding criteria. There is no overarching federal plan or lead federal agency to address the fact that many of the residents of these villages are becoming climate change refugees.

Everything is changing so quickly. Lakes are drying; new insects are appearing; permafrost is melting; berries are disappearing; storms are fiercer; animal populations are changing; our fish are rotting on drying racks; and polar bears are drowning. Because of massive, record-breaking forest fires, our youth and elders are having trouble breathing. Our ice is so much thinner or entirely gone. And our coastlines are eroding, washing away ancient artifacts from our ancestors as well as modern infrastructure.

Inland Impacts

In the summer of 2009, the interior of Alaska had the driest July in 104 years. Two point nine million acres of forest recently burned, and the salmon run was the weakest in recent memory. A warming climate contributed to increased forest fires in 2004 and 2005 as well, devastating more than 11 million acres of Alaska's interior forests.

Many lakes and ponds on the tundra are rapidly drying up as a result of warmer temperatures. Melting permafrost compounds climate change by further releasing additional CO_2 and methane into the air. The loss of permafrost also reduces habitat and increases energetic demands on migrating wildlife.

Warming events have altered the route and time of migration for the Porcupine Caribou Herd, thereby impacting the subsistence lifestyle of such peoples as the Gwich'in Athabascan. The forage habitat of caribou is shrinking with increased forest fires and shifting tundra. The increased frequency of freezing rain due to rising temperatures has resulted in a crust covering lichen, which has diminished the caribou's ability to forage for their primary food source.

Larger Patterns

Throughout the nation in Indian Country, traditional foods are declining, local landscapes are changing, rural infrastructure is being challenged, soils are drying, and lake and river levels are declining. Tribes are experiencing droughts, loss of forests, fishery problems, and increased health risks from heat strokes and from diseases that thrive in warmer temperatures.

If global warming is not addressed, the impacts on Alaska Natives and American Indians will be immense. Models and the best scientific data and traditional knowledge indicate that if we do not reduce greenhouse gas emissions, the entire Arctic ice cap will melt, endangering the culture and subsistence needs of America's Inuit people. Furthermore, flooding, sea level rise, storm surges, and greater storms will endanger my people, the Yupiaqs, as well as tribes in Florida and elsewhere.

Hotter temperatures threaten all American Indians, but especially in the southwest and Florida, where we often do not have adequate means of escaping the heat. Increased global warming will also endanger salmon in the Pacific Northwest, which are crucial to tribes there, as well as in Alaska. Finally, on almost all tribal lands, enhanced global warming will threaten our sacred waters, essential to our physical and cultural survival. Clearly, global warming presents one of the greatest threats to our future, and must be addressed by Congress as soon as possible.

Low-Carbon Opportunities and Initiatives

There are many economic opportunities for Alaska Natives and American Indians in a low-carbon future, especially with respect to renewable energy. Tribes offer some of the greatest resources for helping the nation with renewable energy development, particularly wind, solar power, biomass, and geothermal power.

In Alaska, for example, we are installing wind power in very remote communities, such as Tooksok Bay, St. Paul Island, and Kotzebue. Wind power has also been installed on the Rosebud Sioux Indian Reservation. Port Graham Village is assessing construction of a biomass facility using forestry waste. The Confederated Tribes of the Warm Springs Reservation has analyzed the viability of a commercial geothermal power plant. Also, Native Sun Solar, which provides installation, maintenance, and technical support for photovoltaic systems, has installed hundreds of systems on the Navajo and Hopi reservations.

To achieve Indian Country's and Alaska's renewable energy potential, however, we need investment capital, infrastructure, and technical capacity. Any renewable energy program must include opportunities and incentives for tribes. Also, with training, American Indian and Alaska Native youth and adults can be actively engaged in renewable energy jobs, from engineering, to manufacturing, to installation.

There are also economic opportunities associated with energy conservation. We would welcome tribal-based initiatives to better insulate our homes, to convert our lighting, and to educate our members regarding energy efficiency practices. We want jobs that save us money and reduce our carbon footprint. In general, we believe that a low-carbon economy will provide multiple local benefits by decreasing air pollution, creating jobs, reducing energy use, and saving money.

Actions Needed

In recognition of the tremendously serious impacts that global warming poses to American Indians and Alaska Natives, our most important organizations have passed urgent resolutions outlining problems, threats, and needed action by Congress. These organizations include the Alaska Inter-Tribal Council, the Alaska Federation of Natives, RuralCAP, and the National Congress of American Indians, as well as 150 resolutions from Alaska Native entities (tribes, Native corporations, and regional non-profits).

With respect to adaptation, communities like Newtok, Alaska, are already taking action to move from dangerous sites to higher ground. It is important for Congress to recognize that the adaptation needs are very great. We require planning assistance, federal coordination, and significant financial resources to execute these

Mike Williams mushing in Alaska.

crucial relocations and to fund other adaptation needs. In all instances, it is important that our traditional knowledge be incorporated and respected, that we be consulted, and that our values and needs be honored. Alaska Native villages and Indian tribes as a whole have borne the disproportionate and negative impacts of climate change. I would implore you to consider the circumstances unique to our villages by mandating in climate change legislation that federal agencies develop, fund, and implement a strategic plan that addresses the climate change impacts on Alaska Native villages.

This plan would need to be developed in consultation with Alaska Native villages with their free, prior, and informed consent and include the prioritization and coordination of assistance to Alaska Native villages; the permanent relocation of qualified Alaska Native villages in a manner that obtains their free, prior, and informed consent in the planning and implementation of such relocations (and the removal of barriers for accessing federal funds for such efforts), and also include the mitigation of climate change impacts upon the traditional and subsistence practices of Alaska Natives.

In addition, Alaska Native villages should be provided with a one-percent apportionment of the allowances made available under the bill for domestic adaptation purposes, preferably in the form of monies, to help these villages prepare to adapt to the impacts of climate change facing them on a daily basis. Monies are important in that few if any Alaska Native villages have the capacity and/or expertise to manage the conversion of allowances to monies.

Conclusion

Much of Indian Country and Alaska's 229 Native villages are being seriously threatened by climate change and its impacts. It is therefore imperative that Congress take action to protect the nation's many tribal communities against such impacts and help protect and preserve our planet for the current generation and those to come. In closing, I'd like to share my own personal experience with climate change and its impact to one of my great loves. Alaska Natives have used dogs for transportation for thousands of years. Since 1992, I have had the good fortune of participating in the Iditarod, widely referred to as the "Last Great Race." As a participant, I have seen the race change in a number of ways, most notably the lack of consistent snow cover. Because of an absence of snow in recent years, the race has been forced to move on a number of occasions from more southerly Wasilla to more northerly Willow and even Fairbanks. To keep the dogs cool, since the days are too warm, we have to mush mostly by night now. And, we also mush more on land and less on the frozen rivers because of thawing.

There is so much as stake. I implore you to take meaningful action to address climate change now and to help assure that the traditions of Alaska Native villages and Indian tribes, which have withstood the test of time, continue for generations into the future.

SHARING ONE SKIN
Jeannette Armstrong

Jeannette Armstrong (Okanagan Syilx) is executive director of the En'owkin Centre, Penticton, British Columbia, and assistant professor in Indigenous Studies at the University of British Columbia–Okanagan. She has an interdisciplinary Ph.D. in Environmental Ethics and Syilx Indigenous Literatures from the University of Griefswald, Germany.

Editors' note: This article was adapted from *Paradigm Wars: Indigenous Peoples' Resistance to Economic Globalization*, edited by Jerry Mander and Victoria Tauli-Corpuz and published by the International Forum on Globalization in 2006.

I am from the Okanagan, a part of British Columbia that is very dry and hot. Around my birthplace are two rocky mountain ranges: the Cascades on one side and the Selkirks on the other. The main river that flows through our lands is the Columbia.

My mother is a river Indian. The Kettle River people are in charge of the fisheries in the northern parts of the Columbia River system in our territories.

My father's people are mountain people. They occupied the northern part of British Columbia, known as the Okanagan Valley. My father's people were hunters. My name is passed on from my father's side of the family and is my great-grandmother's name.

I am associated with my father's side, but I have a right and a responsibility to the river through my mother's birth and my family education.

So that is who I am.

When I introduce myself to my own people in my own language, I describe these things because it tells them what my responsibilities are and what my goal is, what I need to carry with me, what I project, what I teach and what I think about, what I must do and what I can't do.

The way we talk about ourselves as Okanagan people is difficult to replicate in English. When we say the Okanagan word for ourselves, we are actually saying "the ones who are dream and land together." That is our original identity. Before anything else, we are the living, dreaming Earth pieces. Dream is the closest word that approximates the Okanagan. But our word doesn't precisely mean dream. It actually means "the unseen part of our existence as human beings." It may be the mind or the spirit or the intellect. We are mind as well as matter. We are dream, memory, and imagination.

Another part of the word means that if you take a number of strands of hair, or twine, place them together, and then rub your hands and bind them together, they become one strand. You use this thought symbolically when you make twine, thread, and coiled baskets. This part of the word refers to us being tied into and part of everything else. It refers to the dream parts of ourselves forming our community.

I explain this to try to bring our whole society closer to that kind of understanding, because without that deep connection to the environment, to the Earth, to what we actually are, to what humanity is, we lose our place, and confusion and chaos enter.

When we Okanagans speak of ourselves as individuals, we speak of four main capacities that operate together: the physical self, the emotional self, the thinking-intellectual self, and the spiritual self. The four selves join us to the rest of creation.

Okanagans teach that the body is Earth itself. Our flesh, blood, and bones are Earth-body; in all cycles in which Earth moves, so does our body. We are everything that surrounds us, including the vast forces we only glimpse. If we cannot continue as an individual life form, we dissipate back into the larger self.

As Okanagans we say the body is sacred. It is the core of our being, which permits the rest of the self to be. It is the great gift of our existence. Our word for body literally means "the land-dreaming capacity."

The emotional self is that which connects to other parts of our larger selves around us. We use a word that translates as heart. It is a capacity to form bonds with particular aspects of our surroundings. We say that we as people stay connected to each other, our land, and all things by our hearts.

The thinking-intellectual self has another name in Okanagan. Our word for thinking/logic and storage of information (memory) is difficult to translate into English. The words that come closest in my interpretation mean "the spark that ignites." We use the term that translates as "directed by the ignited spark" to refer to analytical thought. In the Okanagan language, this means the other capacities we engage in when we take action are directed by the spark of memory once it is ignited.

We know that in our traditional Okanagan methods of education, we must be disciplined to work in concert with the other selves to engage ourselves beyond our automatic-response capacity. We know too that unless we always join this thinking capacity to the heart-self, its power can be a destructive force both to ourselves and to the larger selves that surround us. A fire that is not controlled can destroy.

The Okanagan teach that each person is born into a family and a community. No person is born isolated from those two things. As an Okanagan you are automatically a part of the community. You belong. All within family and community are affected by the actions of any one individual. The capacity to bond is critical to individual wellness. Without it the person is said to be "crippled/incapacitated" and "lifeless." Not to have community or family is to be scattered or falling apart.

The Okanagan refer to relationship to others by a word that means "our one skin." This means that we share more than a place; we share a physical tie that is uniquely human. It also means that the bond of community and family includes the history of the many who came before us and the many ahead of us who share our flesh. We are tied together by those who brought us here and gave us blood and gave us place. Our most serious teaching is that community comes first in our choices, then family, and then ourselves as individuals, because without community and family we are truly not human.

Language of the Land

The Okanagan word for "our place on the land" and "our language" is the same. We think of our language as the language of the land. The way we survived is to speak the language that the land offered us as its teachings. To know all the plants, animals, seasons, and geography is to construct language for them.

We also refer to the land and our bodies with the same root syllable. The soil, the water, the air, and all the other life forms contributed parts to be our flesh. We are our land/place. Not to know and to celebrate this is to be without language and without land. It is to be displaced.

As Okanagan, our most essential responsibility is to bond our whole individual and communal selves to the land. Many of our ceremonies have been constructed for this. We join with the larger self and with the land, and rejoice in all that we are.

The discord that we see around us, to my view from inside my Okanagan community, is at a level that is not endurable. A suicidal coldness is seeping into and permeating all levels of interaction. I am not implying that we no longer suffer for each other but rather that such suffering is felt deeply and continuously and cannot be withstood, so feeling must be shut off.

I think of the Okanagan word used by my father to describe this condition, and I understand it better. An interpretation in English might be "people without hearts."

Okanagans say that "heart" is where community and land come into our beings and become part of us because they are as essential to our survival as our own skin.

When the phrase "people without hearts" is used, it refers to collective disharmony and alienation from land. It refers to those who are blind to self-destruction, whose emotion is narrowly focused on their individual sense of well-being without regard to the well-being of others in the collective.

The results of this dispassion are now being displayed as nation-states continuously reconfigure economic boundaries into a world economic disorder to cater to big business. This is causing a tidal flow of refugees from environmental and social disasters, compounded by disease and famine as people are displaced in the expanding worldwide chaos. War itself becomes continuous as dispossession, privatization of lands, exploitation of resources, and a cheap labor force become the mission of "peace-keeping." The goal of finding new markets is the justification for the westernization of "undeveloped" cultures.

Indigenous people, not long removed from our cooperative, self-sustaining lifestyles on our lands, do not survive well in this atmosphere of aggression and dispassion. I know that we experience it as a destructive force, because I personally experience it so. Without being whole in our community, on our land, with the protection it has as a reservation, I could not survive.

The Way of Creating Compassion for...

The customs of extended families in community are carried out through communing rather than communicating. Communing signifies sharing and bonding. Communicating signifies the transfer and exchange of information. The Okanagan word close in meaning to communing is "the way of creating compassion for." We use it to mean the physical acts we perform to create the internal capacity to bond.

In a healthy whole community, the people interact with each other in shared emotional response. They move together emotionally to respond to crisis or celebration. They "commune" in the everyday act of living.

Being a part of such a communing is to be fully alive. To be without community in this way is to be alive only in the flesh, to be alone, to be lost to being human. It is then possible to violate and destroy others and their property without remorse.

With these things in mind, I see how a market economy subverts community to where whole cities are made up of total strangers on the move from one job to another. This is unimaginable to us.

I do see that having to move continuously just to live is painful and that close emotional ties are best avoided in such an economy. I do not see how one remains human, for community to me is feeling the warm security of familiar people like a blanket wrapped around you, keeping out the frost. The word we use to mean community loosely translates to "having one covering," as in a blanket.

I see how family is subverted by the scattering of members over the face of the globe. I cannot imagine how this could be family, and I ask what replaces it if the generations do not anchor to each other. I see that my being is present in this generation and in our future ones, just as the generations of the past speak to me through stories. I know that community is made up of extended families moving together over the landscape of time, through generations converging and dividing like a cell while remaining essentially the same as community. I see that in sustainable societies, extended family and community are inseparable.

The Okanagan word we have for extended family is translated as "sharing one skin." The concept refers to blood ties within community and the instinct to protect our individual selves extended to all who share the same skin. I know how powerful the solidarity is of peoples bound together by land, blood, and love. This is the largest threat to those interests wanting to secure control of lands and resources that have been passed on in a healthy condition from generation to generation of families.

Land bonding is not possible in the kind of economy surrounding us, because land must be seen as real estate to be "used" and parted with if necessary. I see the separation is accelerated by the concept that "wilderness" needs to be tamed by "development" and that this is used to justify displacement of peoples and unwanted species.

I know what it feels like to be an endangered species on my land, to see the land dying with us. It is my body that is being torn, deforested, and poisoned by "development." Every fish, plant, insect, bird, and animal that disappears is part of me dying. I know all their names, and I touch them with my spirit. I feel it every day, as my grandmother and my father did.

I am pessimistic about changes happening, but I have learned that crisis can help build community so that it can face the crisis itself.

I do know that people must come to community on the land. The transiency of people crisscrossing the land must halt, and people must commune together on the land to protect it and all our future generations. Self-sustaining indigenous peoples still on the land are already doing this. They present an opportunity to relearn and reinstitute the rights we all have as humans.

Indigenous rights must be protected, for we are the protectors of Earth. I know that being Okanagan helps me have the capacity to bond with everything and every person I encounter. I try always to personalize everything. I try not to be "objective" about anything. I fear those who are unemotional, and I solicit emotional response whenever I can. I do not stand silently by. I stand with you against the disorder.

WHERE WORDS TOUCH THE EARTH

Tribal Students Produce a Climate Change Video on the Coast Salish Moons

Greg Mahle and Lexie Tom

My name is Greg Mahle. I am an enrolled member of the Upper Skagit Indian Community located in Sedro-Woolley, Washington. I am a graduate of Northwest Indian College at the Coast Salish Institute as a culture curriculum developer. The mission of the Coast Salish Institute is to revitalize the Coast Salish language and cultures. We do this by teaming up our young staff with elders from the Coast Salish territory so the young staff members learn the stories of our ancestors and their ways of life. Our goals are to bring those ways of life into modern times and attempt to bridge the gap between traditional and contemporary society.

My name is Lexie Tom and I am from the Lummi Tribe. I graduated from Northwest Indian College (NWIC) in 2007 and am currently working toward my degree at Western Washington University. I am also an employee at the Coast Salish Institute, which focuses on revitalizing Coast Salish language and cul-

ture. I truly believe the work we do is very important to the survival of our community.

Recently we were given the great opportunity to work with the Where Words Touch the Earth project, coordinated by Dr. David Adamec of the National Aeronautics and Space Administration (NASA). In spring 2009, NASA approached six tribal colleges (including NWIC) about the project, whose goal was to create twelve- to fifteen-minute documentaries that would shed light on the Native American view of the causes and impacts of climate change.

Director Sharon Kinley of the Coast Salish Institute agreed to the project and asked us to be a part of creating the video. At first it felt as if this project would be similar to our previous cultural videos, which were based on our tribes' traditional stories and knowledge. Dr. Adamec helped us realize the global reach of this project when he came to meet with us at NWIC. After an incredible amount of work and dedication, our video segments were posted on the Teacher's Domain digital media website in May 2010: http://www.teachersdomain.org/special/nasawords

In the Beginning

We made a decision to focus the video on the moons (Coast Salish months), with a major emphasis on the salmon. Traditionally, salmon was our staple food, and we as Indigenous people knew that for us to survive we must respect and appreciate the life of the salmon, for it is the salmon who feeds us. We took a historical approach to the project by researching what it was our ancestors saw and did during each moon. We contrasted the seasonal cycles of the moons to the natural resource cycles we see in our region today, under conditions of climate change. We decided to start our video production in the Deep Snow Moon, because it was the time when the world was coming out of hibernation (see facing page).

We began the project by training students how to use the equipment. We took random shots of the environment to allow them to practice, after which we would play the video on the television to show the students the aspects that went well and the aspects that they needed to improve on. We would then return to the place where we took the original shots and take a new set. Those shots would also be compared to previous shots so students could learn the difference.

As a group we created a list of elders who held a wealth of traditional knowledge whom we could interview. We developed the interview questions and trained students in proper interview techniques. During the interviews, the elders told us many of the locations where the seasonal activities occurred, what they hunted, what they fished, what they gathered, and how they knew it was time to begin these activities.

We created our location shot list after conducting the elder interviews. Throughout the interviews the elders talked about a common theme—salmon. The elders talked about their experiences with salmon fishing and how extreme the changes have been over the years. Several elders talked about the effects of the changing earth on the return of the salmon, and others talked about how the warming of the water (caused by the change in climate) not only allowed a harmful fungus to grow on fish but also forced the fish into colder, deeper water.

In order to enhance the visual images, we incorporated traditional music and the Lummi language in our video. We met with Coast Salish artist Alfred Charles, Jr., to describe our needs, and he produced a beautiful Coast Salish spindle whorl design for us, incorporating images of the brant and salmon. We appreciate his work. For the traditional music we were able to record the George Family (Lummi), Leonard Dixon (Lummi), and Edmond "Budz" Mathias (Upper Skagit). The traditional Lummi language was courtesy of Lucas Washington and the voice of Lexie Tom.

Editing the Video

After all of our location shots and elder interviews were complete, we began the time-consuming and detailed process of importing the video from the digital videotapes onto our Apple G-5 editing machine. A good rule of thumb for creating a video is that one should have one hour of footage for every minute of video one intends to produce. This turned out to be a gross underestimate in our case. In the end we had compiled over sixty-seven hours of raw footage for a fifteen-minute video. We also had completed a total of twenty-two interviews with eleven different elders.

Editing was perhaps the most difficult piece of the project. With so many hours of footage and the significant interviews, we had to decide not only what to put into the video but what to leave out. Of the twenty-two hours of interview footage, we were only able to utilize eight minutes for our video.

The Final Product

The agreement with NASA was to finish the video by March 2010. We were informed later that in September 2009 NASA had scheduled a climate change conference for November to be held at the Shakopee Mdewakanton Community near Minneapolis, and they would be grateful if we could show our progress. Beginning early September, we worked overtime on the project trying to get it substantially completed. Fred Lane (Lummi), who worked with us on the project, and Greg Mahle took the final product to the conference. We met great people who shared a common understanding that climate change is actually occurring and has real impacts on our environment. We showed our video to a full audience and got a very positive reception for the work.

The Experience

The experience that we had while working on the climate change video created great and everlasting memories. With those memories comes a strong feeling of responsibility to our ancestors and children. This is a responsibility to understand the ways of our old people and to pass the traditional knowledge on to the children so that they too can feel and understand their ties to this land, their ties to the salmon, and their responsibility to pass the knowledge of the moons to future generations.

We would like to thank the Northwest Indian College, the Coast Salish Institute, and NASA for allowing us the great opportunity to work on the project. We would like to thank all the elders—for without their stories, our message would not be heard. We would also like to thank the traditional musicians, artists, and language speakers. Lastly, and perhaps most importantly, we would like to thank our ancestors for leaving us the tools and knowledge to continue on as a people. *Hyshqe.*

The Moons

Naka... Slhqalch

*In the Deep Snow Moon the world
is beginning to warm up.*

*The men are finished carving their fishing canoes
that will be needed later on in the year.*

We, the Lummi people, have names in our language for the different seasons. These names come from natural events that occur on our land. *Naka Slhqalch* is the Deep Snow Moon (which comes about the time of February). All winter long the men in the Lummi village were carving dugout-style canoes that would be used in the warmer months.

Wekes... Slhqalch

*In the Little Frogs Moon, Lummi is coming
out of hibernation after a long winter.*

The little frogs begin to sing to the people.

Paddles are being carved, fishing season is almost here.

Nature is the place the Lummis turned to for our calendar. We knew it was time to hunt when the depth of the snow dropped on the mountains. We knew it was time to fish when the salmon berries started to bloom. In the Little Frogs Moon, the Lummi people were preparing for warmer weather. Their whole year was about survival, and the salmon was their lifeline. The Lummi people followed a circular cycle of life.

Qwechwayechten... Slhqalch

When the Onion Moon comes about, the sap
begins to run on the sacred cedar tree.

The people will make baskets and clothing
from this bark in the winter to come.

The Onion Moon is a busy time. The sap is running on the cedar trees, telling the people it is time to strip the bark. Cedar has always been the most important tree to the Lummis. We used cedar for everything we needed— such as canoes, baskets that were used for cooking and storing, houses, clothing, and many other things. Cedar trees were to be treated with respect, and when the weaver stripped the bark, the weaver was taught to strip it in such a way that the tree can grow it back. Preservation was very important, because if we strained our resources of cedar, then what would we have left? Our whole lives were made from cedar. "It's a very important tree," as our elders would say. Once harvested, the bark was dried and stored for the winter to come.

St'et'ale... Slhqalch

In the Little Fawns Moon

the women are finished weaving the fishing nets.

The salmon berries are beginning to bloom.

The camas are blossoming.

Nature has signaled us that the salmon
people are preparing to return.

Nature always had a way of telling the Lummi people of events to come. Camas bulbs were an important resource for us as well. The blooming of the beautiful blue flowers in the Little Fawns Moon signaled the people that spring was in the air. Some say that long ago there were huge camas fields stretching for miles. Today, there are some wild camas beds on the San Juan Islands, but nowhere else in the area.

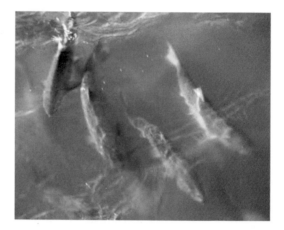

Yomach... Slhqalch

In the Spring Salmon Moon

all of our hard work is about to be completed.

Families start preparing for the first salmon ceremony.

Our elder brother is returning home.

Villagers are gathering up all of their belongings

the huge planks from their longhouses, and

move their families to the San Juan
Islands for the fishing season.

The Lummi people lived in longhouses that were made of flat removable planks. They could strip their houses of these planks, leaving only the house posts, and take them to their temporary fishing ground villages. This is what they did in the Spring Salmon Moon, when they moved to the San Juan Islands, which were the people's ancestral fishing grounds. The Creator gave these specific fishing grounds to us so that we would never go hungry. These fishing sites were ours since the beginning of time, always to be protected. The people prepared for the first salmon ceremony. The first salmon ceremony was a time of celebration, when the Lummi people could honor the salmon people with song and dance, in the hopes of having continuous returns in the years to come.

Chen seqi... Slhqalch

The people fished for five months out of the year.

In the Sockeye Moon the people fished sockeye salmon,

gathered many plants, which they ate
fresh... and dried for winter.

Very rarely was there ever a lack of food in the villages.

So long as the Lummi people

respected the salmon people,

the salmon people then returned.

Salmon is the most important resource for us traditionally. Salmon made up about 90 percent of our diet in precontact times. The Lummis knew where their customary fishing grounds were, and they spent the whole year carving fishing canoes, weaving fishing nets out of nettles, and weaving storage baskets out of cedar. When the time came in Sockeye Moon, they set out to feed their families. They never caught more salmon than each family would need to get through the year, but they had to make sure they caught enough for their families to store for the long winter.

Chenq'echs... Slhqalch

*The Silver Salmon Moon marks the
return of the silver salmon*

and the peak of the salmon fishing season.

The Lummi people are already preparing for the winter.

Families move back to their permanent villages and begin hunting deer.

When the Lummi people moved back to their permanent villages, they mainly fished in the rivers. The salmon were going up river to spawn in plentiful numbers. The Lummi people were starting to prepare for the long winter months by storing what food they needed to survive during those months when food was scarce. They caught enough fish in the Silver Salmon Moon and gathered enough berries and bulbs over the spring and summer to dry and store in their longhouses.

Kwel exw... Slhqalch

The Dog Salmon Moon marks

the last of the salmon people traveling up river.

It is time for them to spawn,

replenishing the population for the seasons to come.

The dog salmon are the last to spawn each year. When these fish started to go up river in the Dog Salmon Moon, the people knew it was almost time to retire their nets and fishing boats until the first salmon ceremony of the following year. These signs that nature gives to the people are considered sacred. They are not to be taken for granted.

The Lummi people had a life cycle that they followed very closely. At the turn of the twentieth century when their world was changing drastically, they knew what was important to them. And this was the salmon.

We still think the salmon are most important. At the turn of the twenty-first century, we believe that the balance of our life cycle has been interrupted to the point of almost no return.

WATCHING FOR THE SIGNS
Chief Willie Charlie

Vice President, Union of British Columbia Indian Chiefs (UBCIC); Chief of Chehalis (Sts'Ailes) First Nation, along the Harrison River, east of Vancouver, British Columbia.

Editors' note: This interview was conducted by Zoltán Grossman on May 17, 2010.

We've been noticing the huge impacts of climate change in intricate ways. The different things we are noticing around us include the snow-capped mountains. At one time our elders would predict the following winter according to the snow packs. There was one mountain close to us; its traditional name in Hul'qumi'num Salish is Theeth-uhl-kay. Everyone knows it today as Mount Cheam. My grandfather used to predict the following winter according to that mountain. It would never go bare, but today quite frequently it does go bare.

We're so dependent on fish. We're the river people, and to us, food is fish. In these past few years, multiple impacts have diminished the returns of the salmon to us. This includes overfishing on the ocean and at the entrance of the mouth of the Fraser River. With everything that's dumped into the river, it's so polluted now, and the fish have to swim through that. The water has become so warm and shallow now because of climate change. The fish are strained from swimming in polluted, warm water, so they can't make it to their destination, or if they do they're stressed.

It's the same with overfishing, on the sandbars and the riverbanks that are lined up with sportfishermen who supposedly catch-and-release. What strain and stress is that putting on the fish? It's not just one thing—not just the snow pack diminishing, not just the water warming up and being polluted, not just the overfishing—it's all of them together.

Seasonal People

Then there is the timing of the animals, birds, and plants. The times of the year that you would normally

Theeth-uhl-kay (Mount Cheam).

get some of the plants are all changed. So we used to go picking cedar or "swamp tea" or different medicines, but those times are changing. We normally would start going out for the fish in late December to January, and it would be at the end of January we would get the early-time chinooks. Now we're not getting those chinooks until April or May. My grandfather used to be able to tell what fish were in the water just by the plants and by the animals that were around. Now some of the animals are not even returning or they're showing up later. So climate change has had some impact for us.

Not knowing the connections of all these things, someone could quite easily say, "If it's not there, it's not there," or "if it's not there yet, just wait." But we're connected to all living things, and we have ceremonies and ways of life that revolve around the different animals and different plants and different fish that return to us. We're very, very seasonal people. Things trigger off where we go to from one season to the next.

In the early time, when the chinooks arrive, it's a time to get ready to harvest. You go out and get your fish, and that's when the plants are coming back, so your winter ceremonial season comes to an end, and it's time to gather things. Once the gathering season is completed, you're waiting for the next run of the sockeye, and the sockeye tells you it's the summertime. That's the time you're out on the water and you're preserving and salting your fish. Then when the weather changes again for the fall time, you get your first snow on the mountain, it's a time to trigger the hunters. Once it gets cold enough, the hunters go up and gather their meat and put the meat away and preserve it for the winter. Once you're finished gathering your meat, then you're into ceremonial season again.

So we're still today very seasonal people, we still do those activities. We still gather for ceremonies, we still gather for that first-time fish, we still fish and preserve sockeye in the summer, and we still are very much into hunting. You can almost tell the seasons by the people who do those things. So at fishing time you can go around the community and feel the buzz of all the fishermen. At hunting time you can go around the community and feel the buzz of all the hunters. The people are anxious to eat that fish, people are anxious to have that first deer. We are still very seasonal people, so if something is not there, then what does that do for everything that is connected, what does that do for us?

Taking Care of All Living Things

We knew that there were things that were going to be impacted by humans, by man. We knew the impacts were going to be great, the impacts that we as human beings have on Mother Earth. A lot of the stories talk about the evolution, talk about the beginning of time, and bring us through all the different eras. In this history, at one time human beings were very powerful. We had a lot of abilities and a lot of gifts, with not just our mind but physically, spiritually, and emotionally. We could do things that we can't do today—like shape-shifting, like out-of-body travel experience, like moving things. We have stories that talk about those times. The way that my uncle described it, he said that we've lost all of those things because of our humanness. People started using these things for the wrong reasons, for the wrong purpose, and therefore we lost them. We understood it would be the same thing with Mother Earth, that [if] we keep taking and taking and not paying attention that there is going to be some impact.

We were supposed to take care of all living things because there was an agreement in time, in history. All living things were the same, and it was through time and evolution that they took different shapes and different forms. But there was an agreement at one point in time that all our relations, all living things—the winged, the four-legged, the plant people, the ones that swim in the rivers and the oceans—they all agreed that they would give themselves to us as humans. All that they asked for in return was to be respected, to be remembered, and to only take what you need. If you take more than what you need, you're supposed to share with those that are less fortunate. There are other parts to the agreement, where we're not supposed to harvest at certain times of the year out of respect for them, so that they can replenish or reproduce. But today it seems like we've forgotten some of those agreements with living things, and we either exploit, overhunt or overfish, or overgather berries and medicines. And we've forgotten how to share with those that are less fortunate.

I think we're fortunate enough still today that we have many knowledgeable elders that understand our connection to the land. They understand the spiritual connection, the emotional, physical, and mental connection to all living things. I think that the traditional knowledge, along with the scientific knowledge, the educational knowledge of today, need to be somehow combined and have respect for each other and work together. But you can't have the impact you need without buy-in from all the different interest areas, starting with First Nations, other governments, and industry. I think they need to have a roundtable, a sit-down with all those interests, to better understand the impacts on Mother Earth, from a scientific point of view and also understanding some of the traditional knowledge of First Nations people.

Listening to Traditional Knowledge

For so long, we had no voice, or the voice was taken away and it wasn't heard. Slowly we're getting the attention here from the government. But in the past they would always use that Western knowledge against you, like "prove it!" So we would tell them about the impacts on the land or the fish or the animals, and people would come and say, "Well, prove it, show us the report." Now, by working with educational institutions, we're able to do some of that and have somewhat of a voice, or least a little louder of a voice.

We have examples of situations here, in our workings with the government agencies, where we told them the impacts, and they carried on with the mitigation. At the end, they said "Well, it's everything that you told us." An example of that was three years ago, when we had a huge landslide at Chehalis Lake. A 2,000-meter-high mountainside came down with enough debris to fill up five football stadiums. It created a tsunami, a big tidal wave up the lake. By the time we heard about it, the government was up there with TV crews, talking about the little First Nation that was endangered. But at no point did anyone from the government call us. The airplane pilot that saw the landslide took it upon himself to phone us, so we hired our own helicopter and went up there. Immediately upon returning to our office, we phoned the government officials and asked, "Why didn't anyone phone us? We're the ones who are here."

We made enough noise that Ministry of the Environment and the Ministry of the Forest said the cleanup would be a tripartite, coordinated effort, which would include us in the management, the planning, and the remediation work. They were trying to understand the full impacts of the landslide. The 30- to 50-foot tidal wave washed back and forth—like how in a tub the water would swish back and forth. It washed away the trees, soil, and vegetation 30 feet up the bank, right into the bedrock. There was so much debris on the lake; trees were 30 feet high, stuck on other trees. It was an amazing impact; half the lake was debris of toothpicks and shattered splinters of wood. The fear was if all that came down the river, it would dam the river up.

Once we told our community members here, I spoke with my dad and a number of my uncles, and they said, "This is what you need to do. This is how you clean the lake up, this is where the danger areas are going to be, and this is how you get by those danger areas." So a few days later, I flew up to the site with the government. They flew in a hydrologist, they flew in a geologist, and we went up with the coordinator from the Ministry of the Environment. I was asking how much it cost them to rent the helicopter for the day, what it cost for the professionals to be there, and what their assessments were. After returning, we had a debrief for a half-hour, and the hydrologist and geologist gave their opinion on what to do. After they said what they did, I just smiled at them, and said, "Is that it? Because that was so common sense to the Indian. Everything that you said now is exactly what our elders said—without even visiting the site." The thing is, our guys were up there hunting and fishing for years, or logging for years. So they know the land, they know how the wind goes.

I went up about two weeks after that, and the debris had blown from the bottom end to the top end of the lake. I was in awe, and so were some of the government people. So I went back and told my dad, "It was so weird about the debris," and he said, "Oh yeah, it does that." That's the traditional knowledge. One of the government officials thought I was just being sarcastic but called me up two days later to ask me what I meant about the elders. It's those stories that acknowledge our intimate knowledge of the land that our people have from living there for hundreds or thousands of years.

Mother Earth Is Crying

My grandfather and my uncles say, "The laws of life are real simple. It's our humanness that complicates life." If we go back to respecting all our relations, having respect for all living things, if we can go back to the collaborative mindset, I think we can change things around before they get too bad or go too far. (Or get worse than they are now, because things are pretty bad now.) My grandmother used to say twenty to thirty years ago, "Mother Earth is crying, and we need to pay attention to what she's saying." While we were watching the TV news, she would talk about the signs, the different things that were happening in the world. Everything happens for a reason, but somehow we have to watch for the signs.

DIFFERENT WAYS OF LOOKING AT THINGS

Larry Merculieff

Ilarion (Larry) Merculieff was born and raised on St. Paul Island, one of the Pribilof Islands off southwestern Alaska in the Bering Sea. His ancient Aleut name, Kuuyux, which means "extension" (like an arm extending out), was passed to him by an Aleut spiritual leader when he was four years old.

Editors' note: This interview was conducted by Zoltán Grossman during an American Indian and Alaska Native Climate Change Working Group meeting at the Northwest Indian College in the Lummi Nation (Washington) on November 8, 2010.

Where I come from, the elders are always saying look at the root causes for anything. It's fine to see the symptoms and deal with the symptoms, but you must keep in mind you are going to address the root causes for any issue. My elders say it's reversed in Western society, where they look at the symptoms and not the root causes at all.

For example, on St. Paul Island, we're called Qawalangin Unangan, which means "People of the Sea Lion." Sea lion are to us what the bison are to the Plains Indians, or like the bowhead whale are to the Iñupiat. In the past thirty years the sea lions declined in population by 80 percent. What that did effectively was to sever the link between the young men and the older, experienced hunters, who didn't go out hunting anymore because there were no sea lions. Aleut hunters, like all other indigenous hunters, are self-regulating. It left the young men out by themselves. In one generation of that happening—in the mid-eighties to mid-nineties—we lost 70 percent of our young men, due to accidental death due to alcohol, suicide, murder, and incarceration for felony crimes. That never happened to our young men before, but nobody from the outside would study what occurred, so there's no documentation. This is some indication again of where science priorities are; they're collecting data to see what the climate is doing rather than also assessing the human impacts on St. Paul Island.

Our people were the first Indigenous people (I don't know about North America, but certainly in Alaska)—who, in the 1970s, first noted ecosystem-wide changes that were anomalous, that indicated a trend. We knew that something was going wrong in the environment and it was ecosystem-wide in the Bering Sea. Specifically what we noted were adult birds with chest bones sticking out, chest muscles caved in, chicks falling off of cliff ledges in large numbers, and sea lions chasing after seal pups with greater frequency than ever in living memory. The first seal pelts were so thin that we could see light through them after we took the fat off of them. We'd never see that before.

Disconnection from Self

So the elders were saying something big is happening, not just around the islands but around the whole Bering Sea, and maybe bigger. They understand that *everything is connected*. This is according to my Aleut elders in the Pribilofs and the elders everywhere throughout Alaska, and elders throughout the world that I have come into contact with. They include Mapuche elders in southern Argentina, the Mayan elders (including Don Alejandro, the keeper of the day calendar and head of the Grand Mayan Council), Lakota elder Chief Arvol Looking Horse, the Cree elders up in Canada, the Iroquois or Haudenosaunee elders, the Yupik elders in Alaska.

All of them are saying basically the same thing. They say the root cause is the disconnection that we have *from ourselves*. When we extrapolate that out, the elders say nothing is created outside until it is created inside first. When we disconnect from ourselves, it's easy to disconnect from anyone on the outside. We're trashing the environment outside because we're trashing the environment inside. We're in conflict outside because we're in conflict inside. Nothing is created outside until it's created inside first. So this dilemma that we're finding in the human contribution to climate change ultimately is caused by a profound disconnection with self. And no one is looking at that issue. And when you're looking at what is being proposed for solutions, they're all external. You know: "Let's use technology, let's use hybrid cars, let's do energy efficiency," and all these kinds of things.

Al Gore had asked me to be part of his team, after *An Inconvenient Truth* came out. I said sure, as long as I can bring in the Indigenous view. And the response was "no, you've got to use our PowerPoint." Right? I had to say I disagreed with this PowerPoint, because the movie has a profound subscript to it, which no one addresses. And that is: "Let's keep our current lifestyle." This whole thing about sustainability is "we don't want to sacrifice anything." Our elders are saying, no, we *have* to simplify. We have to, there is no choice, given that the climate change effects are getting worse and worse and worse. The only thing we can really do right now is to simplify as much as we can and become as self-sufficient as we can in the context of real community. No more importing foods from outside our region and consuming like there is no tomorrow.

The elders talk about how our people had extensive protocols and agreements with places as far away as South America, where each party to the agreement agreed to help each other in times of need and hardship. There were sophisticated trading routes throughout all of what they call Turtle Island. The elders in Alaska said that when we had really hard times, we had to depend on people in South America to bring up healing herbs and healing songs. They went to Canada to get food, for barter exchanges. They're saying we have to go back to those again.

But ultimately they said we have to go back to the Real Human Being—that is, getting back in touch with who we are. My generation was the last generation that grew up this way. I cannot describe to you how profound this connection is, and how we garner information from the environment to direct what we do, how we actually adapt. Everybody's talking about adapting, but no one has a clue about what it really means. So basically we're all fishing in the dark, trying to figure out "what the heck do we do?"

This often touted quote of Einstein, "You cannot solve the problem with the same consciousness that created the problem," shows you've got to be able to move on a conscious level to a different box, a different paradigm, a different worldview. But it's basically the worldview that we've always had. It's been the foundation of all cultures around the entire world, according to all the stories. The template of the original instructions are the same for every culture in the world, reflected differently only by their cultural patterns, but the underlying principles are exactly the same. The elders around the entire world are saying we must get back to these original teachings, if we're going to have any hope whatsoever. When you come from a mindset of an embodied connection with the Earth, you get information about what's happening and also get information to some degree about what we need to do or where we need to move. What is it that we need to adjust?

We also have contemporary leadership (and this is the hard part) that is all well-intended and focused on these issues, but it's mostly mental. The spiritual component, the heart component, is missing. The true solutions, the elders say, are not going to be coming out of any logical construct. It's going to be coming out of another place outside of logic and the rational.

Traditional Youth Camps

These are the ideas we're trying to reimmerse our young people in now. So on St. Paul Island, we started a Stewardship Camp many years ago. We started experiential learning processes, so the kids can get back in touch with the land, with the environment, with the animals. This is with the guidance of stories and storytelling with the elders present, bringing songs back to restore the intimate and profound connection that people had with the environment.

How do we instill these ideas in youth exposed to TV, gadgets, and commercial culture? You've got to take them out of it. Take them away from that situation—to start with for three to four weeks, but build on it quickly, because we don't have much time. I don't mean giving these students instructions. What the adults' responsibility is, based on all these traditions, is to allow these young people the space to really learn experientially, expose them to opportunities to learn so that what is learned is the young person's responsibility. This is a self-empowering way. Learning by rote and direction of authority is disempowering.

For the first two to three days, they're restless. They're used to their TV, their telephone, but they can't take any of those to camp. The addictive aspects fade away in nature, if you provide attention to the young people after about three days, on the average. You start to bring out their own innate intelligence, which comes into play. For example, one of the games we play seems to be a simple game, but it is profound. We find a sandy beach, and we ask the young to close their eyes and walk for ten minutes, then fifteen minutes, up to a half hour. These young people are not used to using their senses at all, except the visual. If they use their hearing, it's for music. But to be able to use all of them on a singular task, having their senses guide them in an environment, starts to put them back in touch. The innate characteristics of the human being start to come out. Just like a blind man or woman takes advantage of their loss of sight by expanding their senses in other areas. The Real Human Being always had that.

We help the students get back in touch, and pretty soon they get a kick out of it: "Look what I can do: I can walk down a quarter mile of beach and not open up my eyes." We do the same thing at night, when it's pitch black. We strategically locate the young people

Coastal erosion at Shishmaref, Alaska, forcing the Native village to move inland.

away from the camp, and we say everybody is going to be totally silent. You've got to find your way back to the camp. There will be no fire, there are no reference points, and you can't see lights in the distance. The kids would usually pair up, encouraging them to help each other, but at the same time to use their senses, and again they get a kick out if it. The elders would then tell stories around the meaning of what those skills are and how those skills were actually used.

Getting the student out of their normal milieu, to be able to function at a higher level of consciousness, is going to be very important in these coming times. It has to happen organically. If you say to students that you have to listen to the elders, it's like another command. The traditional way is you don't give commands; you simply create the opportunity for learning through exposure. So the elder sits down around the campfire, and they're cooking wieners or whatever, and slowly the elder starts talking about something to a young person. Another young person comes, then another, and the elder is still talking, and pretty soon they're all totally glued. They're hearing what that elder is saying. They may be scary stories, but even the scary stories are in the context of a relationship to self and a relation to others. So they learn without even knowing they're learning.

It's a fascinating process, but it's a natural one. The more you structure it, the more you take away from it. What's important about this for the young people is that all situations they are exposed to are self-empowering. The Western system disempowers young people: "You

sit down, you listen to me, I'm the authority, blah, blah, blah." And they get that their entire life. We come to a conference, and we have experts from certain agencies and scientific groups; they're the new authority. Subconsciously we just let go of our own self-authority. This self-authority is really important in terms of critical thinking, being able to think creatively outside the box. So the students are not given any instructions, and they're never scolded or disciplined at all. When that happens, the unbridled exuberance of young people comes out. After a few days, they're just totally engaged. The light is back in their eyes, they are in just observing one starfish. But if you said, "Today we're going to study starfish," it wouldn't work half as well.

All bits and pieces of wisdom are used in our camp. When we started our camp, it was the only one in the whole state. Now there are twenty-seven around Alaska, and they're all doing it in their own culturally relevant way, which is really good. Next year I think it's going to be thirty-five camps, each in a different cultural area.

Functioning from the Heart

One of the sad things is that we, the older generation, are telling the younger generation to "clean up our mess." You're not going to have a future unless you engage in looking at climate change. But everyone who is saying that is coming from the old paradigm, and they're using old paradigms about what it is we need to do—like "we need to do recycling." Recycling by itself, that's fine, but if it's reflective of a mindset, then it's dangerous for the young people—let's recycle because everyone says it is good. We're taking them down the road of a paradigm that is pushing all life support systems over the edge, and we don't even know it. We're in the matrix, and we can't see it. If you work with the inherent intelligence of the young person and understand what that means, you can create opportunities for learning where naturally they are going to get concerned about things that are affecting the Earth.

One of the issues I've got with the focus on climate change is that we've forgotten about all the other environmental issues—deforestation, air pollution, plastics on the planet, all the poisons we're spewing out. The whole focus right now is climate change. That's good, but everything's connected. We have no idea what it really means when we compound the climate issues with all the other issues. Where are we headed?

The Real Human Being never functions out of fear, because when you function from the heart there is no fear. If you function from fear, the elders said, what you choose to focus on becomes a reality. Are we choosing to focus on what we're trying to move away from, or what we're moving toward? Process is more important than the goal; how we get where we're going is more important than getting there.

Meshing Western and Native Research

When teachers teach science and they go into the classroom, they think that they need to teach only one way—using Cartesian-based science. But what we decided to do at the outset is to take our traditional way and Western science and be very conscious that they're equal in their presentation and exposure to students. We wanted to make sure they were equal, without a value judgment about one way being better than the other. The whole idea is that these ideas can complement each other. So we put together a science project that involved local hunters and Russian scientists. We knew that we wanted to bring these two together and they could work together. The outcome would be an organic outcome of our process.

So that's the way we built a project to observe birds on the bird cliffs. We have 2.5 million birds in the Pribilof Islands in the Bering Sea. We call ourselves the Galápagos of the North, because we have the highest number of different species in a single location in North America, on one island. About 227 different species of birds are found throughout the year on the island. We're a bird mecca. So we took three students at random

out of the high school and asked if they'd be willing to participate in this project doing Western science and Aleut ways. Part of our way of doing things is observation skills, being really visually acute. The Western way is documenting. So we took federal Fish and Wildlife plots that they have had for twenty years, documenting number of eggs laid, number of feeding trips, etcetera, so the students learned how to collect the quantitative data. We told them, after you've done that documentation, we want you to add to that anything you observed, not just what you were asked to count.

After a couple of weeks, these three students made simple observations of the migratory sea birds on the cliff ledges. One of the students pointed out that on one plot there was no ledge anymore, because the water got between the cracks, froze in the winter, expanded, and broke off the rock. She noted that the rock had not been there for seven years, but that the plot data had shown no birds in the plot—with no documentation that the ledge wasn't there. That skewed the whole data for the Fish and Wildlife Service.

We took the students' documentation and their abstract and presented it to the regional director for Fish and Wildlife Service, and asked them to critique it. We didn't tell them who did this, but after a couple of weeks the regional director came back and told us, "This is the best graduate-level research we have ever seen, who did it?" When we told him it was high school students, he could not believe it. But as a result of the research, they changed all the whole research protocol for all of Alaska. That was a lot of fun.

We asked the students to come up with a performance—storytelling and/or dancing, or any other way that is culturally relevant that they could use to report on this research project. (We didn't get any support at all from the scientific community to do the traditional report.) The students made up a song in Aleut, and they did sign dancing along with it. It was fantastic. The whole community got together and these students performed it to a standing ovation. They were so surprised that these young people could actually make up a song and sing it in Aleut, and perform it at the same time. The whole community just ate it up, which really made these students proud. Their performance was par excellence.

One of the things that the students were doing in their traditional report was actually Visual Math. If we combine math with visual acuity, it meshes with our cultural strength. It so happened that we had a young math teacher and young biology teacher, both fresh out of teachers' school. They were willing to try new things, so we said why don't we help you design a Visual Math program, using the visual acuity of our students

Muir Glacier in August 1941 (left) and receded in August 2004 (right), in Alaska's Glacier Bay National Park and Preserve.

in watching a fur seal herd. We have seven rookeries out there, and at that time there were about 800,000 seals. Out of that, our people are still able to look at these rookeries of teeming seals and tell what's going on. If something is wrong, we can tell it immediately, no matter how many seals are out there. If there are fewer females or too many breeding males, we know that immediately.

We had the students look at that, and their report was just fantastic. We showed it to the math teacher, who said, "Sure, we could do that, we can combine the qualitative with the numerical." He designed some observation project using math, using the process of the Western way and the traditional way. That's why we got selected as one of two places in the U.S. (along with the Warm Springs Reservation in Oregon), to test out a just-emerging Visual Math textbook.

Silencing the Indigenous Voice

There needs to be an approach in working with Indigenous people that is not ad hoc. We've got a lot of damn good scientists out there who really do want to work with Indigenous people, but they've approached it ad hoc. There's no support in the system for doing that. We need a critical dialogue between scientists and go-betweens like myself to explore what has worked, what hasn't worked, why hasn't it worked. What are we not accessing because of the process that's there? What are the institutional barriers, and how do we deal with them and get something going? We need it.

I've been talking about these kinds of things and about thinking outside the box for about twenty years. Some people get it. I didn't have too much expectation, so to see that some people picked it up was great. We really need to start pursuing those routes, and in order to do that we need the support of people who are not Native. We need non-Native allies. And we need them

to give attribution. With all the incredible things that our people in Alaska have offered Western science over the last thirty years since the advent of climate change, none of it was given attribution except in a few rare cases.

For example, the Iñupiat were the first ones to tell the scientists that we're having problems with ice. That was never given attribution. Then they told the scientists that the ice is going to melt faster than you think, basically because of tipping points. The scientists originally had 2050, and the elders said 2050 is way too out there. So the scientists gave some credence to the elders (since *Iñupiats know ice*) and revised it to 2035. "No," the elders replied, "it's going to be faster than that." Then they devised 2020—"Nope, faster than that!" Now it's at 2015, but in all those revisions there was very little attribution to the source. They give attribution to their own scientific ways of getting the information out there. The same thing has happened over and over again throughout all of Alaska. It is improving, but too slowly.

As long as that kind of thing happens, the Indigenous voice is silenced and marginalized and given less credence because it's not right up front there with all of the popular reports. The Western science system mandates that all scientists have to use the "best available science," particularly if it is going to be directing the policies that are developed. "Best available science" by definition does not include Indigenous ways of knowing, so they're pushed out of the system. This is a monocultural understanding of what is happening on the planet at a time when we've got daunting issues and we need different ways of looking at things.

PART II. EFFECTS OF THE CLIMATE CRISIS

EFFECTS OF THE CLIMATE CRISIS

The effects of the climate crisis are now being felt around the world. In the media, we hear about disastrous effects in remote, unknown corners of the globe, such as Greenland or Tuvalu. We also read of dire predictions about the future, of warming temperatures, rising sea levels, and increasing frequency of extreme weather. But the climate crisis is no longer a far-off concept in time and place. It is *here* and it is *now*, a reality being felt today in our local communities and ecosystems along the Pacific Rim.

The climate crisis is not merely another environmental issue. It is an unprecedented challenge to the economies, cultures, and social fabric of both Native and non-Native peoples, and particularly of Indigenous communities on the front lines of the crisis. They are using every tool at their disposal to understand the changes taking place and how these changes are beginning to shape our everyday lives. As Gregory Cajete observes, tribes, First Nations, and other Indigenous peoples draw on the "Native science" of traditional ecological knowledge, but Indigenous peoples also employ Western scientists to document the effects of the climate crisis. Western scientific researchers gather data and attempt to motivate decision makers to pay attention to the crisis, but they also are beginning to learn from Native science, which has an intimate and detailed relationship with local natural processes.

Indigenous nations cannot avoid or evade the impacts of climate change. Non-Natives can move away to a safer distant place with little thought or effort, but Native peoples will not abandon ancestral burial sites or leave behind clan-identified species without taking a stand. With nowhere else to go, Indigenous communities are generally less willing than most non-Native communities to compromise in the defense of Mother Earth. The former chair of the Inuit Circumpolar Council, Sheila Watt-Cloutier, said in a February 2005 interview in *Satya*:

> We have been witnessing an awful lot of change in our environment. It's so unpredictable these days—the weather patterns and climate and so on…. It goes way beyond environmental concerns for us. It is health first and foremost…. My hope is that we can have the world understand through our story and our challenges that the planet and its people are one….
>
> Our cultural survival is at stake as well…. We have never lost that strong, strong connection to our way of life even in terms of all the changes we've been going through in becoming wage earners and institutionalized. We are trying to maintain our way of life that has sustained us for millennia. We have never yet depleted one species of an animal in that millennia, so we know a little bit about sustainability….
>
> Adaptation, as we know, has its limitations, and what are we to do as a people if one of the most powerful countries in the world who is emitting the largest amount of greenhouse gases is not going to look at these issues? We have to be able to stand up for our human rights. We are Inuit who live and thrive on ice and snow. We thrive on it being frozen. That is what our culture depends upon. In essence we are fighting for our right to be cold.

The January 2012 ice storm wreaked havoc throughout the Pacific Northwest, by downing trees and transmission lines, and causing power blackouts that lasted for many days.

CLIMATE THREATS TO PACIFIC NORTHWEST TRIBES AND THE GREAT ECOLOGICAL REMOVAL : KEEPING TRADITIONS ALIVE

Terry Williams and Preston Hardison

Fisheries and Natural Resources Department of the Tulalip Tribes (Marysville, Washington)

Introduction to Indigenous Peoples and Climate Change

Indigenous peoples are rich in traditional knowledge inherited from the wisdom of tribal ancestors. This knowledge has guided them through many difficult episodes in the past when the Earth has brought forth natural catastrophes. The pulse of life that has sustained tribal cultures has ebbed and flowed. Indigenous peoples developed extensive networks of alliances and trade that helped them to survive environmental changes and upheavals. Many tribes moved with the changes of the waters and lands.

The great encounter of Native peoples and settlers in the Pacific Northwest brought great changes to all sides, and to the environment. Much of the law relating to water and the environment was brought to this continent through European settlers, who saw these lands primarily through the lens of English common law and sensibilities. One hallmark of this worldview was that the world was seen primarily as static and unchanging; that while change might come and go, it cycled around a relatively fixed state. When the new United States signed treaties with the Indian tribes, the common phrase "as long as the rivers run" was used to describe the permanent relationship between the new society and the first inhabitants. By this, it was understood by all that the resources and the land base could forever be assumed to exist in a relatively fixed state and provide abundant and sufficient resources for all.

Standing at the beginning of a new millennium, we now see that this worldview was overly optimistic. In 1992, over fifteen hundred world scientists, including a large number of Nobel Laureates, issued the "World Scientists' Warning to Humanity," which began:

> Human beings and the natural world are on a collision course. Human activities inflict harsh and often irreversible damage on the environment and on critical resources. If not checked,

many of our current practices put at serious risk the future that we wish for human society and the plant and animal kingdoms, and may so alter the living world that it will be unable to sustain life in the manner that we know. Fundamental changes are urgent if we are to avoid the collision our present course will bring about.

One of the gravest threats that faces humanity as a whole, and Indigenous nations as culturally distinct peoples, is global climate change. Climate change presents significant challenges to Indigenous peoples, and is expected to put Native peoples' communities and cultural survival at risk. Some of the threats come from direct exposure to the impacts of climate change, such as extreme temperatures, floods, fires, droughts, and storms. Other threats are more indirect, including the effects of climate change on spreading diseases, significant cultural resources, and the transmission of traditional knowledge.

The involvement of Indigenous peoples in climate change issues can be divided into three rough categories—Indigenous peoples as observers of climate change; Indigenous peoples as victims of both climate change and the actions taken to stop it (mitigation) and to respond to change that cannot be avoided (adaptation); and Indigenous peoples as mitigators of and adapters to climate change.

Some of the earliest reporting on the relationship of Indigenous peoples to climate change emphasized their role as "canaries in the mineshaft," referring to an old practice of coal miners to use canaries to detect the buildup of dangerous gases, such as carbon monoxide and methane. Indigenous peoples, because they live closely related to the land and its cycles and patterns, can provide an early warning system for climate change.

Social and environmental justice and human rights issues are involved with both climate change mitigation and adaptation. Most concerns have addressed mitigation measures to reduce greenhouse gas (GHG) emissions from the use of fossil fuels through fuel efficiency, alternative energy sources, and clean technologies. Mitigation also includes measures to sequester (store) greenhouse gases by conserving existing natural forests or planting new ones.

One of these mitigation measures is the use of biofuels to replace fossil fuels, resulting in the expansion of agricultural areas. Biofuel crops, such as soybeans and corn, can cause the expansion of large-scale agriculture into or adjacent to Indigenous territories and cause the loss of natural areas, the introduction of genetically modified organisms (GMOs), and an increase in the runoff of agricultural pollutants into the environment. The introduction of biofuel crops can also compete for water that could be used for traditional crops and increase the price of foods, particularly subsistence staples such as corn. There is also evidence that, in some cases, biofuel crops do not actually achieve their goal of reducing fossil fuel emissions. Techniques known as life cycle or footprint assessments not only account for the relative carbon emissions from the use of fossil fuels and biofuels themselves, but also analyze the fossil fuels used in biofuels' production, transportation, use, and disposal in the environment. In some circumstances, these analyses demonstrate that some kinds of biofuel production may result in carbon emissions that are as high as or higher than some kinds of fossil fuels (Fargione et al. 2008).

Other major mitigation measures for climate change focus on sequestration of carbon to decrease or prevent the release of carbon dioxide, the primary greenhouse gas, either through the planting of trees (carbon plantations) or by preventing deforestation. One major initiative, United Nations Collaborative Programme on Reducing Emissions from Deforestation and Forest Degradation in Developing Countries (REDD+), seeks to increase carbon sequestration and to conserve forest ecosystems in developing tropical countries. There are some significant issues involved in this program related to the land rights and territorial integrity of Indigenous communities.

This chapter will concentrate on the effects of climate change on tribal natural resources in the Pacific Northwest and on adaptation to those effects, after focusing on the effects of climate change on tribal natural resources in the Pacific Northwest. Indigenous peoples are not just victims of climate change, but are now empowering themselves to take the necessary steps to sustain their cultures in a dynamically changing world.

Some Terminology to Get Started

The detailed dynamics of the climate system—its natural variability, and its responses to changes in solar radiation, anthropogenic (human-made) greenhouse gas emissions, and feedbacks from its interactions with the oceans, vegetation, and other biogeochemical systems—are complex, and will not be covered in detail here. References are given in Part V (under "Native Climate Change Resources") to allow readers to find the details of models and evidence related to climate change. Scientists now refer to "climate change" rather than "global warming," because although a mean global rise in temperature is one of the leading outcomes of changes in the climate system, this is not evenly distributed around the globe and there are other manifestations of climate change. Climate models are moving from sketching the broad warming trend to understanding how the factors promoting change (climate-forcing factors) interact with details of local topography, landscape and vegetation structure, wind, and other features to determine local outcomes—a process known as downscaling. Local factors can determine whether changes in the climate system make local environments wetter or drier, colder or hotter. Climate change is also characterized by changes in extremes. A region may, for example, experience the same amount of rainfall, but concentrated in more intense bursts that generate flooding.

Carbon capture and storage, or carbon sequestration, employs several mechanisms to remove carbon from the atmosphere, most commonly through the creation of plantation forests to sequester carbon, or the protection of existing natural forests. However, much of the carbon and other greenhouse gases cannot be easily recovered, and, over a period of ten to thirty years, gases emitted at the ground level make their way into the upper atmospheric levels, such as the troposphere and stratosphere. Once in the atmosphere, the greenhouse gases have different warming potential, a measure of their heat-trapping potential—or amount of infrared absorption—and their lifetime in the atmosphere. In addition, there is a time lag due to the thermal inertia of the oceans and their response to increased carbon dioxide in the atmosphere. Because water is 500 times denser that the atmosphere, it absorbs and releases atmospheric heat slowly, and the thermal inertia of the oceans is estimated to be between twenty-five and fifty years (NRC 2003). Feedbacks related to the melting of glaciers, ice caps, snowfields, and other ice formations, and to the thawing of permafrost in northern tundra, also cause complex changes in the amount of reflectance of infrared radiation from the land and sea surface, and the release of powerful greenhouse gases such as methane. Thermal expansion of the oceans due to warming will give inertia to sea-level rise, and a warmed ocean will be less effective at absorbing heat from the atmosphere. These delayed feedbacks and inertial effects all work to ensure that climate change will continue to intensify over the next several decades and last for over a millen¬nium (Solomon et al. 2010).

Since the Industrial Revolution, carbon dioxide has

The Anderson Glacier in Washington's Olympic Mountains, in 1978 (left) and receded in 2006 (right).

increased from 280 parts per million (ppm) to 391 ppm by 2011, rising about 2 ppm per year. During the same period, average temperature of the Earth has risen by 0.7°C (1.26°F) over pre-industrial levels. Even if a magic wand were used to stop all greenhouse gas emissions today, it is expected that the world is committed at least to an increase of 1.3°C (2.34°F) (Richardson et al. 2011).

However, the global community is not yet attempting to immediately limit or reverse greenhouse gases in the atmosphere, with a 450 ppm target suggested by the last IPCC report in 2007. And there is troubling evidence that the capacity of natural carbon sinks, such as the oceans, is declining. A number of converging climate models suggest that, if we continue with "business as usual" and fail to drastically reduce emissions, the average global temperature is expected to rise by 4°–4.5°C (7.2°–8.1°F) between 2065 and the end of the century (Richardson et al. 2011). The IPCC expert assessment for business as usual estimates the global average temperature rise to be between 2.9 and 6.9°C (5.22°–12.42°F) by 2090. Regional temperature changes may be significantly higher, in some models reaching 15°C (27°F) in parts of the Arctic and 10°C (18°F) in parts of western and southern Africa in worst case scenarios.

These projected means and extremes of temperature raise serious concerns, because temperature is a master driver of a considerable number of abiotic (non-living, physical) and biotic (biological) phenomena, including patterns of intensity, frequency, and duration of precipitation, droughts, floods, storms, and wildfires; the survival, abundance, and distribution of species; the spread of diseases, invasive species, and pollutants; and the biogeochemical cycles that underpin the productive capacity of the wild and domesticated animals and plants that contribute to the well-being of humankind. These changes are not occurring in a vacuum, but in the context of other global environmental changes,

population growth, and consumption patterns that are presenting great challenges to meeting human and nature's needs.

Regional Climate Change Impacts for the Pacific Northwest

The accounts below are extremely brief, and meant to illustrate in a very general way the major impacts of climate change in the Pacific Northwest as a background to tribal adaptation options. In any detailed analysis, scenarios should be treated with great care. Climate scenarios, with their impacts, are usually presented as a range of possible outcomes related to different model assumptions that include factors such as the volume and growth rate of greenhouse emissions, population growth, technological innovations, and climate feedbacks. The climate impacts presented below are not treated in such a comprehensive manner, for which primary research and assessment reports should be consulted.

Regional Warming Trends

The average temperature in the Pacific Northwest is projected to have increased by 2.0°F by 2020, 3.2°F by 2040 and 5.3°F by 2080 relative to 1970–1999 averages (Littell et al. 2009). These figures represent the mean of a range of projections taken from IPCC low- and medium impact models, with a range of 1.1°–9.7°F. The average rate of increase is projected to be 0.5°F per decade. All seasons are expected to get warmer, with the greatest increase in the summer (Littell et al. 2009).

Precipitation and Snowpack

While models only predict a slight change in the amount of precipitation for the region (1–2 percent), the seasonal and spatial patterns of precipitation are expected to trend toward wetter autumns and winters and drier summers (Littell et al. 2009). Some areas of western Washington are expected to have extremely high increases in rainfall, while less precipitation is expected to fall as snow, decreasing snowpack. The effects of increased winter precipitation on snowpack will depend on whether the affected river basins are dominated by rainfall, by snow, or are transient (affected by mixed rainfall and snowmelt) (Littell et al. 2009).

Hydrologic Impacts

Streamflow hydrology is strongly affected by patterns of rainfall and snowmelt in the Pacific Northwest, and the degree of climate change impact is a function of the kind of river basin (rainfall-dominated, snowmelt-dominated, or transient). Climate change is expected to cause snowmelt-dominated river basins to become transient, with their peak flows occurring in the winter rather than the spring, and with reduced spring and summer flows (Littell et al. 2009, Whitley Binder 2009). Transient basins will also likely shift to becoming rainfall dominated (Littell et al. 2009). Rainfall-dominated basins are expected to be less affected. While precipitation overall is not expected to be affected on average for the region (an increase in 3.8 percent by 2080, with models projecting a range from -10 percent to +20 percent), the pattern is projected to be significantly affected (Littell et al. 2009). With warmer winters and more precipitation falling as rain rather than snow, peak flows are shifting toward the winter, delaying the onset of snow season, decreasing snowpack, and causing winter flooding. Snowmelt is occurring earlier, and released in a few strong pulses rather than many smaller pulses, as spring temperatures rise more steeply and continuously.

Impacts on Salmon and Other Salmonid Species

These changes in hydrology are having, and will increasingly have, effects on salmon and trout and the human communities that depend on them. The degree of impact will to some extent depend on whether the species are resident in streams or are anadromous (ocean-going). Salmonids have complex life histories, with spawning and egg deposition occurring in spawning beds (egg nests, or redds) in the upper watersheds of rivers and streams. The eggs remain in the gravel for one to three months, after which the salmonids emerge as alevins, which live in the gravel for a further one to five months. Following these stages, the fry emerge in the spring or summer and mature as juveniles (known as parr) that live in stream habitats from a few days to four years. After this, ocean-going salmonids undergo

a process of smoltification, in which they change their external appearance (morphology) and physiology to adapt to ocean living. These fish then spend one to four years feeding in the ocean, after which they return to the stream in which they were born to renew the cycle (Climate Impacts Group 2009).

Projected climate impacts are expected to affect salmon and trout at all stages of their life cycle. Winter floods over time can destroy salmon habitat and redds, cause stream and river channelization (which straightens and deepens channels), and cause higher sediment loads and siltation. The high flows also may damage streambanks and destroy streamside vegetation. This can lead to more erosion and weaken streambanks, which brings more sediment from slumps, decreases the ability of the riparian areas to filter nutrients and toxins in runoff, and can decrease the transfer of important stream nutrients (such as from salmon carcasses) to the forest riparian zone.

These outcomes can in turn harm salmon eggs, alevins, fry, and parr. Habitats, particularly side-channel habitat, can be degraded and lost through these peak flows, and the loss of pools and side channels can reduce water infiltration into aquifers. The higher flows may also push smolts more quickly to the ocean, causing them to expend more energy to remain in streams and rivers and move them into the ocean before they complete smoltification.

The low flows in spring and summer can cause streams and rivers to warm in two ways. The lower volumes of water warm more quickly from the warmer air temperature and this is often accompanied by increased solar irradiation due to loss of cover over streams. The faster winter runoff and related loss of habitat to slow water movement can also reduce the amount of water infiltrated into groundwater, which is an important contributor to natural water storage and the seepage of cool groundwater into streams. The warmer water induces whirling disease in salmon, and increases their susceptibility to other aquatic diseases.

Many northern aquatic species are dependent on cold winter temperatures to activate their eggs, and scientists have documented dramatic declines in many temperate aquatic invertebrates over recent decades.

Higher flows may also be associated with higher delivery of sediment, contaminants, and organic nutrients (such as phosphorous and nitrogen) into estuaries and nearshore environments that are important to salmon as nurseries. Terrestrially derived organic carbon can contribute to ocean acidification, which can harm habitat and the food webs that support growing salmon (see below). Other changes in ocean chemistry

A mudslide blocked this highway at the Skokomish Reservation by Hood Canal, after a December 2006 storm hit Washington.

from increased freshwater fluxes and the deposition of other emissions-related pollutants into the ocean also affect salmon survival at sea. Little research has been performed on climate-related impacts on salmon sensory and communication systems, but climate-related changes in water chemistry, acoustic properties, and visibility could also affect such aspects as predator avoidance, prey location, and mating behavior. Climate-related changes in species interactions (biotic interactions) are implicated in the decline of trout species predicted under climate change, with the habitat of cutthroat trout decreasing by up to 58 percent, rainbow trout to 35 percent, brown trout to 48 percent, and brook trout to 77 percent (Wenger et al. 2011).

Heat Waves

Since the 1950s, the number of extremely hot days in the western U.S. has increased by about thirty days (five per decade) (NRDC 2008). These hot days are also often accompanied by excessively hot nights, which keep daytime temperatures high by increasing morning base temperatures. These long heat waves can be deadly to humans, livestock, and crops. One model for the city of Seattle estimates that by 2025, there will be approximately 100 more deaths during heat events among persons 45 and above, and this is projected to rise to over 150 by 2045 (Littell et al. 2009).

Pests and Diseases

Higher temperatures may also favor increased damages from diseases and pests. The generation times of many pathogens and pests decreases with increasing temperatures, which may increase the extent of their

harmful effects. Since pests and pathogens and hosts all have longer active seasons under increased temperatures, this may allow pests and pathogens to build up to epidemic proportions. Forests where trees have been weakened from such attacks have been shown to be more susceptible to forest fires, which further reduces ground cover and increases erosion, reducing the overall infiltration capacity of the soils.

Invasive Species

Climate change is known to foster the intrusion of invasive alien species, which commonly invade disturbed ecosystems. Disturbed riparian corridors provide an invasion pathway into new uncolonized areas for a number of invasive plants. Invasive vegetation often provides poor habitat to non-plant native species, and can reduce burrowing and other activities that turn and aerate soils, reducing infiltration capacity and nutrient turnover. Invasive plants can alter stream hydrology by narrowing stream flood channels and trapping sediment, leading to extreme overbank flooding and further erosion of the streambank. Some invasive plants may have different root systems and provide less cover than native vegetation, causing erosion.

Invasive species may alter habitats important for the survival of traditionally used species, replacing native species of importance to Indigenous peoples. They may not only be important economically, but have great significance in religious, social, and cultural practices.

Species Range Shifts and Ecological Mismatches

Temperature is one of the master drivers of biological processes. Species are generally adapted to live within a certain range of temperature, moisture, nutrient, and other conditions known as their ecological niches. The thermal niche of species is often referred to as their "bioclimatic envelope" (Hannah 2011). Some species are tolerant enough in their existing characteristics to adapt to a range of climates without much change. Others may adapt through their ability to change their characteristics—by changing their shape, physiology, or behavior to meet the demands of a changing bioclimatic envelope. Others may adapt through genetic change through natural selection. If climate change is too abrupt, or restricts available habitat, resources, or niche space too greatly, species are likely to go extinct. For example, during the 1997–1998 El Niño warming cycle in the Pacific Ocean more than 10 percent of all corals died from a phenomenon known as coral bleaching. In the Indian Ocean, coral mortality was as high as 46 percent (Hannah 2011).

One common response to climate change is migration to stay within a preferred bioclimatic envelope; this is known as a species range shift (Hannah 2011). Many species are moving toward the poles, both on the land and in the sea, and vertically up mountain slopes to follow shifting climate zones. One recent study estimates the rate of migration at approximately to two to three times previous estimates (Chen et al. 2011). Species are on average moving vertically up slopes about 11.0 m (36 ft.) and toward the poles about 16.9 km (10.5 mi) per decade.

There are risks associated with range shifts. Species on mountains or other "pocket" habitats might become ecologically trapped and unable to migrate across bioclimatic barriers. Species migrating over terrestrial landscapes require corridors, or permeable landscapes, and can become trapped or hindered by human-made barriers such as highways, fences, and urban infrastructure. Although there is evidence of species migrations in response to climate change in the past, current landscapes are likely to be more hazardous, putting migrants under greater stress and risk.

Mismatches between biological phenomena or relationships also occur. Species are adapted so that events in their life history are timed to occur in relationship to other natural events. The seasonal timing of life-history events—such as flowering, mating, dormancy, and reproduction—are known as phenology. These events can be disrupted by mis-timings (temporal mismatch) or mis-occurrences (spatial mismatch). Predators time their reproduction or emergence to ensure their young are born or emerge during times of prey abundance, so shifts in the reproductive timing of their emergence can cause predator-prey mismatch. Flowers that rely on pollinators for reproduction are timed to bud and open when their pollinators are present, so shifts in pollinators can cause mutualism mismatch. There was an earlier example of mismatches between water availability and salmon ecology and life history, called an ecohydrological mismatch. Spatial mismatches can occur when species involved in a relationship migrate at different rates such that their life-history events occur at the right time, but in the wrong places.

Species range shifts and ecological mismatches can pose serious threats to the cultural sustainability of tribes. The living world is a core part of tribal identity, as animals, plants, fungi, and even microbes are often involved in the expression of tribal culture. They are embedded in tribal stories, are a co-equal part of their kinship relations with their human kin, and are used for subsistence, economy, medicines, architecture, clothing, artistic expression—in practically every aspect of tribal culture. Non-human species are fundamental to spiritual beliefs and important for ritual and ceremonial practices. Many yearly rounds and ceremonies, such as sowing, harvesting, gathering, fishing, and hunting, are linked to the phenology of species.

Species range shifts pose a special burden, because tribes in the United States, as is largely true for indigenous peoples everywhere, are confined to reservations with legally defined boundaries. When signing their treaties, the tribes generally reserved some rights (known as "usufruct rights") to gather, hunt, trap, and fish off their legal reservations on what were at the time "open and unclaimed lands,." These lands have dwindled over the years, and currently tribes find much of their treaty access limited to federal lands within thickets of regulations. Climate change threatens to move culturally important species away from both their legal territories and lands to which they have usufruct rights to use resources. This raises significant questions regarding treaty rights and federal trust obligations to ensure tribal access to their treaty-reserved cultural resources. Aside from the legal questions are the bottom line of cultural continuity and survival.

Emerging Diseases and Pests

Temperature alterations have been mentioned as one cause of the increase in diseases that affect the health of forests, streams, and other ecosystems. Hydrological changes can also increase disease incidence in ecosystems in a number of ways. Increases in nutrient load can favor the growth of algae or phytoplankton that creates conditions for hypoxia (oxygen starvation), eutrophication (dense growth of plant life), and red tides (a type of toxic bloom), resulting in fish kills, the proliferation of toxic substances (e.g., paralytic shellfish toxins), and the growth of pathogenic microorganisms and harmful algal blooms. In Puget Sound, the frequency, duration, and geographic extent of shellfish toxicity (associated with harmful algal blooms of the dinoglagellate *Alexandrium catenella)* has increased in 1976 and 2007 (Moore et al. 2011). Even a moderate IPCC emission scenario predicts that the window of opportunity for climate-related harmful algal blooms for this species will increase thirteen days by 2100. The conditions benefiting the algae include warmer air, low streamflows into Puget Sound, low winds, and low tidal variability. The overall period during which harmful algal blooms may occur are predicted to start two months earlier and persist one month later than has been true. Even if species are not killed by such climate-related events, they can be rendered inedible, which blocks tribal access to culturally and economically critical resources.

Wildfire in Alaska exacerbated by the spruce bark beetle infestation.

Increased surface-water and storm-water runoff from compacted soils and impervious surfaces can also increase the delivery of pathogens and toxins into aquatic ecosystems. The loss of riparian ecosystem services (such as riparian vegetation that prevents erosion and filters diseases and toxins from runoff) also increases disease burdens in aquatic ecosystems. Stirring up sediments can stir up trouble, as pathogens are pushed into water columns and move downstream. Pathogens such as aquatic viruses in this way sometimes are able to meet and exchange genes to produce more virulent strains.

Invasive species can bring their foreign pathogens with them, and cause "virgin water epidemics" in aquatic organisms that have never been exposed to them, and virgin soil epidemics in riparian zone species. Similarly, species range shifts can move disease-carrying plants and animals into new, often disturbed, environments that provide opportunities for the pathogens to infect new hosts. These "emerging diseases" may have existed for some time, but are becoming new and prevalent threats to the health of the public, wildlife, livestock, crops, and other vegetation because of the rapid changes caused by climate change.

Climate-related changes also leave many ecosystems more vulnerable to pests and diseases. Warmer winters mean greater survival rates for many, so that pests can build up greater levels of infestation and diseases can infect more people, often with increased virulence. Many Western forests, for example, are being attacked by wood-boring and bark beetles, such as the mountain pine beetle. These beetles have in the past been controlled naturally by the diversity of tree species and ages that reduce their attack success rates, and killing temperatures that reduce their numbers (Han-

nah 2011). These beetles are surviving better through warmer winters, allowing their numbers to build. Coupled with a history of poor forestry practices that have created even-aged forests with little tree species diversity, the greater numbers of beetles are attacking millions of trees in the Pacific Northwest and Western Canada. This leaves the trees valueless for commercial use and highly susceptible to very hot fires that destroy the capacity of the land rather than rejuvenate it.

Sea Level Rise

Sea level is expected to rise between two and 13 inches. In some extreme models, which assume the complete melting of the Greenland Ice Sheet, the level could reach 50 inches by 2100 (Littell et al. 2009). The relative level of rise will vary by location, because absolute sea level rise can be offset by marine basin topography, land subsidence, upward movements of land masses, and local rates of thermal expansion of the ocean. But any amount of sea level rise is threatening to coastal tribes, because their land base is set by treaty. Sea level rise puts coastal populations at risk from coastal erosion and inundation by storm surges. A February 2006 storm surge caused significant tidal inundation and property damage on the Swinomish Indian Reservation in LaConner, Washington. The Swinomish Tribe has identified zones at risk of inundation, which threaten up to 15 percent of coastal reservation lands (Swinomish Indian Tribal Community 2009).

Ocean Acidification

Carbon dioxide in the atmosphere, when absorbed in seawater, makes it more acidic. The acidity of the ocean is now thought to be at its highest concentration for the past three hundred thousand years. The impacts of this acidification are unclear. It is known that acidification lowers the concentration of calcite and aragonite, elements used in building the calcium shells and skeletons of marine organisms. Many of these organisms, such as shellfish and crab, are critical for Indigenous livelihoods. Calcium carbonate provides the building material for coral reefs, which are also already suffering from seawater temperature rise. It also affects the abundance of plankton, which form the basis for marine food webs. Their degradation or loss could lead to a widespread decline in fisheries production, leading to loss of livelihoods.

Exacerbating Impacts and Climate Change Feedbacks

Climate change co-occurs with numerous existing adverse impacts, and many of its effects cause feedbacks to the climate system. Alone and together, these pressures and feedbacks can significantly amplify the impacts of climate change.

Although precipitation is not expected to increase greatly, more is projected to fall as rain on a landscape already impacted by development. Much of the surface is hardened, or armored, from various processes such as paving, soil compaction in forestry and farming areas, rooftops, and soil drying. Open canopies and riverside soil compaction also reduce the survival of soil mycorrhiza (a fungus that grows in association with plant roots), which are important for nutrient cycling in forest ecosystems. The mycorrhizae are also important in water cycling, slowing and holding water in the soil leading to greater groundwater infiltration. Soil compaction, drying, and fire-related mycorrhizal mortality decrease soil's ability to support native vegetation and decreases groundwater recharge. When soils can no longer slow down the rain and runoff to prevent erosion and promote infiltration, these effects add to the direct impacts of climate change on precipitation and flooding, adding to storm-water volume and contaminant loads.

On the coasts, sea level rise is exacerbated by the building of settlements up to the coastal fringe that prevents the landward migration of habitat (also called coastal squeeze). Dredging and pollution have taken their toll on hundreds of miles of sea-grasses in Puget Sound. Sea grasses normally sequester carbon in the ocean, absorb wave energy, and act as biological barriers against sea level rise. They also hold sediment and keep it from being transported by long-shore currents, which normally would potentially allow them to build sediment platforms to keep pace as sea level rises.

The loss of snow and ice cover also creates feedbacks, decreasing the albedo (reflectance) of solar radiation, as more of the energy is absorbed as heat by darker surfaces. This is causing more warming of the land and ocean surface, which generates more melting under glaciers, snowpack, and sea ice, reducing the extent of snow and ice even more.

Climate change impacts are additional to existing impacts such as from landscape and habitat fragmentation, population growth, urban sprawl, pollution, overharvesting, and overconsumption, which already impose inequitable burdens on Indigenous peoples in many places. Non-climate impacts and feedbacks

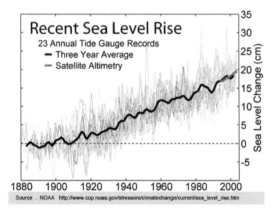

Global sea-level rise between 1880 and 2000.

must be assessed and addressed in any climate impact assessment and planning.

The Political Context

Tribes have a unique relationship to their ancestral homelands and to the United States. Their identity is deeply rooted in their lands—places from which they emerged, where their ancestors dwell, to which their stories and language refer, and to which they have continuing spiritual and collective obligations. Their status in the United States is commonly misunderstood. American Indians are citizens of the United States, and yet they are more than that—they also belong to their tribal nations, which are not simply a kind of stakeholder or special interest group, as is sometimes believed. And the treaties they made with the U.S. government are not faded relics of a nostalgic and romanticized past, but represent the highest and most solemn agreements among different peoples that were intended to be as permanent as the agreement that established the Constitution.

When the founders of the United States became independent from Britain, they were sensitive to the abuses of colonial rule and power. The Constitution was established to put the new country under the rule of law. Article 6, Section 2 establishes that the Constitution and "all Treaties made, or which shall be made, under the Authority of the United States, shall be the supreme Law of the Land." The founders recognized the legitimacy of treaties with tribal nations prior to independence, and bestowed on the president the power to make new treaties with the tribes, provided support by two-thirds of the Senate.

The question of the political status of the Indian tribes was decided by the Supreme Court when it considered a series of cases between 1823 and 1832 known as the Cherokee Cases or "Marshall Trilogy," after Chief Justice John Marshall, who handed down the majority opinions. In the final case, *Worcester v. Georgia* (1832), Marshall held "The words 'treaty' and 'nation' are words of our own language, selected in our diplomatic and legislative proceedings, by ourselves, having each a definite and well understood meaning. We have applied them to Indians, as we have applied them to the other nations of the earth. They are applied to all in the same sense." In other words, the tribes have retained sovereign powers that are on an equal footing with those of other foreign nations. In a later decision, *United States v. Winans* (1905), the Supreme Court further clarified that the Indian treaties are "not a grant of rights to the Indians, but a grant of rights from them—a reservation of those not granted." The political status of the tribes arises from their inherent right of self-determination as peoples, on the same level as any nation, and not from a grant of power by the United States. The treaties were not about grants of privileges by the U.S. government, but about cessions of sovereign jurisdiction by the various tribes. The tribes are not foreign nations in the sense of the Constitution, but they are domestic nations that retain their sovereign rights to govern themselves, with the United States making a pledge to be their guardian or trustee.

This accords with how the tribes interpret the treaties, and the federal "canons of construction," which affirm that treaties should be interpreted as the tribes understood them, and any ambiguities should be resolved in a way most favorable to the rights and interests of the tribes. What the tribes understood of the treaties was that they were reserving their homelands forever. They generally reserved sovereign rights both to their reservation lands and access to off-reservation resources purposefully, because their leaders knew they required more than reservation land bases to sustain their place-based cultures.

This political context is critical for understanding tribal issues related to climate change. Our brief presentation of climate change impacts only scratches the surface of the ways in which climate change affects tribal lifeways and puts tribal economies, practices, and values in harm's way. These impacts affect not only tribal members as individual citizens or stakeholders. Climate change also affects collective rights reserved by treaty to species that are culturally important to Native peoples, species that the United States has assumed a trust responsibility to protect. Without action, the situation will lead to "ecological removal" or "ecological dispossession" of these species. Rather than removing

Native peoples to reservations, as happened during the first waves of removal, the changing climate is pulling the living carpet out from under tribal nations.

Tribal Engagement and Biocultural Adaptation

Most policy attention related to climate change has focused on mitigation measures to reduce carbon dioxide and other greenhouse gases in the atmosphere to avoid, as much as possible, altering climate on a planetary scale. Hundreds of books exist on lifestyle and consumption choices that can move everyone towards "zero emissions," "decarbonization," or a "low-carbon future." Legislation has been attempted, but failed, in the U.S. that put the greatest emphasis on incentives to limit emissions and invest in cleaner, alternative energy sources. Tribes have engaged by investing in wind energy, biomass energy, energy-efficient traditional construction such as straw-bale housing, and other green technologies. Some are also taking steps, similar to those of many municipalities and governments, to promote low-carbon lifestyles and green operations by tribal governments on reservations. Global society as a whole will never be able to get climate change under control until everyone learns to control their greenhouse gas appetites.

But mitigation falls short. Tribes must address adaptation to unavoidable climate change. This change is not a specter of the future, but is already present, and will increase its grip on everyone in the coming years. This is guaranteed by lag times in the Earth's systems. We must continue to strive to personally take all possible measures we can, and to move politicians to commit us all to actions necessary to avoid climate tipping points and dangerous climate change. Adaptation cannot be use as an excuse to avoid responsibilities to all peoples and future generations and curtail climate change.

Actions need to be taken now in anticipation of climate futures on the order of decades to centuries. There must be accelerated and increased investment in adaptation measures to address the many impacts of climate change that will not be avoided. As with mitigation policies, most adaptation policies look at options related to energy and water conservation, changes in production sectors (such as breeding crops resistant to drought and heat), and supply chain and public procurement systems that make businesses, consumers, and citizens more resilient in the face of climate change.

Nature needs to be taken into account in adaptation. This need arises from our obligations as nature's fiduciaries, guardians, and stewards. It also follows from the condition that ecosystems also have their thresholds and tipping points, and their own lag times in response

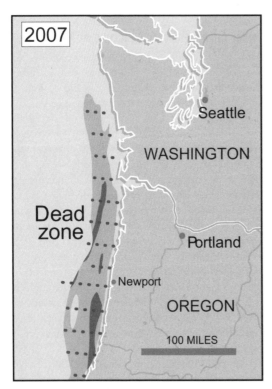

"Dead Zone" area affected by hypoxia (oxygen starvation) off the Pacific Northwest Coast in 2007.

to interventions, leading to ecological debt and ecological change in the pipeline. At stake are not only increasing future costs of restoration and resilience. In the face of ecological thresholds and potential irreversible changes to ecosystems, delays in making ecosystems more resilient and resistant to climate change can make restoration infeasible. The fate of species is tied to ecosystems, and many species possess traits that make them particularly vulnerable to climate-induced ecosystem changes, deterioration, and collapse. Such traits include narrow food or habitat specializations, low dispersal ability, small population size, or dependence on other species (mutualisms) that are themselves in decline from climate change. The consequences of ecosystem collapse would place great burdens on society and fall most heavily on Indigenous peoples and local communities that still have close connections to the land.

Indigenous peoples have had much experience with climate change and environmental extremes in the past, which forms a part of what some have called "traditional ecological knowledge" (TEK). The explanation for this traditional knowledge differs depending on who is asked. In textbooks and research papers it is often described as "a cumulative body of knowledge, practice,

and belief, evolving by adaptive processes and handed down through generations by cultural transmission. [It concerns] the relationship of living beings (including human) with one another and with their environment" (Berkes 2008). Tribal elders will often have a different explanation, seeing their traditional knowledge as a gift from the Creator, the ancestors, acquired through dreams, or in direct conversations with the spirit world, the plant people, the animal people, or other spiritual origins. Indigenous peoples regard their traditional knowledge as both practical and spiritual.

There is much traditional knowledge related directly to climate change, or providing indirect benefits. The Nuu-chah-nulth of Vancouver Island were reported to have transported salmon eggs in wet moss between streams and around landslides in the 1860s, making them some of the first recorded peoples to have practiced "assisted migration" (Campbell and Butler 2010). There exists traditional knowledge related to architecture (e.g. straw-bale and adobe house construction) and water harvesting in arid landscapes that could aid in adapting to climate-induced drought and warmer temperatures (Corum 2005, Mays and Gorokhovich 2010).

There are several ways to think about the relationship of traditional ecological knowledge to climate change. The examples above illustrated some of the direct benefits of applying TEK to climate change adaptation. One way to accomplish this is to adjust the environment, such as providing microenvironments, burning to create ecological succession, and so on. Creating traditional irrigation works could better capture, store, and release water from snowpack runoff (slowing springtime floods and increasing low summertime flows). Another path is to change exposure or sensitivity to climate change, such as by selecting drought-tolerant traditional seed varieties, changing the start of hunting, harvesting, or gathering, their duration, or the life-history stages or species harvested, guided by deep knowledge of landscapes, ecosystems, and species.

The second framework is to consider the impacts of climate change on traditional knowledge, practices, and beliefs, and all of the scales of the environment on which they depend—genes, ecosystems, landscapes, biomes—and find ways to protect the traditional knowledge systems and their relationships to the land and life. There are a number of traditional ways of being—such as plants used in rituals—that may not be directly useful to adaptation, but which the tribes wish to preserve.

A third framework is to be mindful of the potential impacts of climate change adaptation and mitigation on traditional ecological knowledge and its basis in the land. The potential impacts of biofuels production and monocrop tree plantations mentioned earlier are well known. Using pesticides, which could get sprayed on traditional basketry grasses in gathering areas, to control climate-related weed invasions, is one potential impact of an adaptation on Indigenous peoples. One common adaptation recommendation is to provide corridors to aid in species migration, so that they may better find their ways to areas that can support them under climate change. Corridors have characteristics that can impinge on culturally important resources—they cross the landscape, and may be adjacent to or run through tribal territories. Corridors can have the effect of encouraging movement out of an area, to the detriment of ecosystem management measures to help keep (at least some) populations in the legal areas of tribal use rights. They can also deposit species in target areas that cause disruptions of ecosystems there, or bring in invasive species as well as native ones.

U.S. tribes have consistently called for including their traditional knowledge (or Native science) along with Western science in addressing climate change issues. The Department of Interior Secretarial Order 3289 - Addressing the Impacts of Climate Change on America's Water, Land, and Other Natural and Cultural Resources, issued on February 22, 2010, recognizes and affirms U.S. trust responsibilities toward the tribes, and the goal of protecting cultural resources from climate change. Section 5 on American Indians and Alaska Natives states that:

Climate change may disproportionately affect tribes and their lands because they are heavily dependent on their natural resources for economic and cultural identity. As the Department has the primary trust responsibility for the Federal government for American Indians, Alaska Natives, and tribal lands and resources, the Department will ensure consistent and in-depth government-to-government consultation with tribes and Alaska Natives on the Department's climate change initiatives. Tribal values are critical to determining what is to be protected, why, and how to protect the interests of their communities. The Department will support the use of the best available science, including traditional ecological knowledge, in formulating policy pertaining to climate change. The Department will also support substantive participation by tribes in deliberations on climate-related mechanisms, agreements, rules, and regulations.

The tribes need to be brought into the planning process at all levels of governance and in all issues that touch upon their cultural identity and heritage, and on their rights reserved by treaty. This will require some sensitivity and mutual respect. One common way of expressing tribal participation has been to refer to the "integration" of Western science and traditional knowledge. It might be thought that the tribes will bring their knowledge to the table to be shared with scientists, the government, and the public, where it can be studied, analyzed, and disseminated for the greater good. But considerations related to the spiritual nature of traditional knowledge, the complex ways in which the traditional knowledge is circulated within Native cultures, and the desire for cultural privacy are all factors that will make the conversation between worldviews a challenging exercise.

There is some concern about the potential exploitation of traditional knowledge, stemming from experiences by Indigenous peoples in the exploitation of their knowledge of biodiversity in the development of drugs and foods, without their consent or compensation, and in ways that go against their traditional values. There are many tribes willing to share at least some of their traditional knowledge to solve climate issues, only asking that it be used respectfully, with acknowledgement, and in a way consistent with their beliefs. But in other cases, the benefits of traditional knowledge may not come through sharing so much as through the benefits of tribal self-management of their own climate adaptation and mitigation processes on their own lands.

Like the Swinomish mentioned above, tribes are beginning to develop their own tribal adaptation and mitigation plans, incorporating indicators and values that reflect their priorities and make sense to themselves, while also incorporating best-science practices to better understand the future threats facing them. Tribes and others should recognize the unprecedented nature of the threat that climate change poses, and the necessity to anticipate future climate impacts, particularly those that threaten keystone cultural resources and the ecosystems on which Native nations depend. There are many uncertainties ahead, and tribes will need greater resilience to cope with the changes and find the incentive and resources to preserve, restore, and enhance the landscape to keep as many cultural values in place as possible. Despite the uncertainties, one thing is clear. Tribes are bound to their homelands by kinship with the land and the beings that dwell there, by their ancestors, by their stories, and the law. Their solutions will not be found in flight from climate change, but in caring for, healing, and defending their place, and securing their homelands to the generations to come.

Further Reading

Ad hoc Technical Expert Group on Biological Diversity and Climate Change. 2003. *Interlinkages between Biological Diversity and Climate Change: Advice on the Integration of Biodiversity Considerations into the Implementation of the United Nations Framework Convention on Climate Change and its Kyoto Protocol.* Montréal, Canada: CBD Technical Series 10. Secretariat of the Convention on Biological Diversity (SCBD). www.biodiv.org/doc/publications /cbd-ts-10.pdf

Allan, J. David. 2004. Landscapes and Riverscapes: The Influence of Land Use on Stream Ecosystems. *Annual Review of Ecology, Evolution, and Systematics* 35: 257-284. http://ecolsys.annualreviews.org/

Anderson, James J. 2001. Decadal Climate Cycles and Declining Columbia River Salmon. In: Knudsen, E. Eric, Cleveland R. Steward, Donald D. McDonald, Jack E. Williams, and Dudley W. Reiser (eds.): *Sustainable Fisheries Management: Pacific Salmon.* Boca Raton, Florida: Lewis Publishers Inc. Pp. 467-84. www.crcpress.com/

Bergkamp, Ger, Brett Orlando, and Ian Burton. 2003. Change: Adaptation of Water Resources Management to Climate Change. Gland, Switzerland: IUCN - The World Conservation Union. www.iucn.org/themes/wani/pub/ Brochure-UICN-Change.pdf

Berkes, Fikret. 2008. *Sacred Ecology: Traditional Ecological Knowledge and Resource Management.* 2nd ed. London: Routledge.

Bernhardt, Emily S., Gene E. Likens, Robert O. Hall, Don C. Buso, Stuart G. Fisher, Thomas M. Burton, Judy L. Meyer, et al. 2005. Can't See the Forest for the Stream? In-stream Processing and Terrestrial Nitrogen Exports. *BioScience* 55(3): 219-230. www.aibs.org/bioscience/

Betts, Richard A., Matthew Collins, Deborah L. Hemming, Chris D. Jones, Jason A. Lowe, and Michael Sanderson. 2009. When Could Global Warming Reach 4°C? Hadley Centre Technical Note 80. Exeter, Devon, UK: Met Office Hadley Centre for Climate Change. 22 pp.

Burton, Ian, Kristie L. Ebi, and Joel Smith. 2004. *Integration of Public Health with Adaptation to Climate Change: Lessons Learned and New Directions.* London: Taylor & Francis Ltd.

Campbell, Sarah K., and Virginia L. Butler. 2010. Archaeological Evidence for Resilience of Pacific Northwest Salmon Populations and the Socioecological System over the last ~7,500 Years.

Ecology and Society 15(1): 17 [online]. www. ecologyandsociety.org

The Climate Impacts Group. 2009. The Washington Climate Change Impacts Assessment: Evaluating Washington's Future in a Changing Climate. Executive Summary. Seattle: University of Washington. 20 pp.

Chen, I-Ching, Jane K. Hill, Ralf Ohlemüller, David B. Roy, and Chris D. Thomas. 2011. Rapid Range Shifts of Species Associated with High Levels of Climate Warming. *Science* 333(6405): 1024-26.

Corum, Nathaniel. 2005. *Building a Straw Bale House: The Red Feather Construction Handbook.* Princeton, New Jersey: Princeton Architectural Press.

Costa, John E., Andrew J. Miller, Kenneth W. Potter, and Peter R. Wilcock (eds.). 1995. Natural and Anthropogenic Influences in Fluvial Geomorphology. Geophysical Monograph 89. Washington, DC: American Geophysical Union. https://www.agu.org/

Dialogue on Water and Climate. 2003. Climate Changes the Water Rules: How Water Managers Can Cope with Today's Climate Variability and Tomorrow's Climate Change. Delft, The Netherlands. www.wac.ihe.nl/report.htm

Diaz, Henry F., and Barbara J. Morehouse (eds.). 2003. Climate and Water: Transboundary Challenges in the Americas. *Advances in Global Change Research* 16. Dordrecht, The Netherlands: Kluwer Academic Publishers. springeronline.com/

Dorava, Joseph M., David R. Montgomery, Betty B. Palcsak, and Faith A. Fitzpatrick (eds.). 2001. Geomorphic Processes and Riverine Habitat. *Water Science and Application* 4. Washington, DC: American Geophysical Union. https://www.agu.org/

Elsner, Marketa M., Lan Cuo, Nathalie Voisin, Jeffrey S. Deems, Alan F. Hamlet, Julie A. Vano, et al. 2010. Implications of 21st Century Climate Change for the Hydrology of Washington State. *Climatic Change* 102: 225-260.

Epstein, Paul R., and Dan Ferber. 2011. *Changing Planet, Changing Health: How the Climate Crisis Threatens Our Health and What We can Do About It.* Berkeley: University of California Press.

Fargione, Joseph, Jason Hill, David Tilman, Stephen Polasky, and Peter Hawthorne. 2008. Land Clearing and the Biofuel Carbon Debt. *Science* 319 (5867): 1235.

Few, Roger, Mike Ahern, Franziska Matthies, and Sari Kovats. 2004. Floods, Health and Climate Change:

A Strategic Review. Tyndall Centre Working Paper 63. Norwich, U.K.: University of East Anglia, Tyndall Centre for Climate Change Research. http://tyndall.e-collaboration.co.uk/publications/publications.shtml

Field, John C., Donald F. Boesch, Donald Scavia, Robert Buddemeier, Virginia R. Burkett, Daniel Cayan, Michael Fogarty, et al. 2001. Potential Consequences of Climate Variability and Change on Coastal Areas and Marine Resources. In National Assessment Synthesis Team (eds.): *Climate Change Impacts on the United States: The Potential Consequences of Climate Variability and Change: Foundation Report.* Washington, DC: United States Global Change Research Program, National Assessment Synthesis Team. www.usgcrp.gov/usgcrp/Library/nationalassessment/foundation.htm

Hamlet, Alan F., with David L. Fluharty, Dennis P. Lettenmaier, Nathan J. Mantua, Edward L. Miles, Philip W. Mote, and L. Whitely Binder. 2001. *Effects of Climate Change on Water Resources in the Pacific Northwest: Impacts and Policy Implications.* Preparatory White Paper / CIG Publication 145. Seattle: University of Washington, Climate Impacts Group. http://jisao.washington.edu/PNWimpacts/Publications/Pub144a.htm

Hannah, Lee, Guy F. Midgley, and Dinah Millar. 2002. Climate Change-integrated Conservation Strategies. *Global Ecology and Biogeography* 11(6): 485-496. www.blackwellpublishing.com/journals/geb/

Hannah, Lee. 2011. *Climate Change Biology.* ,San Diego, California: Academic Press.

Harvell, C. Drew, Charles E. Mitchell, Jessica R. Ward, Sonia Altizer, Andrew P. Dobson, Richard S. Ostfeld, and Michael D. Samuel. 2002. Climate Warming and Disease Risks for Terrestrial and Marine Biota. *Science* (Washington) 296(5576): 2158-62.

Hilborn, Ray H., Thomas P. Quinn, Daniel E. Schindler, and Donald E. Rogers. 2003. Biocomplexity and Fisheries Sustainability. *Proceedings of the National Academy of Sciences of the United States of America* 100(11): 6564-68. www.pnas.org

Jensen, Mari N. 2004. Climate Warming Shakes up Species. *BioScience* 54(8): 722-729. www.aibs.org/bioscience/

Kabat, Pavel, Martin Claussen, Paul A. Dirmeyer, John H. C. Gash, Lelys Bravo de Guenni, Michel Meybeck, Roger A. Pielke, Sr., et al. 2004. *Vegetation, Water, Humans and the Climate: A New Perspective on an Interactive System.* New York: Springer-Verlag. www.springeronline.com/sgw/cda/frontpage/0,11855,4-0-0-0-0,00.html

Lackey, Robert T. 2003. Pacific Northwest Salmon: Forecasting their Status in 2100. Reviews in *Fisheries Science* 11(1): 35-88.

Lewis, William M., Jr. (ed.) 2003. *Water and Climate in the Western United States.* Boulder: University Press of Colorado. www.upcolorado.com/

Lim, Bo, Erika Spanger-Siegfried, Ian Burton, Elizabeth Malone, and Saleemuk Huq. 2004. *Adaptation Policy Frameworks for Climate Change: Developing Strategies, Policies and Measures.* Cambridge, UK: Cambridge University Press. www.cup.org/

Littell, Jeremy S., Marketa McGuire Elsner, Laura C. Whitely Binder, and Amy K. Snover (eds). 2009. *The Washington Climate Change Impacts Assessment: Evaluating Washington's Future in a Changing Climate.* Seattle: University of Washington, Climate Impacts Group. 414 pp. http://cses.washington.edu/cig/res/ia/waccia.shtml

Lovejoy, Thomas E. 1995. Will Expectedly the Top Blow Off? *BioScience* 45(6): S3-S6 (Supplement: Science and Biodiversity Policy). www.aibs.org/bioscience/

Lovejoy, Thomas E., and Lee Jay Hannah (eds). 2005. *Climate Change and Biodiversity.* New Haven, Connecticut: Yale University Press. http://yalepress.yale.edu/yupbooks/

Mantua, Nathan, and Robert C. Francis. 2004. Natural Climate Insurance for Pacific Northwest Salmon and Salmon Fisheries: Finding Our Way through the Entangled Bank. In: Knudsen, E. Eric, Donald D. MacDonald, and Yvonne K. Muirhead (eds.): *Sustainable Management of North American Fisheries.* Baltimore, Maryland: American Fisheries Society. www.fisheries.org/

Martens, Willem Jozef Meine (Mertens, Pim), and Anthony J. McMichael (eds.). 2002. *Environmental Change, Climate and Health: Issues and Research Methods.* Cambridge, UK: Cambridge University Press. www.cup.org/

Mays, Larry W., and Yuri Gorokhovich. 2010. Water Technology in the Ancient American Societies. In: Mays, Larry W. (ed.). *Ancient Water Technologies.* New York: Springer. Pp. 171-200.

McCally, Michael (ed.). 2002. *Life Support: The Environment and Human Health.* Cambridge, Massachusetts: MIT Press. http://mitpress.mit.edu/

McGinn, Nature A. (ed.). 2002. *Fisheries in a Changing Climate*. Bethesda, Maryland: American Fisheries Society. www.fisheries.org/publications/catbooks/x54032.shtml

Mendelsohn, Robert, and James E. Neumann (eds). 2004. *The Impact of Climate Change on the United States Economy*. New York: Cambridge University Press. www.cup.org/

Miles, Edward L. 2001. *Global Climate Change and Marine Policy: Planning for Impacts, Adaptation, and Vulnerability*. CIG Publication / JISAO Contribution 176 / 835. Seattle: University of Washington, Climate Impacts Group. http://tao.atmos.washington.edu/PNWimpacts/Publications/ Pub176a.htm

Miles, Edward L., and Amy K. Snover (eds.). 2005. *Rhythms of Change: Climate Impacts on the Pacific Northwest*. Seattle: University of Washington, Climate Impacts Group. www.cses.washington.edu/cig/

Miles, Edward L., Amy K. Snover, Alan F. Hamlet, Bridget Callahan, and David L. Fluharty. 2000. Pacific Northwest Regional Assessment: The Impacts of Climate Variability and Climate Change on the Water Resources of the Columbia River Basins. *Journal of the American Water Resources Association* 36(2): 399-420 (Special issue on Water Resources and Climate Change) www.awra.org/jawra/

Moore, Stephanie K., Nathan J. Mantua, and Eric P. Salathé, Jr. 2011. Past Trends and Future Scenarios for Environmental Conditions Favoring the Accumulation of Paralytic Shellfish Toxins in Puget Sound Shellfish. *Harmful Algae* 10(5): 521-529.

Morrison, Jason, and Peter Gleick. 2004. *Freshwater Resources: Managing the Risks Facing the Private Sector*. Oakland, California: Pacific Institute / Trillium Asset Management Corporation. Pacific Institute. www.pacinst.org/reports/business_risks_of_water/index.htm

Mote, Philip W. 2004. How and Why Is Northwest Climate Changing? In: Peterson, David L., John L. Innes, and Kelly O'Brian (eds.). *Climate Change, Carbon, and Forestry in Northwestern North America: Proceedings of a Workshop, November 14 - 15, 2001, Orcas Island, Washington*. USFS General Technical Report PNW-GTR-614. United States Department of Agriculture, United States Forest Service, Pacific Northwest Research Station, Portland, Oregon. Pp. 11-22. www.fs.fed.us/pnw/publications/index.shtml

Mote, Philip W., D. Canning, D. Fluharty, R. Francis, J. Franklin, J., A. Hamlet, M. Hershman, et al. 1999. *Impacts of Climate Variability and Change in the Pacific Northwest*. Seattle: University of Washington, Climate Impacts Group. www.cses.washington.edu/cig/

Mote, Philip W., Alan F. Hamlet, Martyn P. Clark, and Dennis P. Lettenmaier. 2005. Declining Mountain Snowpack in Western North America. *Bulletin of the American Meteorological Society* 86(1): 39-49.

Mote, Philip W., Edward A. Parson, Alan F. Hamlet, William S. Keeton, Dennis Lettenmaier, Nathan J. Mantua, Edward L. Miles, et al. 2003. Preparing for Climatic Change: The Water, Salmon, and Forests of the Pacific Northwest. *Climatic Change* 61(1-2): 45-88.

National Research Council. 2002. Abrupt Climate Change: Inevitable Surprises. (National Academy of Sciences, National Research Council, Committee on Abrupt Climate Change.) Washington, DC: National Academy Press. www.nap.edu/

National Research Council. 2003. Understanding Climate Change Feedbacks. (National Academy of Sciences, National Research Council, Panel on Climate Change Feedbacks.) Washington, DC: National Academy Press. www.nap.edu/

National Wildlife Federation. 2005. *Fish out of Water: A Guide to Global Warming and Pacific Northwest Rivers*. Washington, DC: National Wildlife Federation. www.nwf.org/nwfwebadmin/binaryVault/Fish%20Out%20of%20Water%2020051.pdf

National Wildlife Federation. 2011. *Facing the Storm: Indian Tribes, Climate-Induced Weather Extremes, and the Future for Indian Country*. Washington, DC: National Wildlife Federation. 27 pp.

Odum, William E. 1982. Environmental Degradation and the Tyranny of Small Decisions. *BioScience* 32(9): 728-29. www.aibs.org/bioscience/

Petersen, J. H., and J. F. Kitchell. 2001. Climate Regimes and Water Temperature Changes in the Columbia River: Bioenergetic Implications for Predators of Juvenile Salmon. *Canadian Journal of Fisheries and Aquatic Sciences / Journal canadien des sciences halieutiques et aquatiques* 58: 1831-41. www.nrc.ca/cgi-bin/cisti/journals/rp/rp2_cont_e?cjfas

Pimentel, David (ed.). 2002. *Biological Invasions: Economic and Environmental Costs of Alien Plant, Animal, and Microbe Species*. Boca Raton, Florida: CRC Press Inc. www.crcpress.com/

Pimentel, David, Bonnie Berger, David Filiberto, Michelle Newton, Benjamin Wolfe, Elizabeth Karabinakis, Steven Clark, et al. 2004. Water

Resources: Agricultural and Environmental Issues. *BioScience* 54(10): 909-918. www.aibs.org/bioscience/

Pimentel, David, Rodolfa Zuniga, and Doug Morrison. 2005. Update on the Environmental and Economic Costs Associated with Alien-invasive Species in the United States. *Ecological Economics* 52(3): 273-288. www.elsevier.com/wps/find/journaldescription.cws_home/503305/description

Pollack, Henry N. 2003. *Uncertain Science . . . Uncertain World.* New York: Cambridge University Press. www.cup.org/

Richardson, Katherine, Will Steffen, and Diana Liverman (eds.). 2008. *Climate Change: Global Risks, Challenges and Decisions.* Cambridge, UK: Cambridge University Press.

Roe, Dilys (ed.). 2004. The Millennium Development Goals and Conservation: Managing Nature's Wealth for Society's Health. London: International Institute for Environment and Development. www.iied.org/docs/mdg/MDG2.pdf

Root, Terry L., Dena P. Macmynowski, Michael D. Mastrandrea, and Stephen H. Schneider. 2005. Human-modified Temperatures Induce Species Changes: Joint Attribution. *Proceedings of the National Academy of Sciences of the United States of America* 102(20): 7465-69. www.pnas.org/

Root, Terry L., Jeff T. Price, Kimberly R. Hall, Stephen H. Schneider, Cynthia Rosenzweig, and J. Alan Pounds. 2003. Fingerprints of Global Warming on Wild Animals and Plants. *Nature* (London) 421(6918): 57-60. www.nature.com/

Rosenberg, Norman J., and James A. Edmonds (eds.). 2005. Climate Change Impacts for the Conterminous USA: An Integrated Assessment. *Climatic Change* 69(1). www.kluweronline.com/issn/0165-0009

Saunders, Stephen, Charles Montgomery, and Tom Easley, with Theo Spencer. 2008. Hotter and Drier: The West's Changed Climate. Boulder, Colorado, and Washington, DC: The Rocky Mountain Climate Organization and Natural Resources Defense Council. 54 pp.

Schneider, Stephen H., and Terry L. Root (eds.). 2002. *Wildlife Responses to Climate Change: North American Case Studies.* Washington, DC: Island Press. www.islandpress.com/

Solomon, Susan, Gian-Kasper Plattner, Reto Knutti, and Pierre Friedlingstein. 2009. Irreversible Climate Change Due to Carbon Dioxide Emissions. *Proceedings of the National Academy of Sciences of the USA* 106(6): 1704-9.

Solomon, Susan, John S. Daniel, Todd J. Sanford, Daniel M. Murphy, Gian-Kasper Plattner, Reto Knutti, and Pierre Friedlingstein. 2010. Persistence of Climate Changes Due to a Range of Greenhouse Gases. *Proceedings of the National Academy of Sciences of the USA* 107(43): 18354-59.

Steffen, Will L., Jill Jäger, David L. Carson, and Clark Bradshaw (eds.). 2003. *Challenges of a Changing Earth. Proceedings of the Global Change Open Science Conference, Amsterdam, the Netherlands, 10-13 July 2001.* Global Change - The IGBP Series. International Geosphere-Biosphere Programme. New York: Springer-Verlag. www.springeronline.com/sgw/cda/frontpage/0,0,0-0-0-0-WEST,0.html

Swinomish Indian Tribal Community. 2009. *Swinomish Climate Change Initiative Impact Assessment Technical Report.* La Conner, Washington: Swinomish Indian Tribal Community, Office of Planning and Community Development. 141 pp.

Tribal Climate Change Project. 2011. A Guide for Tribal Leaders on U.S. Climate Change Programs. Draft. Eugene, Oregon: Tribal Climate Change Project, University of Oregon. http://tribalclimate.uoregon.edu/.

United States Global Change Research Program, Pacific Northwest Regional Assessment Group. 2000. *Preparing for a Changing Climate: The Potential Consequences of Climate Variability and Change: Pacific Northwest.* Seattle: Climate Impacts Group, University of Washington. http://jisao.washington.edu/PNWimpacts/

Valigura, Richard A., Richard B. Alexander, Mark S. Castro, Tilden P. Meyers, Hans W. Paerl, Paul E. Stacey, and R. Eugene Turner (eds.). 2001. *Nitrogen Loading in Coastal Water Bodies: An Atmospheric Perspective.* Washington, DC: American Geophysical Union. www.agu.org/

Ward, Robert C., Roger Pielke, Sr., and Jose Salas (eds.). 2004. Is Global Climate Change Research Relevant to Day-to-Day Water Resources Management? *Water Resources Update* 124. Carbondale, Illinois: Universities Council on Water Resources. http://ucowr.siu.edu/

Wenger, Seth J., Daniel J. Isaak, Charles H. Luce, Helen M. Neville, Kurt D. Fausch, Jason B. Dunham, Daniel C. Dauwalter, et al. (2011). Flow Regime, Temperature, and Biotic Interactions Drive Differential Declines of Trout Species under Climate

Change. *Proceedings of the National Academy of Sciences of the USA* 108(34): 14175-80.

Whitely Binder, Lara C., Jennifer Krencicki Barcelos, Derek B. Booth, Meriel Darzen, Marketa McGuire Elsner, Richard Fenske, Thomas F. Graham, et al. 2009. Preparing for Climate Change in Washington State. *Climatic Change* 102(1-2): 351-376.

Wohlforth, Charles. 2004. *The Whale and the Supercomputer: On the Northern Front of Climate Change.* New York: North Point Press. www.fsgbooks.com/nothpointpress.htm

World Scientists' Warning to Humanity: www.worldtrans.org/whole/warning.html

CLIMATE CHANGE IN THE QUILEUTE AND HOH NATIONS OF COASTAL WASHINGTON

Chelsie Papiez

Master in Environmental Studies, The Evergreen State College, Olympia, Washington

The Native peoples living along the Washington state coast have an intimate connection to the land and ocean and have adapted to many previous environmental and social changes, from the receding glaciers of the last ice age to European American colonization. Yet they may be facing unprecedented disruptions to their coastal way of life because of climate change.[1] Washington tribes are limited to sovereign land bases, essentially fixed political islands within the landscape, without space to migrate away from the effects of climate change, such as sea level rise. The tribes limited to small land bases on the coastline, whether on the shores of the Salish Sea (including Puget Sound) or the Pacific Ocean, are most at risk. The Hoh and Quileute reservations on the Pacific coast of Washington each have only one square mile or less, and therefore their residents have very little land to which they can retreat from increased storm surges and sea level rise. The effects that marine and

---

[1] This chapter is based on the author's 2009 thesis, which can be downloaded in its entirety at http://archives.evergreen.edu/masterstheses/Accession86-10MES/Accession86-10E-Theses.htm

terrestrial ecosystem disruptions will have on tribes' way of life will disproportionately affect the viability of these low-lying oceanside reservations, making it difficult for people to maintain traditional and non-traditional place-based practices. This is especially true for the Quileute and Hoh tribes, whose reservations are bordered on three sides by the Olympic National Park. Not only are their reservation lands close to sea level, but their livelihoods rely on the availability of natural resources such as fisheries.

It is important to document current changes being felt by the Hoh and Quileute peoples so that their intimate knowledge of the coastal landscape can be shared with other communities that will soon experience the impacts from climate change. Their responses could help serve as a model for communities to begin to mitigate and adapt to the changes to come. In the project from which this chapter is derived, I documented the current environmental changes that the tribes are experiencing and the responses both communities have made, are making, and are considering. I used an interview process that relies on the traditional ecological knowledge (TEK) that these communities, like many Indigenous communities, still possess. In a multidisciplinary approach, I aligned TEK with Western scientific findings and predictions. TEK provides the earliest warnings of ecological changes, while Western scientific research projects must often wait years for funding, publication, and review. We no longer have years to meet the challenges of climate change, so TEK is crucial in understanding the current crisis. The final objective of the project was to present a model that can be used by other Native communities and non-Native coastal communities (on methods, see Papiez 2008).

There is no universally accepted definition of TEK. A commonly cited definition identifies it as "a cumulative body of knowledge, practice, and belief, evolving by adaptive processes and handed down through generations by cultural transmission, about the relationships of living beings with one another and their environment" (Berkes et al. 2000). Cajete provides a working definition of what he terms "Indigenous science":

Indigenous science is that body of traditional environmental and cultural knowledge that is unique to a group of people and that has served to sustain those people through generations of living within a distinct bio-region. This is founded on a body of practical environmental knowledge learned and transformed through generations through a form of environmental

Huge driftwood logs piled up next to the Quileute tribal school in La Push, after a storm surge in February 2006.

and cultural education unique to them. Indigenous science may also be termed "traditional environmental knowledge" (TEK), since a large proportion of this knowledge served to sustain Indigenous communities and ensure their survival within the environmental contexts in which they were situated. (Cajete 2000)

Quileute and Hoh Tribes

The Quileute and Hoh peoples have lived in the coastal rainforest of the Olympic Peninsula for thousands of years. They speak a language unrelated to any other in the world and were historically known for their accomplished seal and whale hunting. Life for the Quileute and Hoh peoples has not always been peaceful. When European colonization threatened their traditional land base, legal rights, and culture, they fought for their traditional rights on the land and sea.

Historically, the people of the Quileute Tribe used some 900 square miles from the coast to the Olympic Mountain Range (C. Morganroth interview; Pettitt 1950).[2] Chris Morganroth III, a Quileute elder, noted that they shared camps with other coastal tribes, moving with the seasonal weather cycles from the coast to inland areas.

For the past 100 years or so, since reservations were set up, we were not allowed to migrate

anymore. We used to move with the weather…, moving to different camps with the season from the ocean to further inland. We were not nomads, but rather we moved with the weather to the best camps. There were nine tribes on the Olympic Peninsula; many had communal homes called longhouses. The Quileute people had several villages that were kept open by fire. (C. Morganroth interview)

The Hoh people had a similar pattern. Historically, the Hoh village was characterized by a "busy waterway with 7 settlements along its course and a resident population of 110 or more" (Powell 1999). There was always seasonal movement (before the reservation was established) to hunting and gathering grounds and camps along the river. Jay Powell noted, "the entire watershed was utilized in traditional times. The Old Peoples' cognitive maps of the river were dotted with place names, the boundaries of hunting grounds, and the sites associated with mythic narratives, spiritual beliefs, ritual sites, burial locales, tribal historic events, and favorite foraging spots" (Powell 1999).

During negotiations with Governor Isaac Stevens of Washington Territory and President James Buchanan, the Quileute signed the Treaty of Quinault River in 1855, followed by the Treaty of Olympia in 1856 (Buchanan 1855 and 1856), which resulted in the tribe giving up large tracts of lands for a guarantee of continued access to the traditional fishing grounds "that had

---

[2] All interviews were conducted by Chelsie Papiez as part of research for a master's thesis from June to December 2008.

long sustained their people" (Ralston 2008). Yet their negotiations with the state did not leave them with their "usual and accustomed places" to fish, as identified in the treaties, but instead forced them to move south to a reservation located on the lands of the Quinault people, their "traditional enemy" (Ralston 2008). The Quinault Reservation was an attempt by the U.S. government to consolidate a land base for "the different tribes and bands of the Quinaielt and Quillchute Indians" and eventually for the Southwest Washington tribes, too, after their treaty talks collapsed (Buchanan 1855 and 1856). The Quinault Reservation is approximately 115 miles south of La Push on the current road system. Many Quileute people walked to the Quinault Reservation never to return to their homeland, while some grew homesick and returned to La Push.

After thirty-three years of waiting, the Quileute finally gained a reservation recognized by the United States. On February 22, 1889, President Benjamin Harrison created a separate reservation for the Quileute people, one square mile of coastal land bordered by the Quillayute River and the Pacific Ocean. Currently, their lands are limited to 814-acre reservation, the original reservation with the addition of some private land holdings—a huge reduction from their traditional 900-square mile territory. The reservation is home to approximately 450 members of the Quileute Tribe, with an additional 256 members living elsewhere (Ralston 2008).

The Hoh Tribe gained its reservation four years later when President Grover Cleveland signed an executive order creating it on September 11, 1893. This one-square-mile reservation lay along the Hoh River within the bounds of their traditional lands (Wray 2002). Today this land comprises 443 acres bordered by the ocean, the Hoh River, and Olympic National Park. The tribe has approximately "186 registered Hoh tribal members, 94 of whom live on the reservation" (Wray 2002). Ninety percent of the Hoh land is in the 100-year flood plain, which seems to be flooding more frequently in recent years. In addition, 100 percent of the 443 acres is within a high-risk tsunami zone (Berry interview; M. Riebe interview).

Following the United States' recognition of their sovereignty, the Quileute and the Hoh faced another challenge with the creation of the Olympic National Park in 1938 and the 1953 addition of coastal lands that directly border both reservations (ONPS 2008). The National Park Service has a preservation mission that does not always include Indigenous people, making negotiations for resources especially difficult. Having a

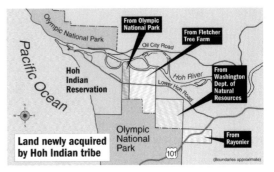

Hoh Reservation tsunami inundation zone, and lands acquired to move housing and government structures to higher ground.

River watershed. Both tribes also re-established ocean fishing rights (Boldt 1974).

To this day the members of the Hoh Tribe make their livelihood primarily from fishing, with some individuals also selling traditional crafts such as cedar woven baskets, cedar dugout canoes, and other carvings (Hoh Tribe 2008). Both the Quileute and Hoh people still gather many of their resources from the tidelands, including razor and butter clams, boots (chitons), and slippers (mussels), and they harvest the forest for cedar, grasses, and berries.

The Landscape and the Communities Today

The Quillayute watershed is fed by rainfall, snowfields, and glacial lakes, making its rivers slow-moving and susceptible to warmer temperatures and lower flows during crucial summer and fall months, when Chinook and chum salmon return to the river system (Hook 2004; Moon interview). In addition to susceptibility to lower flows, the Quillayute sub-basin has an alarming problem with invasive plant species that does not bode well for salmon habitat (Hook 2004).

The Hoh River watershed receives an astounding average of 160 inches of rain annually (by comparison, the city of Seattle receives an average of 36 inches annually). The river, which the Hoh call Cha'laK'at'sit (Southern River), flows from Mount Olympus down to the Pacific Ocean, descending 7,000 feet in a 50-mile stretch before it reaches the ocean (Powell 1999). This heavy rainfall and the steep descent help maintain the river's flow, along with glacier and snow pack run-off that feeds the river system with cool water during the summer months (Golder Associates 2005), though this combination also makes the river more susceptible to extreme high flow events, flooding, and erosion during the winter months.

The Quileute Reservation has two villages. The first is a lower village where the tribal government offices and tribal elementary and middle school sit just above current sea level and are bordered by the Quillayute River. The upper village, where there are more offices, housing, and the high school, is less vulnerable to the rising sea level, but it is still susceptible to large tsunami waves. The Hoh Reservation sits along the mouth of the Hoh River, 90 percent of which is within the historic flood plain and 100 percent within tsunami risk zone (Berry interview; M. Riebe interview). Like La Push, the Hoh Reservation also has lower and upper villages. The lower village was abandoned when the mouth of the river changed and flooding increased in the 1970s (Sampson interview). Family homes were moved to a higher elevation on the reservation near where the current tribal government offices sit, all of which are still in high-risk zones for tsunamis and river channel migration.

The 1964 Alaskan earthquake, the largest ever recorded in North America at magnitude 9.2, triggered a tsunami that smashed into the Pacific Coast from Alaska to California. This is a very vivid memory for many people living on the Washington coast. In La Push boats were damaged, but fortunately no lives were lost. Following a major storm in February 2006 that pushed huge pieces of driftwood up to the school playground (and the December 2004 tsunami that hit South and Southeast Asia), the children of La Push held a mock funeral to protest the dangerous location of their school in the lower village (Kowal 2006). The 2011 Japanese earthquake triggered a tsunami that caused 1.7-foot waves at La Push, rattling nerves but causing no serious damage. Tsunamis are not the only danger; the effects of climate change and sea level rise could also be devastating. Nearly half of the current Quileute Reservation land base is in the river flood zone, making the options slim for moving any of the lower village to safety within the designated reservation. As James Jaime, former executive director of the Quileute Tribe, noted, "If a tsunami warning were issued for the coast, people would have just minutes to travel nearly a mile down the only road out of the reservation" (Crawford 2004).

Currently, the children at the tribal school, along with the entire lower village, organize tsunami drills in which everyone must evacuate up to the highest part of the reservation, where the current high school sits, the A-Ka-Lat. In 2004 the Quileute community successfully evacuated everyone (about 250 people) from the lower village in nine minutes, within the amount of time tribal leaders believe necessary to avoid loss of life (Crawford 2004). Out of fear of tsunamis and increased flooding

during the winter months, the Quileute Tribe began to seek national park lands on higher ground.

Changes on the Coast

I have categorized the changes observed by the Quileute and Hoh communities and their responses to them in six realms: storms, fisheries, diseases and pests, population changes, gathering, and land acquisitions. I have aligned this traditional knowledge with Western science predictions as a way to document the changes that people are experiencing. Although many of the observed changes may very well be caused by the increasingly variable climate, it is difficult to attribute them directly to climate change. Whatever the cause of these environmental processes, the tribal knowledge of and responses to them are unique and have value for other coastal communities, where the impacts of climate change will be felt more strongly in the future.

Storms

Many people on the Quileute Reservation report that storms are increasing in intensity and frequency during the winter (interviews with M. Black; R. Black; Conlow; Hobson; Loudon; C. Morganroth; C. Penn, Jr.; E. Penn; Ratliff; M. Schumack). They report that these changes in storm patterns have only begun to intensify in the last few years, when "superstorms" have hit the area with fierce winds and violent rains. In 2006, waves whipped up by a superstorm threw giant driftwood up to the elementary and middle school playgrounds in the lower Quileute village, sparking the demonstration and evacuation drills mentioned above (interviews with Hobson and Loudon). DeAnna Hobson noted that it was the first time in her lifetime that had happened (Hobson interview).

Quileute elders commented extensively on recent storms. About the 2006 storm, the late elder Chris Penn, Jr., said: "For the past two to three years we've seen an increase in storms…[and] four years ago there was a superstorm that almost took the school away. The lower village was evacuated for the day. Strong winds and currents were coming in from the northwest. There was also high water. The break wall was broken right out front of the school and near DeAnna's house" (C. Penn, Jr., interview). "The ocean almost broke through there [in front of the school] with big storms," Beverly Loudon recalled. "The waves, they get so big that they come crashing through there, knocking everything around. Last year or the year before it was all flooded through there [the school in the lower village]" (Loudon interview).

On the Hoh Reservation storms are also increasing, leading to more floods. In the 1970s the lower village was abandoned because of flooding and erosion at the river's changing mouth. Now the lower village is farther inland from the ocean waves, though still within the tsunami and river flood zones. Testimony from Hoh elders coincides with that of their Quileute neighbors to the north. "More trees are falling. They [storms] pack more wallop now than they ever did. It just costs more money to have these terrible storms, that's what it amounts to" (V. Riebe interview). Local people in the surrounding area also believe that the storms have increased within the last few years (interviews with Dickerson; B. Johnson; Lien; Northcut).

Unable to revive the seasonal migrations of their ancestors, the Quileute and the Hoh must endure the increasing storms without retreating inland (C. Morganroth interview). This is a tough situation to face, because both tribes recognize that the storms are eroding away their recognized sovereign lands, both literally and figuratively. Marie Riebe, a member of the Hoh Tribal Council, asked, "How are we going to save what land base we have left, because you can't create more land" (M. Riebe interview)?

For Quileute crab fishermen, increased winter storms mean little hope of getting out of the narrow La Push Harbor to practice their livelihoods. One interviewee said, "The last two or three years it has almost been unfishable. We had five days that were 17-foot or greater [swells], then we had one day with a 33-foot combined swell, and there's a lot of days where you'll have the 40-knot winds. It's crucial [economically], that [crab fishery] makes or breaks the fishermen for the year" (Northcut interview).

In response to increased storm activity, the Quileute Tribe negotiated a "Fair Weather Agreement" with Washington state that allows the Quileute fishermen equal access to fishing grounds with additional days added to the season when the weather is unremittingly too rough for fishing vessels to leave port. Fishermen from the north or the south of La Push can depart from their ports more easily than the residents of La Push, so this agreement helps the Quileute catch their quota. "In the last two seasons there have been very violent storms that have stayed non-stop in the region for thirty days or greater, so we negotiated a twenty-day guaranteed fair weather agreement," the Director of Quileute Natural Resources told me (Moon interview). For many tribal fishermen, the new agreement brings them hope of maintaining their income if the crab is abundant—

but this species, too, faces environmental challenges, as I discuss shortly.[3]

Precipitation

Along with more severe storms comes more rainfall. It may be that more precipitation is falling in the form of rain instead of snow, increasing the river flows and flooding in the winter months. Looking back to her childhood and recalling walking to her uncle's house for get-togethers during the winter, the late Pearl Conlow remarked, "I guess I just didn't notice it [rain] when I was small, when I was younger. There didn't seem to be as much rain, because a lot of our winters were spent in get-together." Many people on the Quileute Reservation agree that there is more rain falling now than ever before (interviews with M. Black; R. Black; Conlow; Hobson; Loudon; Matson; C. Morganroth; C. Penn, Jr.; E. Penn; Penn-Charles). "Rains have increased during the months of November through February. Seven-to-ten-day floods have made the river breach its banks and cause flooding in the lower village" (Hobson interview). The records kept by Jerry King in Forks (1907 to present) coincide with the elders' testimony. The highest annual rainfall occurred in 1997 with 162.14 inches, followed by 160 inches in 1999, whereas on average the Forks area receives 118.33 inches annually (King 2008). These high rainfall years span El Niño Southern Oscillation (ENSO) variations on the El Niño (1997–1998), and La Niña (1998–2000) weather phases, which may indicate that something is occurring outside of these "normal" climatic cycles (NOAA 2008a). Typically, El Niño years tend to bring drier and warmer winters, while La Niña years tend to bring cooler and wetter winters (CIG 2008).

Changes are occurring in the timing of rainfall as well. This is very important for many natural cycles, but for the Quileute and Hoh people, timing is everything

---

[3] In the summertime, thunderstorms have been in decline. I received this report late in the project, after most interviews were complete, so I didn't ask anyone specifically about summer thunderstorms. However, this is an important observation that should not go unmentioned. "I guess the thing I'm worried about most is the loss of thunder[storms]. I remember having them a lot when I was a boy, my dad would tell me it was Thunderbird and Whale fighting again. We used to have twenty to thirty thunderstorms a summer, this year [2008] we had two" (C. Penn, Jr., interview). More people should be interviewed about thunderstorms, because they play an important role in oral history and signal a change in summer weather fronts.

for productive salmon runs. In recent years the timing of rainfall has not been consistent (interviews with Leitka; C. Morganroth; Penn-Charles). As Mary Leitka (Hoh) noted, "Really it's a change in the weather. Before it used to be, when August would come, it would rain so hard…. That's when the run of the salmon would come in. Now that's changing and we don't get rain until later on, so the run is changing its cycle in time" (Leitka interview). Salmon waiting for fall freshets at the river's mouth have to spawn in the mainstem of the river where water levels are flowing adequately. When heavy rains follow mainstem spawning, the incubating eggs can be devastated by high flows, as fast-moving waters scour out the gravel beds where the eggs rest.

Conversely, precipitation in the form of snow has declined on the coast. Winters are now milder in temperature with fewer robust cold fronts (interviews with Conlow; Jackson; Wallerstedt) to keep snow on the ground for more than a day. On both the Quileute and Hoh reservations, people indicated that there has not been snowfall in recent years as there used to be (interviews with M. Black; Jackson; Matson; E. Penn; Sampson; Wallerstedt). "We used to look forward to the snow," Bertha Wallerstedt, a Quileute elder, remembered, "because we had all four seasons then. We haven't had that for a long time. Winters are mild; it seems like they're not hard like they used to be, they're not real cold like they used to be" (Wallerstedt interview). The late Quileute elder Pearl Conlow, said, "Well there seems to be a lot of difference now. It seems to be a lot warmer now than it used to be" (Conlow interview). Gene Sampson (Hoh), reported that "my parents would talk about the abundance of everything…[and] that was because they were getting regular [events], they got five feet of snow every year" (Sampson interview). Moreover, snowstorms and the flooding that used to follow are no longer predictable on the Hoh Reservation.

In this case, local ecological knowledge (LEK) agrees with TEK that snowfall around both reservations and in the Olympic Mountains is not what it used to be (interviews with B. Johnson and Peterson). LEK is a knowledge base that has a shorter timescale than TEK and is held by people who do not have historical origins in the area but who have observed natural patterns during their lifetime and/or passed their knowledge down through several generations of their families and other resource users. These TEK and LEK responses align with predictions by the Climate Impacts Group (CIG) at the University of Washington (and others in the scientific community) that the amount of precipitation falling as snow will decline as the climate warms, and there is evidence in the Cascade Mountains that it is already happening (CIG 2008).

Moreover, precipitation as rain is only supposed to increase with the onset of a warmer, wetter climate in the Pacific Northwest (CIG 2008). This may already be occurring in the Hoh and Quillayute River systems. Increased river flow and flooding has, in recent years, profoundly affected both the Hoh and Quileute communities, initiating both tribes to seek higher ground out of the channel migration zone (CMZ).

For example, during a December 3, 2007, rainstorm, high flows occurred in both the Quillayute and Hoh Rivers, adversely affecting the Chinook salmon runs that had recently spawned in the river. In the last few years flows on the Hoh River reached more than 60,000 cubic feet per second (cfs), whereas the average winter monthly flow rate is about 4,000 cfs (USGS 2007).[4] On the Quileute Reservation, flooding has been increasing because of both high river water levels and an increase in ocean storm surges. The lower village is extremely susceptible to flooding after heavy rains that cause the river to breach its banks, coinciding with a high tide and wind-driven waves. Bertha Wallerstedt, a Quileute elder, described a recent flood: "Year before [the winter of 2006-2007] it flooded and the waves came in up there at Lonesome Creek [Store, in lower Quileute village], and they came all the way up to the cabins over here, it was unusually high, high water" (Wallerstedt interview).

With four rivers—the Sol Duc, Dickey, Bogachiel, and Calawah—feeding into the Quillayute, increased rainfall and water flow exacerbate erosion along the river's edge and make traditional resources more difficult to obtain. Cedar logs were traditionally captured along the Quillayute River during the winter months, so that the roots of the cedar tree could be harvested and used to weave traditional burden baskets. Elder weavers recall gathering the roots with their grandmothers when they were young, but now flooding has eroded away all of the banks where logs were harvested (interviews with L. Morganroth and M. Schumack). Moreover, the heavy rains have saturated the soil to the point that trees are falling over with even the slightest gusts of wind. "Rainfall has increased," Miss Ann Penn-Charles (Quileute), observed. "The soil has become soft or saturated so much that any little wind blows over huge trees…two or more people around" (Penn-Charles interview).

---

[4] In October 2004 and November 2006 Hoh River flows reached 62,100 and 60,700 cubic feet per second (cfs). Such extremely high flow peaks had never been seen since recordkeeping started in 1961.

Living on the Quillayute River, DeAnna Hobson watches it rise every winter as the rains come, observing the floods firsthand. She stated that "millions of board feet" have been coming down the river annually, but the quantity and high water flow make it too dangerous to capture the logs as they drift downriver (Hobson interview; see also Ratliff interview). During the winter, mounds of debris can be found on the beach, with nearly every inch of the river covered in downed trees. The increased water flow has affected tribal reforestation efforts. Steve Allison, the Director of the Hoh Natural Resources Department, stated that the 4,000 trees planted in an area called Schmidt Bar on Hoh River Trust Land had been lost in the winters of 2006 and 2007. "I mean, we lost forty acres there alone and we planted probably 4,000 trees" (Allison interview). Elder Roger Jackson on the Quileute Reservation made a similar observation (R. Jackson interview).

Almost every winter now the Quileute Reservation is cut off due to extreme flooding (interviews with Hobson; C. Morganroth; Penn-Charles). In 2004 the flood waters covered the La Push Road and forced people out of their homes: "People were in distress up river…getting rescued by boat" (Hobson interview). When this, the lone highway into La Push, is blocked by high waters, the only way that the Quileute people can get further inland is by traveling rough logging roads (Davis interview).

During November 2006 a severe storm hit the Hoh Reservation, causing the river to spill over its banks and flood into the village and its septic system. Mary Riebe (Hoh) recalled, "Our septic system backed up and toilets became fountains!" (M. Riebe interview). The Hoh received no help from the county government, so U.S. Congressman Norm Dicks urged mobilization of the National Guard, which arrived three hours later, bringing food and supplies to the village (M. Riebe interview). During this particular storm, the Hoh River flow peaked at 60,700 cubic feet per second on November 6, further exacerbating the flooding and erosion on the Hoh Reservation (USGS 2007).

Alexis Berry, former executive director of the Hoh Tribe, said that tribal members have told her that the Hoh River floods used to be on a 10-year cycle but that the frequency has increased; floods are now occurring every year. Everyone that I interviewed responded similarly (interviews with M. Riebe; V. Riebe; Sampson). Hoh elder Viola Riebe noted that "it's flooding more than when I was a little girl. We always have high water when it rains in the fall time" (V. Riebe interview). The lower village of the Hoh Reservation has already relocated once because of flooding near the river mouth.

According to interviewees, the last family to occupy that area on the oceanfront left in the 1970s (interviews with Sampson and V. Riebe). Viola Riebe said, "No one lives down there anymore. They moved up over the hill and then up over the hill. They're having to move from there because that's … major flooding every year" (V. Riebe interview). Permanent sand bags now line most homes and facilities in what is now the lowermost village. The tribal office and community center has been repeatedly flooded in recent years and now has a permanent berm around its perimeter, built with eight feet of imported soil (Sampson interview).

In response to the extreme floods, the Hoh Tribe is pushing to get certified by the Federal Emergency Management Agency (FEMA) so that the tribe can in the future apply for grants. The tribe is currently training its staff in the National Incident Management System (NIMS) so that they are prepared to handle natural disasters. FEMA has already supplied the tribe with two satellite phones, emergency radios for every home, and a defibrillator for every building. "We need to be ready…not wonder, who has it or where is it?" (M. Riebe interview).

Not all of these problems can be blamed on climate change. Past logging practices are one of the major causes of the increased flow and flooding of the Quillayute River (interviews with R. Black; Hobson; C. Morganroth; Penn-Charles; J. Schumack; Williams). Chris Morganroth III, a Quileute elder, pointed out that "the logging caused more runoff that straightened out the river's path so it no longer bends and winds down to the mouth" (C. Morganroth III interview). Without a meandering river bed to slow down the river's flow and with an increase in rainfall, flooding occurs less predictably. As Gene Sampson (Hoh) noted,

> They [elders] knew when to go and to get out…. Now it could be like this here now [it was a hot summer day] and tomorrow we could be floating…. Not even anyone can predict what it does anymore…. The bad thing about it is that we don't know when it's going to change because it used to happen around November, that was the framework we expected to get ready to sand bag…. Last couple of years…all of a sudden when no one's ready [a flood may occur]. (Sampson interview)

Further, although river flooding on both lower reservations has always been a problem to some extent, according to the elders, today with no opportunity to move inland seasonally, the tribes suffer more from the blunt impacts of flooding.

Fisheries

In both the Hoh and Quillayute River systems, the number one factor that has been shown to influence the maximum sustainable harvest (MSH) of salmon is the summer low flows (Northcut interview). Successful spawning is reduced by low flow events during key summer months, lessening the success of the salmon runs in the Quillayute River. This knowledge is extremely important when looking toward future impacts that climate change is projected to have—including warmer temperatures, less runoff from declining snow pack, and changes in the timing of rainfall in key summer months. The likelihood of increasing occurrences of summer low flows will be a crucial factor in salmon survival.

Timing of Runs

One of the most significant runs, the surf smelt, has been seasonally late for the past few years. As part of an annual tradition, the Hoh and Quileute peoples seine-net for smelt in the rivers and on the ocean shore. Communications take place between all tribes along the outer coast in order to report the first arrival of the smelt. "Smelt changes are true up and down the coast," noted Miss Ann Penn-Charles (Quileute), "because they are gathered from Neah Bay to Queets, wherever the first smelt arrive" (Penn-Charles interview). Many people look forward to the smelt harvests, for the fish are a delicacy and a subsistence food of both tribes. The smelt were put up in tribal smokehouses to prolong the food source. But throughout my interviews with Quileute and Hoh tribal members, Natural Resource staff, and local fishermen, I found that although smelt had been seasonally abundant in the past, now few return to spawn (interviews with G. Johnson; Leitka; Lien; E. Penn; Penn-Charles). The recent unprecedentedly low annual returns of smelt are undermining this age-old tradition of seasonal seine-netting.

For the Quileute people the timing of the smelt runs has been a significant part of an annual Elder's Week celebration. Traditional foods are harvested and prepared for the elders. One of the most prized was the "stink eggs" fermented from the smelt spawners returning annually to the Quillayute River mouth. Miss Ann of La Push tells of the shift in timing of the smelt run and the effects of this change:

> Fish are coming in later and later in the summer. In 2007 beach smelts or night smelts came a month late. Normally they come in May, by Elder's Week, [but] that year they came in June.

There have not been stink eggs ready for Elder's Week for the past three to four years. That would require the smelt to come in early May and be fermented for three weeks before Elder's Week. (Penn-Charles interview)

For Hoh people, smelting has been part of their livelihood since the beginning of time. The tribe's oral history tells of a transformation performed by the "Changer" K'wati "who went around the world making things as they are today" (Andrade 1969). At Hoh River he found the "inhabitants of the area were upside down people, who walked on their hands and handled their smelt dipnets clumsily with their feet. They weren't very good at it, so they were famished and skinny. K'wati set them rightside up and showed them how to operate their nets with their hands" (Andrade 1969). K'wati then told the ancestors, "You shall use your feet to walk… Go and fish smelt. You shall catch much fish when you fish smelt. Ever since then there is much smelt at Hoh" (Andrade 1969).

Emphasizing how important smelt are to the Hoh Tribe, Mary Leitka said, "You know, I think that because of the change and now I see this is probably going to be the worst thing of all to see, and that's [the loss of] our smelts." She went on to say that when she visited Sitka, Alaska, people there reported not seeing the silver smelts return to their area, so in response Mary made arrangements to trade smoked Hoh River smelt for herring eggs. "So that was going to be our trade and now we can't do it…. We'd used to be able to get them on our river and we'd smoke them and now there's none left" (Leitka interview). Intertribal trading was always a way to exchange abundant local foods for something rare, but now it is becoming even more important in maintaining traditional food systems, as species ranges shift and/or decline in response to climate change.

Not only is smelting an important tradition, but smelt is an important subsistence food source for many people in the tribe. Gene Sampson talked about the past abundance of the smelt in the summer being so great that "ten years ago you could sit here every day and you could smell them from here [the tribal office], from the beaches." It was such an important food source that anyone hungry "could walk down and get some." Times are getting "scary" for people so dependent on the seasonal production of food sources (Sampson interview).

In response, many are resorting to putting up farmed fish in their smokehouses as a way to get by in the absence of the once abundant annual returns of the smelt. "Smelts we used to smoke, now we have to go to our hatcheries to get fish," Gene Sampson (Hoh)

Sandbags permanently around the Hoh Tribal Offices to guard against river and coastal flooding.

lamented. "Farmed fish they are doing now just to keep people satisfied, you know. We were the smelt capital of the world; now it's…very slim pickings. Just a couple of years ago you'd get so sick of eating them…back in the '70s there'd be smelt.… Those are the things that you heard…[that the fish] will never ever be gone" (Sampson interview).

Population Changes

In addition to the smelt arriving late, there have been shifts in other species populations. Many of these changes are difficult to attribute directly to climate change, because they may be influenced by a number of factors.

In 1999, Jay Powell reported that in tidal areas around the Hoh River mouth there were "plentiful" skatefish, tomcod, and sturgeon. At that time, fisherman Frank Fisher caught a 14-foot-long sturgeon at the river's mouth (Powell 1999). Now, sturgeon are very scarce there (interviews with Leitka and Sampson). Tomcod and flounder have almost disappeared. Gene Sampson blamed the temperature changes in the river for the loss of sturgeon, flounder and tomcod:

> We average 57-degree water, [then it was] 66 or 64 [degrees] for four or five years was way unnormal for the temperature of the waters and when that started happening, we lost a lot of our sturgeon, our flounders. We used to get the tomcods, [but we] lost those in the early '80s, they moved to La Push…and I don't know what year they lost a lot of theirs, it used to be a regular thing catching them in the boat basin, a lot of the kids over there used to catch them and sell them to the long liners as bait, not even the kids can make any money now. (Sampson interview)

The decline is also occurring in the Quillayute River. The tomcods used to be so thick in the water that the whole surface of the water was "caked with them" (R. Black interview). People seining for smelt used to bring up a lot of perch and flounder, but they are not as "plentiful" as they once were (E. Penn interview). A commercial fisherman who has fished out of La Push for twenty-two years also stated that tomcods have declined since his arrival, when they were abundant (G. Johnson interview).

The same is true for salmon populations in both the Quillayute and Hoh River systems. Many practices have threatened salmon populations—overharvesting, logging, development, and so on—but it is important to take note of the abundance that had been present in both river systems. Mary Leitka from the Hoh Tribe recalled a hike she took with an elder upriver to visit areas that were used for seasonal fish camps. The elder told her that the chum had been so thick "we could walk on their backs" but "now we don't see them anymore" (Leitka interview).

Fortunately, the tributary waters of both the Quillayute and the Hoh rivers are within the Olympic National Park boundaries, not on privately owned timberlands. Park ownership has kept the habitat more or less intact since the area has been protected (Lien interview), although the boundary may be contested for gathering, hunting, and fishing rights. Still, salmon runs in both river systems are not what they were historically.

In 2008 the salmon runs were "pretty depressed" for the Hoh River, according to the Hoh Natural Resources Director. Fishing was "shut off for spring Chinook [the prized run] on June 16th and normally it's the last week of July" (Allison interview). The fishery had never been shut down that early in the season, at least since the 1980s when Steve started working for the Hoh Tribe (Allison interview). One Hoh tribal fisherman recalls poling upriver in the 1960s and getting forty to fifty 35-pound salmon in a single drift, but today "it's down to hatchery size, which is very, very little" (Sampson interview).

In response to the lower numbers, both tribes are improving riparian habitats. The Quileute have been actively removing invasive Japanese knotweed in the watershed to help improve natural water conductivity (Northcut interview). The Hoh Tribe has planted trees to help keep the river cool in the summer months (Allison interview). From a fisheries management perspective, if salmon numbers continue to decline, there are few options other than to hope that "the state will further limit their [fishery] and our fishery" to keep the runs going (Lien interview). Such limits were imposed in California and Oregon in 2008. "It affects everybody, not just the Natives," Gene Sampson said, "but the non-Indians too. It…affects everybody one way or another and it's tough for our people" (Sampson interview). Looking towards the future, the late Quileute elder Chris Penn, Jr. (Jiggs), testified during the Boldt Treaty case, "Resolution for all of these problems…[is] that we've got to sit down together, otherwise we're all sitting on the bank. Our resources come first, that's what keeps us alive" (C. Penn, Jr., interview).

Hypoxia

Ocean "dead zones" have been recorded along the Pacific coast from La Push south through Oregon. These dead zones are caused by low oxygen levels—hypoxia—in the ocean, which results in massive crab and fish kills that wash to shore. The recent hypoxia events along the Pacific coast have been called "unprecedented" and "severe" within the California Current System along the West Coast (Grantham et al. 2004). The events are alarming to everyone along the coast who depends on the ocean for their food and livelihood. Changes in ocean upwelling, which brings nutrient-rich cold water to the surface, may be the cause of these pockets of low oxygen. Ocean upwelling contributes to the physical transport of low oxygen water and nutrients to the near-shore productive continental shelves (Chan et al. 2007; Grantham et al. 2004).

According to the Intergovernmental Panel on Climate Change (IPCC), low oxygen levels may occur because of a variety of factors including the following: "biological activity, changes in the physical transport of oxygen, or … a change in temperature and salinity" (IPCC 2007b). The respiration of marine life in these productive shelves can further decrease oxygen availability, resulting in hypoxic conditions along the shelf (Schwing and Mendelssohn 1997). The timing of upwelling depends on decadal cycles in the Pacific Ocean, such as the El Niño Southern Oscillation (ENSO) and the Pacific Decadal Oscillation (PDO) (NOAA 2008b). Many pelagic fish rely on coastal upwelling for their reproductive success. The strength and timing of upwelling is significant in providing the right mixing of water columns to maintain the concentration of food organisms (Cury and Roy 1989). When this system fails, it can result in the major fish and invertebrate kills that have been observed up and down the Washington and Oregon coastline in the first decade of the twenty-first century.

Regional researchers recently made a connection between changes in upwelling patterns and climate change, stating that "delayed early-season upwelling and stronger late-season upwelling are consistent with predictions of the influence of global warming on coastal upwelling regions" (Barth et al. 2007). According to the 2007 IPCC Report, ocean water "freshening [the inflow of fresh water from Arctic melt] is pronounced in the Pacific," and this may be contributing to the changes in upwelling and oxygen transportation being seen in the Pacific (IPCC 2007b).[5]

In the eastern Pacific Ocean off the western coast of the United States, coastal upwelling is wind-driven (Bakun 1990). As a result, changes in wind direction, strength, and timing can also greatly contribute to upwelling, which may then have a significant impact on marine productivity. The Quileute and Hoh elders whom I interviewed are already observing changing patterns in wind direction and timing along the coast of Washington.

For the people of La Push and Hoh River, these changes have affected and continue to affect their subsistence food sources and marine resource economies. Hypoxia events have been the priority issue for many First Nation gatherings and will continue to be because they harm the salmon, the lifeblood of many Northwest tribes, as well as other resources (Lekanof 2008).

Crabbing has been particularly vulnerable to hypoxic events up and down the West Coast, and for Quileute fishermen, crabbing is very economically important. A published report on low-oxygen conditions off of Newport Beach, Oregon, indicated that the key timeframe for hypoxia is from late summer through September, during crab molting season (NOAA 2008b). For tribal fisheries, crab season typically begins in November, after the crab have developed hard shells and their meat content is high. In addition to the crab kills south of La Push and the Hoh River, fishermen reported that crab have been slow to "harden up" in the fall. The state requires that the crabs being harvested have 23 percent meat content with hard shells, but in the last few years their shells have not hardened on time for harvest (interviews with G. Johnson; Northcut; Ratliff; J. Schumack). A member of the Quileute Natural Resources department remarked, "In fact, one of the years I believe it went all of the way into January before the meat content was ready.… In the 2004-2005 season, it [shell hardening] didn't start until week 16 [the

---

[5] *Freshening* is the addition of fresh water to the ocean, lowering the ocean's salinity, particularly at the surface. Fresh water from increased precipitation and the melting of land-based ice is expected to increase ocean freshening in the decades to come (Fedorov et al. 2007).

middle of January] when the crab season [starts] for tribe [October 1 to September]" (Northcut interview). People worried that the water temperatures were too warm and that was causing the delay in the crab's life cycle. As Quileute fisherman Steve Ratliff mentioned, "The crab, kind of a strange cycle for a while…. It was taking way too long for them to harden up, the water temperature was just strange, it was warmer than it should have been [last year]" (Ratliff interview).

The delay in shell hardening is worrisome if it is indeed related to the timing of ocean nutrient upwelling. Dungeness crabs molt every season and must receive the right nutrients to be able to rebuild their shells for the winter months. The delays in the formation of crab shells may be an early sign that something is changing in the oceans. More changes in the marine environment may be seen in the future because of the additional threat of ocean acidification.[6]

Domoic Acid

In addition to hypoxic events, domoic acid outbreaks have become common along the Washington state coastline. A product of a diatom (algae) species called *Pseudo-nitzschia*, domoic acid is absorbed by crab and shellfish and is a human health risk if ingested. According to the Quileute Natural Resources office, domoic acid events on the West coast began in 1991 (interviews with Moon and Northcut) and are now common during the summer. Before the 1990s, it had only been found on the east coast, and we do not know whether it came to the Pacific in ship ballast water or if it arrived as a result of warming coastal waters.

For Quileute and Hoh peoples, domoic acid outbreaks affect their ability to freely gather shellfish because contaminated shellfish may cause short-term memory loss and other neurological problems (Burkholder 1998; Moon interview). Not only does domoic acid pose a human health risk, but it can also adversely affect marine mammals and birds who ingest infected shellfish and crab (Gulland 2000; see also Ramsdell and Zabka 2008).

In 2008, Hoh and Quileute peoples reported that clams were particularly small in size and number on the ocean beaches (interviews with Leitka; V. Riebe; Sampson; Wallerstedt), problems that may be the result of domoic acid. But with more restricted harvests

because of the presence of toxic levels of domoic acid there would logically be more shellfish available. Some studies have indicated that domoic acid from *Pseudonitzschia* causes blood cell abnormalities in the Pacific oyster (Jones et al. 1995), so it could be that the acid may have other physiological effects, such as stunting the growth of coastal clams.

The people I talked to often suggested that hypoxic events and disease were the cause of the problem for shellfish. Mary Leitka, a Hoh elder, told me, "I think that warmth is causing a lot of the disease that is getting into our seafood" (Leitka interview). "The waters don't have enough oxygen," Viola Riebe, another Hoh elder, pointed out. "Oxygen is not plentiful. Last year or the year before our ocean floors were filled with dead sea life. Now we can dig clams at Kalaloch and now they are 1 ½ to 2 inches when before they were 4 inches or bigger. Huge, huge clams" (Riebe interview). "What really hurts all the tribes along the coast is the water conditions [domoic acid and hypoxia]…anything along the coastline is no longer predictable whether we are going to have this or that anymore," Gene Sampson (Hoh) remarked (Sampson interview).

As Gene Sampson made clear, LEK also revealed that changes in the ocean are becoming unpredictable, and the extent of the problem is becoming hard to follow. "You can't just count on the normal trends that were happening in the past," Kris Northcut of Quileute Natural Resources told me. "All of those indicators don't hold a lot of weight nowadays because there are just so many different things taking place out in the ocean, certain upwellings aren't taking place, temperatures aren't happening, dissolved oxygen problems…. I mean it's… things that weren't happening are happening now, so if you're not tracking it closely, you could have some serious problems" (Northcut interview).

In response to the increasing outbreaks of domoic acid along the coast, the Quileute began regular testing at local beaches to ensure that shellfish harvests are safe (interviews with Davis and Moon). The tribe also started a program with the Washington State Department of Ecology to monitor the domoic acid blooms by taking shellfish samples from the beach. The director of Quileute Natural Resources, Mel Moon, explained the tribe's goal and the success of the program: "Our goal was to have rapid assay [toxin] tests so that people could know immediately if there was trouble and not to harvest. Now we have a lab that has the capacity to test them. They are day-after tests and we post the results the next day" (Moon interview).

With rapid testing and the state-operated shellfish hotline, harvesting can continue when levels are safe.

---

[6] Ocean acidification is occurring as the ocean absorbs anthropogenic carbon (since 1750), a process that may have negative impacts on "marine shell-forming organisms" (IPCC 2007a: 9).

In addition to testing, the Quileute have been working with the Northwest Fisheries Science Center (NWFSC) to map the toxic blooms and find out what is triggering them (interviews with Moon and Northcut). It appears that nutrients coming out of the Strait of Juan de Fuca, along with upwelling nutrients, now create the "ideal conditions for *Pseudo-nitzschia*" to bloom (Moon interview). Warming ocean temperatures and changes in ocean upwelling may also be to blame. Many algal blooms have been associated with El Niño events, which bring warmer water, so this warming trend "has led to the suggestion that global climate change and warming trends may also encourage their [the algae's] growth" (Burkholder 1998). The increased activity of algal blooms and their shifting geographic range may also be caused by climate warming trends (Epstein et al. 1993).

Along with being involved in the NWFSC, the Hoh and the Quileute are working with the Intertribal Policy Council (IPC), which brings together coastal tribes, the National Oceanic and Atmospheric Administration (NOAA), and the Olympic National Coast Marine Sanctuary (OCNMS). Joe Gilbertson, the Fisheries Management biologist for the Hoh Tribe, explained,

> The coastal tribes have formed the partnership with IPC…to have a consensus and an agreement to move forward in a unified and collective manner with regard to evaluating these physical and biological … parameters. Partnership with IPC is to get the buoy system running better to record surface temperatures, wind, and swell directions, [and] physical parameters." (Gilbertson interview)

In 2006, the OCNMS installed monitoring stations along the coastline that will bring new data on hypoxia by measuring the temperature, dissolved oxygen, and salinity of the ocean (OCNMS 2006). With this knowledge gained through Western science, combined with TEK, the Quileute and Hoh peoples may be able to start predicting hypoxic and domoic acid events in a changed environment.

Diseases and Pests

Warming temperatures on land and sea enable influxes of invasive and noxious plants into new areas, so pests and diseases are predicted to increase with climate change (CIG 2008). On the Olympic Peninsula some of these invasions are already beginning to take place. Through interviews with the natural resource staff from both the Hoh and Quileute tribes, I gained a sense of the types of issues both nations have begun to face as a result of these changes. New insects and bacteria are causing illness among native species on land, just as *Pseudo-nitzchia* has in the ocean.

Noteworthy among new pests is a new louse that has affected deer and elk populations around the Olympic Peninsula. Commonly termed "hair slip," this non-native species has "further pushed the elk population down" (Geyer interview). How the louse species was introduced or whether it has anything to do with the changing climate is unknown. Personnel at both the Hoh and Quileute Natural Resources offices mentioned this as a new problem that, as of yet, has no solution (interviews with Allison and Geyer).

In addition to the louse, the elk are suffering from a new "hemorrhagic disease," which was reported by the Director of the Hoh Natural Resources. According to Allison, the wasting disease is caused by an anaerobic bacterium that grows in fields with standing water. The elk come in contact with the bacteria and "it's like a hemorrhagic disease that is fatal.…They witnessed that down in Clearwater [a tributary of the Queets River]. That's a function of heavy rains in spring and sunlight that would cause a bloom of the stuff, and they would ingest it and it would cause a hemorrhage" (Allison interview). With increased flooding, the disease could become more of a health problem for the elk, a subsistence food source for the Quileute and Hoh peoples.

Animals are not the only ones to suffer from new pests. The Sitka spruce weevil (*Pissodes strobe*) has begun taking hold of the spruce trees within the Hoh River watershed. In the mid-1980s this insect spread explosively on private timber plantations (Thysell and Allison 2005), and it has been shown to perform better with the warmer temperatures (Heppner and Turner 2006) that monoculture plantations provide. It primarily attacks the Sitka spruce because this tree grows at lower, and therefore warmer, elevation, but Engelmann spruce at higher elevations is also susceptible (Heppner and Turner 2006). A weevil attack causes deformed limbs and branching, along with stunted growth and needle drop (Heppner and Turner 2006). This can be devastating to riparian zones, especially in the temperate rainforest where spruce shading of streams is "a key component of salmon habitat" (Thysell and Allison 2005).

As a way to control the proliferation of the spruce weevil within its watershed, the Hoh Tribe has planted around 40,000 alder trees to keep temperatures cool around groves of spruce trees, essentially shading out the weevil. This is an environmental control mechanism that avoids the use of chemicals in a riparian area, helping to sustain one of the traditionally predominant riparian trees in the watershed by lowering insect activity (Allison interview).

In response to increased pests and diseases, both the Hoh and the Quileute have continued to focus on the ecosystem level to figure out how best to address species declines, because everything is interconnected. As Hoh elder Mary Leitka said,

> I think that we need to look at the connection of the food sources and to what is disappearing, because…it's not just the elk and the deer. There's many…things that depend one upon the other, even clear down to the little spiders and bugs, Mom used to talk about that. I think it's important that we realize that somehow we're going to have to work at protecting many of them. (Leitka interview)

Species Range Shifts

As predicted, marine and terrestrial species have shifted their ranges northward in latitude or upward in elevation in response to the warming climate. "Climate-related changes in fish distribution have been typically characterized as range shifts or displacement away from the center of the home range, as temperatures grew warmer" (Zeidberg and Robison 2007; see also Perry et al. 2005). People on the Quileute and Hoh reservations have already observed some potential range shifts in avian, marine, and terrestrial species.

Among the birds, brown pelicans (*Pelecanus occidentalis*) are a new visitor to the Washington coastline, according to many of my interviews with elders (Jackson; Matson; C. Morganroth) and other holders of LEK (Dickerson 2008; B. Johnson 2008; G. Johnson 2008; Payne 2008), tribal fishermen (Moon 2008; Ratliff 2008; J. Schumack 2008), and natural resource staff (Geyer and Northcut). The consensus was that they had arrived to spend a few summer months in Washington around the mid-1980s (interviews with Moon; C. Morganroth; Payne). Now, the brown pelican are arriving during the second or third week in June and departing the last week in October (interviews with Geyer; Moon; C. Morganroth; C. Penn, Jr.).

The brown pelican's original northern range was limited to California. In the late 1960s, it became a federally listed endangered species after the DDT pesticide crisis left many birds unable to form eggshells (Blus 2007; Wickliffe and Bickham 1998). The California population seemingly recovered and moved north into

Washington, but sources still list the brown pelican's northern range as California (Cornell Lab of Ornithology 2003). Warming waters and the resulting decline in food, as well as recovery from DDE, may all be factors in their migration northward to the Washington coastline.

Among marine species, the Humboldt squid (*Dosidicus gigas*) started frequenting the waters off the coast of Washington in the late 1990s and early 2000s. The Humboldt squid can be more than 2 meters long and weigh up to 110 pounds (Zeidberg and Robison 2007). It is a voracious top predator that may pose a problem for Washington coast fisheries, especially salmon runs (Blumenthal 2008). Sources of TEK (interviews with R. Black; Moon; Ratliff; J. Schumack; Williams) and LEK (interviews with Dickerson; G. Johnson; Northcut; Payne) indicate that this creature originally arrived during an El Niño event but kept returning even during cooler La Niña cycles. Such observations confirm several reports from the scientific community that indeed the Humboldt squid is expanding its original range beyond the warm waters around the equator. During the 1997–1998 El Niño, the squid migrated to central California, and after the 2002 El Niño they moved in to stay (Blumenthal 2008; Cosgrove 2004; Zeidberg and Robison 2007). "This geographic expansion occurred during a period of ocean-scale warming, regional cooling, and the decline of tuna and billfish [previously the top predators] throughout the Pacific" (Zeidberg and Robison 2007; see also Sibert et al. 2006).

Though the change in the distribution does not follow the "normal" range shifts or displacements, because the Humboldt squid is not leaving its original range but rather expanding north and south, the animal's

Brown pelicans have migrated in large numbers to the Washington coast since the 1990s, even though bird guides still show them no farther north than San Francisco Bay. This pod of pelicans was at La Push, Quileute Nation, in 2005.

presence in historically colder waters raises concerns. The waters of the northeastern Pacific may be warming enough to support its habitation. Zeidberg and Robison stated that its range expansion does not strongly correlate with warming sea surface temperatures, but its expansion during El Niño events suggest a strong "warm-water affinity" (2007). However, fishermen in La Push have been seeing an increasing number of squid during La Niña years, as recently as 2008, when ocean waters should be colder. Steve Ratliff (Quileute) remarked, "They're showing up more than usual, yesterday [Sep. 17, 2008] one of the sport fishermen caught a 5 ½-footer. For the last couple of years, few years actually, there's been a Humboldt free-for-all…, just enormous size squid" (Ratliff interview).

To have giant squid still present in 2008 during a La Niña cycle indicates that some environmental factor is driving the Humboldt squid to the north. Whether climate-induced or not, the creature's presence means something is shifting in the food chain off Washington coast, with the arrival of a new top predator. The TEK of Quileute and Hoh fishermen is providing an early warning for commercial and subsistence fisheries off Washington coast that range shifts may already be underway. In the North Sea, scientists have already identified distribution shifts among "nearly two-thirds" of the marine fish species, changes that will likely have "profound impacts on commercial fisheries through continued shifts in distribution and alterations in community interactions" (Perry et al. 2005).

Plants have also been affected by climate change, among other factors. The Quileute and Hoh peoples traditionally gathered many plants for medicinal, cultural, and subsistence purposes, and they still harvest many of them from the forest. However, some of these resources have seen changes in recent years.

For example, cedar bark is an important resource for tribal basketweavers. It is harvested every year during a very specific time before the bark becomes too pitchy for use. According to some elders, the bark has been quite dry recently, when usually the inner pith is very wet. These reports may indicate the impact of warmer summer weather on the cedar trees (E. Penn interview). After the bark is cut from the tree, it is stripped and laid out to dry. The length of the drying process depends on the moisture content of the bark. One elder told me that the bark is drying out faster now (L. Morganroth interview). In addition, because of logging there are no longer old-growth cedar trees available. Agreements with the local logging companies (interviews with Geyer; C. Morganroth; L. Morganroth) have allowed some continued seasonal bark peeling, but

often the bark is gathered from trees already cut down, lessening its quality (L. Morganroth interview).

As a way to respond to these changes in quantity and quality of bark, a few basketweavers reported that they often purchase bark from gatherings or trade with other tribes (interviews with E. Penn and L. Morganroth), another example of intertribal cooperation in response to shifting and changing resources. An elder basketweaver said, "You have to find the other places they [traditional resources] might be, or start trading with other tribes that have the grasses that they travel with to the different pow-wows…. Sometimes there's people from Canada that come down to sell their grasses and their cedar bark….Yellow cedar bark is a nice commodity. It's a prized one" (E. Penn interview).

In recent years, some species of berries (such as huckleberries, salmonberries, blackberries, and thimbleberries) have been late to ripen and relatively scarce (Loudon interview). During the spring and summer of 2008, Native and non-Native harvesters reported that various berries ripened three weeks to two months late (interviews with Hobson; Loudon; Payne; E. Penn; Penn-Charles). Wild strawberries on the beach are normally ready in July or August, but in 2007 they were not ready until late August and early September. Salmonberries used to be ready for Elder's Week in May but are coming on at different times now: "The harvesting of different foods is affected by the weather," as elder Eileen Penn mentioned (E. Penn interview).

Sprouts or first shoots of the blackberry, once harvested in May to be eaten with stink eggs during Elder's Week (a delicacy to many), are increasingly late as well (Penn-Charles interview). Huckleberries in the mountains are also having "hard cycles" in recent years, late and scarce (interviews with Penn-Charles and Sampson). Ba·áts or Indian celery (*Equisetum sp.* or horsetail), used to alleviate springtime allergies, also used to sprout during March and April and by June would be ready for harvest. By June 2008 they had not come out yet. Remembered Miss Ann Penn-Charles (Quileute), "Springtime used to be so busy; it was the time everything was awakening" (Penn-Charles interview).

Climate change is not alone in causing these changes in the seasonal growth of traditional resources. Factors such as logging, land management practices, and illegal harvesting on the Olympic Peninsula have a role in the availability or scarcity of plants, but these activities do not necessarily influence the timing of spring growth.

Responses to Change

Adaptation is a common response for Native peoples because the subsistence harvesting of natural resources is still part of their lives. They have adapted their harvesting practices to new rules and regulations since European American colonization, and in more recent times they have adapted to changes in the timing of fish runs and berry harvests, the declines in the availability of important weaving materials and smelt, and the increased occurrence of unpredictable weather and severe storms.

Further, both the Hoh and Quileute tribes are currently obtaining higher ground in the Olympic National Park. These efforts are a direct response to the tsunamis, storm surges, and flooding that threaten much of the two reservations, as well as a long-term strategy for surviving the sea level rise and increased storm surge events predicted to occur because of climate change.

Obtaining higher ground will enable the tribes to move the lowermost villages to areas safe from flooding and tsunamis. This is no easy undertaking for any community; as Hoh elder Viola Riebe pointed out, "Moving a village is a big task when you have lots of people involved" (V. Riebe interview).

The Quileute Tribe also has a land zoning issue with the National Park Service, adding tension to the negotiations. When the Quileute Reservation was created in 1889, the mouth of the Quillayute River was located north of its current location. In 1910 a storm "caused the river's mouth to close and the river to move southward" (Jaime 2008). The shift of the river mouth left a parcel of land cut off from the Quileute Reservation. Following the storm, a survey done in 1916 (now commonly termed the "erroneous survey of 1916") excluded this parcel of land from the reservation (Jaime 2008). Unfortunately, when the coastal part of Olympic National Park was designated in 1953, its boundaries were based on the erroneous 1916 survey, which failed to recognize the original 1889 reservation boundaries (Jaime 2008). The parcel of land cut off by the river's shift was deemed part of the Olympic National Park and is the current location for the Rialto Beach parking lot. This change in boundaries has added political tension for the tribe's request for higher ground from the National Park Service.

In 2008, the Quileute Tribal Council drafted a bill for Congress requesting additional Quileute Trust Land out of the way of tsunamis, sea level rise, and increased storm surges (Hobson interview). On February 27, 2012, President Barack Obama signed the legislation, House Bill 1162 into public law. H.R. 1162 'To Provide the Quileute Indian Tribe Tsunami and Flood Protection, and for other Purposes'

would authorize the transfer of appropriate tracts of higher elevation land from Olympic

National Park... [and] also settle, by mutual agreement, a longstanding dispute between the Olympic National Park and the tribe over the northern boundary of the reservation. In addition, the bill will guarantee public access to beaches on the Washington coast and designate as wilderness thousands of acres of land currently within the Olympic National Park boundary. (Cantwell 2011)

The Hoh Tribe also pursued higher ground outside the Hoh River channel migration zone and beyond the range of tsunamis. U.S. Congressman Norm Dicks, along with the Hoh Tribal Council, drafted House Bill 1061, the "Hoh Indian Tribe Safe Homelands Act," which became public law on December 22, 2010 with signature from President Barack Obama.[7] Through this legislation 37 acres of previously logged National Park Service land and 434 acres of land owned by the tribe (160 of which were transferred from the Washington Department of Natural Resources) were converted into Hoh trust lands, making higher grounds available for relocation. [7] This addition would provide the tribe with a continuous parcel of land from the ocean to Highway 101, including higher ground where the lower village can be relocated. Councilwoman Marie Riebe reminds us that upon federal approval this bill will initiate the second move for the Hoh people in recent times. The first move occurred in the mid-1970s when the lower village, along the ocean, was abandoned because of flooding (M. Riebe interview).

The Hoh and Quileute tribes' land requests will provide homelands safe from rising sea levels, increased storm surges, and flooding. Global sea levels have already risen 18 centimeters (7.1 inches) in the past century (Pendleton et al. 2004) and are projected to rise an additional 48 centimeters (18 inches) by 2100 (IPCC 2007b). Thus sea-level rise will continue to shape the policies of the Quileute and Hoh tribal councils into the future. But for now the tribes have responded to the most imminent dangers.

Marine Sanctuary

In addition to the land acquired from the National Park Service, there is a new entity on the Washington coastline that the Quileute and Hoh tribes are working with to help safeguard marine resources. The Olympic Coast National Marine Sanctuary was established in 1994 to preserve the unique habitats along the coastline from

---

[7] Hoh Indian Tribe Safe Homelands Act http://www.gpo.gov/fdsys/pkg/PLAW-111publ323/pdf/PLAW-111publ323.pdf

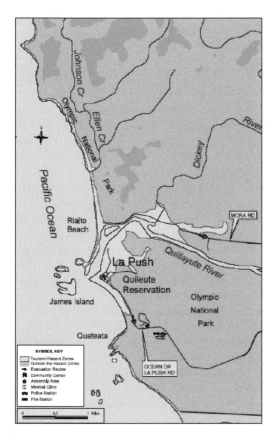

Quileute Reservation tsunami inundation map, showing low-lying coastal zone at La Push threatened by tsunamis and storm surges.

Cape Flattery south to the mouth of the Copalis River. The sanctuary encompasses the usual and accustomed fishing grounds of the Makah, Quileute, Hoh, and Quinault nations, making it of fundamental importance that all tribes work with the sanctuary to ensure their resources are protected and preserved for future generations. Mel Moon, the Director of Quileute Natural Resources, expressed the tribal resource concerns raised by the creation of the sanctuary.

> We were concerned that the MPA [marine protected areas] would create issues with no take zones [these are areas, also know as marine reserves, where human activities such as fishing may be restricted] and wouldn't recognize treaty rights, and would just kind of create a park out there. The sanctuary and the tribes have not had the greatest relationship, but we finally came to a point where we realized we need to change…. We are trying to get a better relationship established through a committee called the Intergovernmental Policy Council, and hopefully we will

have a better relationship in the future. We don't see them being able to do a fisheries management [without] staff. [At] this point in time it's really not there. (Moon interview)

As Mel Moon's words suggest, many Natives regard OCNMS as a threat to Native livelihoods because it could create a park-like area in the ocean (including no-take zones such as marine protected areas) that could negatively affect tribal economies. The uneasiness and fear felt by many tribal members primarily centers on the bans on traditional fishery practices at a time when climate change and population declines are increasingly undermining them. For the Quileute and Hoh peoples, the sanctuary is especially worrisome because their relationships with the National Park Service have not always been positive.

Though the marine sanctuary has caused concern for tribes along the coast, they agree that a level of resource protection is also desirable to address declining salmon runs. Partnering with the Olympic Coast National Marine Sanctuary and having a voice through the Intergovernmental Policy Council (IPC), the Quileute and Hoh tribes (along with the Makah and Quinault Nations) can begin to work together to find solutions for changing ocean conditions. Communications through the IPC will hopefully pool the knowledge of coastal tribes, NOAA, and the state on the domoic acid and dead-zone events, changes in sea-surface temperatures, and shifting species that have been witnessed along the coast. Their collective responses may shape policies that will safeguard resources for all peoples into the future.

Discussion and Conclusion

Native peoples are the world's early warning system that climate change is affecting human communities. With place-based oral histories of their homelands stretching back centuries, Native Americans hold vital knowledge of ecological change in the United States. TEK is crucial in understanding environmental changes affecting natural resources and the entire ecosystem and provides more immediate warning of changes underway than does Western science, allowing more time to prepare responses. Moreover, climate change will severely disrupt Native peoples' place-based rights and way of life. On the northern coast of Washington state, traditional ecological knowledge gathered through in-depth interviews strongly suggests climate change is already affecting the Quileute and Hoh reservations. Both tribes live on low-lying coastline, bordered on three sides by the Olympic National Park, and are susceptible to sea level rise, extreme storm surges, and shoreline erosion.

Quileute and Hoh peoples are already experiencing and responding to increased winter storms and flooding associated with the increased precipitation predicted by climate change models. They are already seeing the results of species range shifts, as new warm-water species, such as the brown pelican and Humboldt squid, are showing up in their waters. Meanwhile, traditional resources in the terrestrial, freshwater, and marine environments have begun to decline, corresponding to an increase in invasive species, hypoxia, and domoic acid events.

The Hoh and Quileute tribes have acted to address these threats, instituting rapid testing for domoic acid, for example. People also increasingly rely upon intertribal cooperation to obtain scarce resources and thus maintain traditional practices. Increased storm severity and flooding prompted the Quileute Natural Resources office to obtain a fair weather fisheries agreement with the state. The same conditions on the Hoh Reservation led the tribe to provide access to emergency radios and satellite phones to all residents. Perhaps the most important step the tribes are taking is acquiring land at higher elevations so that they can relocate villages endangered by rising water and storm surges.

Oral history tells the Quileute and Hoh peoples that they have prepared for and survived major environmental changes in the past. Elder Chris Morganroth III tells the story of the ancestors who were warned by the Great Spirit that change was coming for their people. Through this process, "our people prepared for the ice age," he shared:

> The Great Spirit told people a long period of ice and snow was to come. People stored food away. Leaders [when food didn't last] prayed food would be available. Thunderbird appeared over them. People were scared. Thunderbird flew out to the ocean and got lost. Two to three weeks later Thunderbird came back with whale. It was thundering, and lightening was shooting out of Thunderbird's eyes. Whale was in Thunderbird's mouth [and] it dropped at the feet of the Quileute people. (C. Morganroth interview)

Whale provided the nourishment that enabled the people to make it through the ice age. Like their ancestors before them, the Quileute and Hoh peoples see changes happening in their environment and have begun to take action to prepare for the worst.

Model for Coastal Communities

Through the example of the Hoh and Quileute tribes, Native and non-Native coastal communities can begin to understand the effects of climate change in their own communities and the importance of traditional ecological knowledge and local ecological knowledge in documenting and responding to such changes. Connecting TEK and Western science, as I have done here, is an approach transferrable to other Indigenous communities. Information may be easily found by going to any tribe's natural resources department, which is almost certainly working on issues somehow related to climate change.

Within all communities there are groups of people who meet to share information on particular ways of gathering resources and traditional practices. For example, fishermen meet on a regularly basis to share information on new quotas; during these meetings, they often discuss changes in the environment. Recording the TEK of fishermen during these meetings will help identify some of the changes that are occurring and will help resource managers work with governments on appropriate responses. For changes in the terrestrial environment, weaving circles are a good source of knowledge about traditional resources. These groups will be able to identify changes not only in weaving materials but also plants, animals, and weather cycles that weavers often experience when out gathering.

Recommendations

It is important to continue documenting changes on the coast, such as species range shifts, changes in natural resources, storm surges, and sea-level rise, as well as documenting the continued responses of both the Quileute and Hoh tribes. The OCNMS has been studying the unique habitats of the sanctuary, but more needs to be done to ensure that tribal information about the current and projected climatic changes are included in the sanctuary's master plan.

The Quileute and Hoh tribes must continue to pursue higher ground so that they can retreat from sea level rise and increased storm surges.

People must continue to use intertribal cooperation on local and regional scales to ensure that traditional practices and foods survive for future generations. Strengthening these relationships and continuing to form trading partnerships with First Nations in Canada are crucial. Since species will be moving northward (as well as higher in elevation) as the climate warms, sharing knowledge about plants and animals with neighboring communities to the north must become a common practice.

This is the time when all levels of tribal governance must make strong efforts to communicate with other Indigenous nations and their own tribal members to share plans and advice on strategies for adapting, from education to village relocation. There is much to be shared.

Preparations for and responses to climate change cannot be limited to tribal government and agencies; adaptation must involve all levels of the community to be successful. It is too important to be left up to staff who already have large-scale projects within their job descriptions. Fortunately, many Native groups still have a strong sense of community that can bring everyone together.

Youth are an important part of a community's response to climate change, as they will affect future generations. Youth action is a very powerful force for any community. Youth are already heavily involved in community care and awareness. Committees that join together many age groups, from students to elders, are ideal outlets for developing focused community and individual responses to the effects of climate change. Among the Quileute, youth have already been involved in activism by peacefully demonstrating against the dangerous location of their school. This type of youth action can be extended by forming an environmental health club, which could focus on healthy choices and also empower youth by involving them in community-level planning for environmental changes. Another means of involving students is the Youth Opportunity Program (YOP) offered to tribal high school students.

In addition to these steps, the Quileute and the Hoh tribes must continue to strive to be included in all discussions regarding the designation of protected areas as part of the OCNMS. It is fundamentally important that communications between the tribes and sanctuary managers are strengthened to ensure that the distribution of environmental benefits are equal and treaty rights are upheld for all the coastal tribes. The Sanctuary has begun to show greater interest in hearing feedback from local communities, but more needs to be done, such as a community forum with the sanctuary staff.

Communication between tribal, state, and federal governments is essential in order to safeguard tribal treaty rights and the future of subsistence fishing. Knowledge of change in both the marine and terrestrial ecosystems is available and must be used to respond in a timely way.

Communities around the globe are beginning to come together to plan for adaptation and mitigation strategies on the local level. TEK within the tribal communities reveals that changes are already occurring on the coast. Communities in the Pacific Northwest, such as the Quileute and Hoh tribes, also have access to a leading climate research team, the Climate Impacts Group (CIG) at the University of Washington. To aid in planning large-scale projects (such as village and school relocations), the CIG has a community planning guide. TEK identifies specific impacts, such as the locations of flooding and storm surges. The CIG planning guide can help coastal communities map environmental impacts and at-risk areas. Community mapping can lead to the development of adaptive strategies to current and predicted environmental impacts. As a next step, the designated planner in each tribal office can consider incorporating a long-term plan for climate change adaptation. Much of this is already underway among the Quileute and Hoh tribes.

Community forums and conferences on climate change have been the focus of many tribal gatherings throughout the Pacific Northwest. My final recommendation is that community gatherings on climate change implications, adaptations, and mitigation must be held by all of the Washington coastal tribes. Coastal reservations will see the worst effects of climate change. It is essential that these gatherings for sharing information specific to coastal effects be organized. Traditional knowledge offers many lessons about previous change and adaptation, and it will provide solutions for coastal tribes facing the effects of global climate change.

References

Allison, Steve. 2008. Interview by Chelsie Papiez. Recorded by microcassette, July 18, 2008. Hoh Natural Resource Interviews. Hoh Reservation.

Andrade, Manuel José. 1969. *Quileute Texts*. New York: AMS Press.

Bakun, Andrew. 1990. Global Climate Change and Intensification of Coastal Ocean Upwelling. *Science* 247(4939):198–201.

Barth, John A., Bruce A. Menge, Jane Lubchenco, Francis Chan, John M. Bane, Anthony R. Kirincich, Margaret A. McManus, Karina J. Nielson, Stephen D. Pierce, and Libe Washburn. 2007. Delayed Upwelling Alters Nearshore Coastal Ocean Ecosystems in the Northern California Current. *National Academy of Sciences* 104(10):3719–3724.

Berkes, Fikert, Johan Colding, and Carl Folke. 2000. Rediscovery of Traditional Ecological Knowledge as Adaptive Management. *Ecological Applications* 10(5):1251–1262.

Berry, Alexis. 2008. Interview by Chelsie Papiez. Recorded by microcassette, September 30, 2008. Hoh LEK Interviews. Hoh Reservation.

Black, Margaret. 2008. Interview by Chelsie Papiez. Recorded by microcassette, August 28, 2008. Quileute Interviews. La Push.

Black III, Roy 2008. Interview by Chelsie Papiez. Recorded by microcassette, September 9, 2008. Quileute Interviews. La Push.

Boldt 1974. *United States v. State of Washington*, 384F. Supp. 312.

Blumenthal, Les. 2008. Jumbo Squid Makes New Home in Northwest. *The Olympian*: April 27.

Blus, Lawrence J. 2007. *Contaminants and Wildlife— The Rachel Carson Legacy Lives On*. Paper presented at the Rachel Carson Centennial Birthday Celebration. Silver Spring, MD.

Buchanan, James. 1855 and 1856. *Treaty Between The United States and the Quinaielt and Quillehute Indians*.

Burkholder, JoAnn M. 1998. Implications of Harmful Microalgae and Heterotrophic Dinoflagellates in Management of Sustainable Marine Fisheries. *Ecological Applications* 8(1):S37–S62.

Cajete, Gregory. 2000. *Native Science: Natural Laws of Interdependence*. Santa Fe, NM: Clear Light Publishers.

Cantwell, Maria. 2011. Key Agency Backs Cantwell Bill to Move Quileute Tribe out of Tsunami Zone. Department of Interior Supports Cantwell Plan at Indian Affairs Hearing; Cantwell Urges Approval of Bill. Press release, April 14, 2011, http://cantwell. senate.gov/news/record.cfm?id=332515 .

Chan, F., J. Barth, J. Lubchenco, A. Kirincich, H. Weeks, W. Peterson, and B. Menge. 2007. Emergence of Anoxia in the California Current Large Marine Ecosystem. *Science* 319:920.

Childress, James J., and B. Seibel 1998. Life at Stable Low Oxygen Levels: Adaptations of Animals to Oceanic Oxygen Minimum Layers. *The Journal of Experimental Biology* 201:1223–1232.

CIG (Climate Impacts Group). 2008. University of Washington Climate Impact Group. http://cses. washington.edu/cig/ (accessed October 31, 2008).

Conlow, Pearl. 2008. Interview by Chelsie Papiez. Recorded by microcassette, June 26, 2008. Quileute Interviews. La Push.

Cornell Lab of Ornithology. 2003. All About Birds. http://www.birds.cornell.edu/ AllAboutBirds/ BirdGuide/Brown_Pelican.html (accessed December 15, 2008).

Cosgrove, James A. 2004. *The First Specimens of Humboldt Squid in British Columbia*. Victoria, BC: Natural History Section, Royal British Columbia Museum.

Crawford, George. 2004. *Tsunami Evacuation Case Study: La Push, Washington*. Quileute Reservation. Olympia, WA: Washington Military Department Emergency Management Division.

Cury, Philippe, and C. Roy. 1989. Optimal Environmental Window and Pelagic Fish Recruitment Success in Upwelling Areas. *Canadian Journal of Fisheries and Aquatic Sciences* 46(4):670–680.

Davis, Jack. 2008. Interview by Chelsie Papiez. Recorded by microcassette, September 25, 2008. Quileute Interviews. La Push.

Dickerson, Darby. 2008. Interview by Chelsie Papiez. Recorded by microcassette, July 28, 2008. Commercial Fisherman Interview (LEK). La Push.

Epstein, P., T. Ford, and R. Colwell. 1993. Marine Ecosystems. *The Lancet* 342(8881):1216–1219.

Fedorov, Alexey, Marcelo Barreiro, Giulio Boccaletti, Ronald Pacanowski, and George S. Philander. 2007. The Freshening of Surface Waters in High Latitudes: Effects on the Thermohaline and Wind-Driven Circulations. *Journal of Physical Oceanography* 37:896–907.

Geyer, Frank. 2008. Interview by Chelsie Papiez. Recorded by microcassette, June 13, 2008. Quileute Natural Resources Interview (LEK). La Push.

Gilbertson, Joesph. 2008. Interview by Chelsie Papiez. Recorded by microcassette, July 15, 2008. Hoh Natural Resource Interview (LEK). Hoh Reservation.

Golder Associates, Inc. 2005. *WRIA 20: Watershed Planning Draft Multi-Purpose Storage Assessment Report*. Redmond, WA: Golder Associates.

Grantham, Brian A., F. Chan, K. Nielsen, D. Fox, J. Barth, A. Huyer, J. Lubchenco, and B. Menge. 2004. Upwelling-Driven Nearshore Hypoxia Signals Ecosystem and Oceanographic Changes in the Northeast Pacific. *Nature* 429:749–754.

Gulland, Frances. 2000. *Domoic Acid Toxicity in California Sea Lions (Zalophuscalifornianus) Stranded Along the Central California Coast, May–October 1998.* Report to the National Marine Fisheries Service Working Group on Unusual Marine Mammal Mortality Events. U.S. Department of Commerce 2000, NOAA Technical Memo (NMFS-OPR-17): 45.

Heppner, Don, and Jennifer Turner. 2006. Spruce Weevil and Western Spruce Budworm Forest Health Stand Establishment Decision Aids. *British Columbia Journal of Ecosystems and Management* 7(3):45–49.

Hobson, DeAnna. 2008. Interview by Chelsie Papiez. Recorded by microcassette, June 12, 2008. Quileute Interviews. La Push.

Hoh Tribe. 2008. Chalá•at: People of the Hoh River. http://www.hohtribe-nsn.org (accessed November 12, 2008).

Hook, Abigail. 2004. *WRIA 20: Technical Assessment Level I Water Quality and Habitat.* Forks, WA: University of Washington Olympic Natural Resources Center.

Intergovernmental Panel on Climate Change (IPCC). 2007a. *Climate Change 2007:Synthesis Report Summary for Policy Makers.* Working Group contributions to the Fourth Assessment Report. Valencia, Spain, 12–17 November 2007.

Intergovernmental Panel on Climate Change (IPCC). 2007b. *Climate Change 2007: The Physical Science Basis. Contribution of Working Group I to the Fourth Assessment Report of the Intergovernmental Panel on Climate Change.* S. Solomon, D. Qin, M. Manning, Z. Chen, M. Marquis, K. B. Averyt, M. Tignor, and H. L. Miller, eds. Cambridge: Cambridge University Press.

Jackson, Roger. 2008. Interview by Chelsie Papiez. Recorded by microcassette, September 3, 2008. Quileute Interviews. La Push.

Jaime, James. 2008. Boundary Dispute between the Quileute Indian Tribe and the Olympic National Park. http://www.quileutenation.org (accessed September 12, 2008).

Johnson, Babs. 2008. Interview by Chelsie Papiez. Recorded by microcassette, October 8, 2008. Local Ecological Knowledge Interviews. La Push.

Johnson, Gerry. 2008. Interview by Chelsie Papiez. Recorded by microcassette, July 28, 2008. Local Ecological Knowledge Interviews. La Push.

Jones, T., J. Whyte, N. Ginther, L. Townsend, and G. Iwama. 1995. Haemocyte Changes in the Pacific Oyster, *Crassostrea gigas*, Caused by Exposure to Domoic Acid in the Diatom *Pseudo-nitzschia pungens* f. *multiseries. Toxicon* 33:347–353.

King, Jerry R. 2008. *Jerry King's Weather Service* (Cooperative Station Forks 1E). Forks, WA: John Ten Harkel.

Kowal, Jessica. 2006. In a Bid for Higher Ground, a Low-Lying Indian Tribe Raises the Stakes. *The New York Times,* July 30. http://www.nytimes.com/2006/07/30/us/30beach.html (accessed June 21, 2011).

Leitka, Mary K. 2008. Interview by Chelsie Papiez. Recorded by microcassette, September 22, 2008. Hoh Interviews. Hoh Reservation.

Lekanof, Debra. 2008. Coast Salish Gathering. http://www.coastsalishgathering.com/ (accessed December 17, 2008).

Lien, Roger. 2008. Interview by Chelsie Papiez. Recorded by microcassette, July 24, 2008. Quileute Natural Resources Interviews. La Push.

Loudon, Beverly. 2008. Interview by Chelsie Papiez. Recorded by microcassette, August 26, 2008. Quileute Interviews. La Push.

Matson, Pat. 2008. Interview by Chelsie Papiez. Recorded by microcassette, December 6, 2008. Quileute Interviews. La Push.

Moon, Mel. 2008. Interview by Chelsie Papiez. Recorded by microcassette, June 25, 2008. Quileute Natural Resources Interviews. La Push.

Morganroth III, Chris 2008. Interview by Chelsie Papiez. Recorded by microcassette, June 4, 2008. Quileute Interviews. La Push.

Morganroth, Lela Mae. 2008. Interview by Chelsie Papiez. Recorded by microcassette, September 6, 2008. Quileute Interviews. La Push.

Mote, Philip. 1999. *Impacts of Climate Variability and Change in the Pacific Northwest. The JISAO Climate Impacts Group University of Washington.* A Report of the Pacific Northwest Regional Assessment Group for the U.S. Global Change Program. Seattle, WA: University of Washington. http://www.usgcrp.gov/usgcrp/nacc/pnw.htm (accessed June 21, 2011).

National Oceanic and Atmospheric Administration (NOAA). 2008a. Cold and Warm Episodes by Season. http://www.cpc.noaa.gov/products/analysis_monitoring/ensostuff/ensoyears.shtml (accessed August 4, 2008).

National Oceanic and Atmospheric Administration (NOAA). 2008b. National Oceanic and Atmospheric Administration, United States of America. http://www.nwfsc.noaa.gov/research/divisions/fed/oeip/ca-pdo.cfm (accessed October 1, 2008).

Northcut, Kris. 2008. Interview by Chelsie Papiez. Recorded by microcassette, July 1, 2008. Quileute Natural Resource Interviews (LEK). La Push.

Olympic Coast National Marine Sanctuary (OCNMS). 2006. Hypoxia Monitoring. http://www.nps.gov/olym/historyculture/index.htm (accessed June 18, 2008).

Olympic National Park Service (ONPS). 2008. http://www.nps.gov/olym/historyculture/index.htm (accessed November 4, 2008).

Payne, Sue. 2008. Interview by Chelsie Papiez. Recorded by microcassette, August 1, 2008. Quileute Interviews (LEK). La Push.

Pendleton, Elizabeth A., Erika S. Hammar-Klose, Robert E. Thieler, and Jeffress S. Williams. 2004. *Coastal Vulnerability Assessment of Olympic National Park to Sea-Level Rise*. Washington, DC: United States Geological Survey. http://pubs.usgs.gov/of/2004/1021/images/pdf/olym.pdf

Penn Jr., Chris (Jiggs). 2008. Interview by Chelsie Papiez. Recorded by microcassette, June 11, 2008. Quileute Interviews. La Push.

Penn, Eileen. 2008. Interview by Chelsie Papiez. Recorded by microcassette, August 6, 2008. Quileute Interviews. La Push.

Penn-Charles, Ann L. (Miss Ann). 2008. Interview by Chelsie Papiez. Recorded by microcassette, June 12, 2008. Quileute Interviews. La Push.

Perry, Allison L., Paula J. Low, Jim R. Ellis, and John D. Reynolds. 2005. Climate Change and Distribution Shifts in Marine Fishes. *Science* 308(5730):1912–1915.

Peterson, Gary. 2008. Interview by Chelsie Papiez. Recorded by microcassette, August 13, 2008. Local Ecological Knowledge Interviews. Peak 6 Adventure Store, Hoh River Valley.

Pettitt, George A. 1950. *The Quileute of La Push, 1775–1945*. Berkeley: University of California Press.

Powell, Jay. 1999. Hoh River Watershed Analysis: Cultural Resources Module Part I: Hoh Tribal Cultural Resources. Ongoing research available at http://hohtribe-nsn.org/history2.html

Preston, Debbie. 2007. Monster December Storm Likely Harmed Future Olympic Peninsula Chinook Returns. *Northwest Indian Fisheries Commission*, December 27. http://nwifc.org/2007/12/monster-december-storm-likely-harmed-future-olympic-peninsula-chinook-returns/

Ralston, Larry. 2008. Is Your Tribal Land Secure? http://www.evergreen.edu/tribal/docs/Quileute_Jan_2008.doc (accessed May26, 2008).

Ramsdell, John S., and Tanja S. Zabka. 2008. *In Utero Domoic Acid Toxicity: A Fetal Basis to Adult Disease in the California Sea Lion (Zalophus californicanus)*. *Marine Drugs* 6:262–290.

Ratliff, Steve T. 2008. Interview by Chelsie Papiez. Recorded by microcassette, September 18, 2008. Quileute Interviews. La Push.

Riebe, Marie. 2008. Interview by Chelsie Papiez. Recorded by microcassette, October 8, 2008. Hoh Interviews. Hoh Reservation.

Riebe, Viola. 2008. Interview by Chelsie Papiez. Recorded by microcassette, June 23, 2008. Hoh Interviews. Forks.

Sampson, Gene. 2008. Interview by Chelsie Papiez. Recorded by microcassette, August 14, 2008. Hoh Interviews. Hoh Reservation.

Schumack, John. 2008. Interview by Chelsie Papiez. Recorded by microcassette, August 29, 2008. Quileute Interviews. La Push.

Schumack, Marian. 2008. Interview by Chelsie Papiez. Recorded by microcassette, August 12, 2008. Quileute Interviews. Forks.

Schwing, Franklin B., and Roy Mendelssohn. 1997. Increased Coastal Upwelling in the California Current System. *Journal of Geophysical Research* 102(C2):3421–3438.

Sibert, John, John Hampton, Pierre Klieber, and Mark Maunder. 2006. Biomass, Size, and Trophic Status of Top Predators in the Pacific Ocean. *Science* 314:1773–1776.

Thysell, Rod, and Steve Allison. 2005. Hoh Tribe's Planting Effort Using Science to Combat Insect. *Northwest Indian Fisheries Commission*, March 25. http://nwifc.org/2005/03/hoh-tribes-planting-effort-using-science-to-combat-insect/

USGS (United States Geological Survey). 2007. *Water-Data Report 2007: 12041200 Hoh River at U.S. Highway 101, Near Forks, WA*. Washington, DC: United States Geological Survey.

Wallerstedt, Bertha. 2008. Interview by Chelsie Papiez. Recorded by microcassette, September 16, 2008. Quileute Interviews. La Push.

Washington Military Department (WA MD) and
Washington State Department of Natural Resources
(WA DNR). 2007. Tsunami! Evacuation Map for
La Push and Vicinity. Olympia, WA: WA MD
Emergency Management Division and WA DNR
Division of Geology and Earth Resources.

Washington State Department of Natural Resources
(WA DNR). 2007. Tsunami! Evacuation Map for
the Hoh Reservation. Olympia, WA: Department of
Natural Resources, Division of Geology and Earth
Resources.

Wickliffe, J., and J. Bickham. 1998. Flow Cytometric
Analysis of Hematocytes From Brown Pelicans
(*Pelecanus occidentalis*) Exposed to Planar
Halogenated Hydrocarbons and Heavy Metals.
*Bulletin of Environmental Contamination and
Toxicology* 61:239–246.

Williams, Mark. 2008. Interview by Chelsie Papiez.
Recorded by microcassette, July 30, 2008. Quileute
Interviews. La Push.

Wray, Jacilee, ed. 2002. *Native Peoples of the Olympic
Peninsula: Who We Are*. Norman: University of
Oklahoma Press.

Zeidberg, Louis D., and Bruce H. Robison. 2007.
Invasive Range Expansion by the Humboldt Squid,
Dosidicus gigas, in the Eastern North Pacific. *PNAS*
104(31):12948–12950.

MAORI PERSPECTIVES ON CLIMATE CHANGE
Ata Brett Stephenson

*Ata Brett Stephenson is of Te Kapotai, Ngapuhi,
from Aotearoa/New Zealand.*

The reality of climate change is slowly dawning on people, but public debate is largely confined to heralding the global shifts, potential changes, and in some instances the somewhat alarming possible impacts on existing human resources and behaviors. Global *warning* of global warming seems to be a preoccupation of the socially conscious even though climate change is an anathema to the economy of an industrialized world. Most people appear to place little value in the voices or opinions of Indigenous people in this crisis. However, Indigenous peoples hold a regionally specific,

collectively powerful body of climatological knowledge, holistically interwoven with what we might call the environmental knowledge of a culture [in Maori, *matauranga putaiao*] and strategies for its *kaitiakitanga* [collective guardianship or management].

The opportunity therefore exists for Indigenous peoples to move beyond waiting—the poignant moment of global realization, acceptance, and acknowledgement—and depart from a framework determined by capitalist economic and political expediency. In general we Indigenous peoples have environmental management practices that are less wasteful and more biodiverse than the paradigm that leads to crop monoculturalism and overfishing. Our attitudes and behaviors always express strong regard for the interreliance of species and resources in both current and future generations. We do not have a particular sense of *ownership* to land and resources—we are *part* of those resources, and we identify with the land through *whenua* [umbilical attachment]. It is not a simple thing to destroy the resources with which we are connected.

Maori Ideology and Identity: Pacific Connections

Indigenous people have long, historical collections of natural observations, a detailed and expert knowledge of sea, landscape, and atmospheric phenomena. While the migratory pathway of Maori to Aotearoa/New Zealand from a Pacific homeland is less than a thousand years old, our ancestors did not come from a vacuum, the kind of *terra nullius* suggested by the first British explorers; they had already developed experiential knowledge about the natural world because of their journeys and settlement across the Pacific. Neither can Western science be excused for overwriting (Stephenson 1998) and trivializing the Maori knowledge base of the natural world, an effort that arose as part of a concerted, hegemonic project to foster British settler colonization.

We live on the edges of Pacific Plate boundaries and subduction zones—earthquakes and volcanoes are part of our history. We are sea voyagers, navigators, peoples of first exploration and discovery different from those who came later following predetermined routes. The depth of geomorphological knowledge held by Maori is reflected in a passage of oral history published in *Ko Ngatoro I Rangi raua ko Tia he rangatira no Te Arawa waka* (New Zealand Geographic Board 1990). Two *rangatira* [chiefs] of the Te Arawa *waka* [canoe] brought their canoes from Hawaiki down the Kermadec-Tonga

Trench, guided by their *taniwha* [water being] and left track markers of hot embers along the line of the trench into Te Moana a Toi (the Bay of Plenty in North Island).˙ These two chiefs explored the landscape, lakes, and mountains. From Mount Tongariro one of them, Ngatoro I Rangai, commanded the view and challenged the other, Hape Tu A Rangi (on the desert floor below): "Do not dare to climb this mountain or I will cause darkness to descend upon you."

Hape began his ascent nevertheless, and Ngatoro I Rangi immediately called on Ruamoko (the god of earthquakes and volcanoes) to destroy the trespasser. A massive eruptive force was created from underground, and dense black clouds cast darkness. Snow fell over the desert, and Hape was frozen to death. After destroying his rival, Ngatoro I Rangi continued to climb but encountered snow, sleet, and cold winds. Weakened by his climb, he cried out to his two sisters, Kuiwai and Haungaroa (who resided in Hawaiki), to assist him. They heard him and sent sources of heat with the assistance of fire gods. That trail remains, marked by a succession of volcanoes—Whakaari, Mautohora, Putauki, Rotoehu, Tarawera, and Taranaki—and by a collection of *ara* [talisman stones], one of which Ngatoro I Rangi threw to mark the sacrificial death of his servant and which now stands as Mt. Ngauruhoe.

This commonplace story gives a Maori explanation for the source of volcanism occurring along the western contact boundary of the Australian/Pacific oceanic plates. With the description of "track markers of hot embers," Maori had ascertained the existence of a visible line of undersea volcanoes extending from the land, but Western scientists did not fully contemplate or realize these volcanoes' "reality" until the studies of the past twenty-five years. This example reflects the postcontact, colonial western attitudes to Indigenous (Maori) knowledge.

The migratory seabirds such as the albatross and titi (sooty shearwater or *Puffinis griseus*) follow the trail of the deep sea subduction zone (the Tonga-Kermadec trench), reflecting the continuing close connections between Pacific people and their resources. Between September and mid-May, breeding and pre-breeding titi disperse widely throughout New Zealand waters. Some may move into the Southern Ocean as far south as the outer edges of Antarctic pack ice. Most adults depart on migration to the North Pacific between late March and early May—fledglings leave colonies from mid-April and follow a similar route. The birds arrive on the west coast of North America in April and May

and drift northwards to reach the Gulf of Alaska by June. Breeding adults return south in August, with non-breeders leaving a little later. Tracking the bird migrations in September 2004 found that homecoming birds from San Luis Bay, California, flew about twenty days to reach Taiaroa Head, Otago. Titi traveled at about 25 km/hr, often following a zigzag course, and some evidence suggested that birds travel further in 24-hour spans when the moon is full (Adams 2005). Of interest to Maori navigators in the birds' return passage was the possible avoidance of northern trade head winds between Hawaii and Central America, making it particularly productive to follow the birds' route. Equally interesting, the birds made a second route adjustment to follow the Kermadec-Tonga Trench (which is a magnetic anomaly), the same passage that Maori navigated in our migration across the Pacific. We followed the birds, the stars, and the swells—we did not just drift with winds. There is an intimate relationship between the birds and Maori origins.

Research on the southern New Zealand titi population has been successfully developed by a cooperative venture between Rakiura Maori harvesters of juvenile titi (knowledge holders and *kaitiaki* or guardians of a food resource) and scientists at the University of Otago (a source of Western scientific information). To date, little is known about the potential or actual interbreeding between New Zealand and South American populations and therefore stock size, but these data are crucial for estimating sustainable harvesting levels. Preliminary results indicate that there are no major differences between eastern and western Pacific populations in the first set of gene sequences—but there is more to be discovered. If interbreeding is found to be frequent, then the two populations could have a strong influence on each other with respect to harvesting levels, and the birds' migration patterns could serve as markers to track climate change patterns in the Pacific.

The Ocean around Aotearoa/New Zealand

Throughout most of the last 65 million years, sea surface temperatures around New Zealand have been warmer than they are now. Marine fossils, particularly mollusks, provide excellent evidence of subtropical temperature peaks in the early Eocene (c. 50 Ma BP or million years before present), when seas are thought to have been approximately 9°C warmer, and again in the early Miocene (c. 16 Ma BP), when they were about 8°C warmer. Many warm water species that had earlier dispersed into the region in shallow warm water currents, as planktonic larvae or juveniles, died out in the progressively cooler climate of the late Tertiary (up to 2 Ma BP).

The evidence of sea level changes visible in terraced coastal cliff topographies date from the more recent Pleistocene ice ages (c. 2 Ma BP), where the effects of the worldwide locking of water into ice caps is apparent. While sea levels have periodically risen as much as 15 to 105 m above their current positions, land instability (and small upward movements of land) have also occurred, making it difficult to calculate sea level changes precisely.

Climatic shifts occur naturally, and modern record-keeping provides a progressively growing wealth of accurate and precise information. Mean surface air and sea-surface temperature data are closely related, and the marine evidence shows a warming trend of 0.7ºC over the period from 1871 to 1995. The decade of 1900–1910 was the coldest of the twentieth century, while the most marked warming began around 1950. Additionally, since 1950 nighttime temperatures have warmed more than daytime maximum temperatures, with a marked decline in frost frequency (Bosselmann et al. 2002). Scientists detect a twenty-year cycle in climate shifts for the New Zealand–South Pacific region, driven by a recently described climate feature called the Interdecadal Pacific Oscillation (IPO), which reverses climate features every one to three decades (S. Power et al. 1999). In the positive phase of IPO, southwesterly winds are more frequent; conversely, a negative phase brings periods of more intense northeasterly winds in the New Zealand region.

Our deep sea fauna, at depths of more than 1000 m, is worldwide in its representation because the dispersal and distribution of these fauna rely on the major deep sea current circulation systems. Water carries more heat than air, and it is the oceanic circulation of heat that drives climate patterns. Warm currents of lower density move at the surface from equatorial to polar regions, producing warm climates in nearby landmasses (such as the U.S. eastern seaboard). Conversely, cold currents move cool water towards the equator. Ocean water is stratified by temperature differences, causing the establishment of a permanent *thermocline*, or temperature layer between the surface and deep water. Polar water is both cooler and more saline than warm equatorial water because the high rainfall over equatorial seas effectively dilutes saltwater. As a result, polar water has a higher density, causing it to sink before circulation, and deep water is cold. Deepwater thermohaline circulation is largely separate from surface oceanic circulation, but the deep water retains plenty of oxygen and nutrients— ocean upwelling brings sources of high productivity.

What is at risk from climate change is the alteration of atmosphere–sea surface exchange at any point between equatorial and polar regions to the extent that deep sea circulation is interrupted. By way of example, we might note that the Humboldt deep cool current shifts during the La Niña/El Niño oscillation in the Pacific.

Emerging information suggests that global climate changes may soon (or are already) affecting deep sea circulation. Not least of this evidence is the frequently reported decay of Arctic and Antarctic coastal boundaries because ice sheets and icebergs are melting, confirming that temperatures are warming on both land and sea. A news report in the *New Zealand Herald* (Sep. 8, 2006) addresses compounding issues in "unexpected" greenhouse gas (GHG) emissions. The report cites the director of global ecology at the Carnegie Institute on research that shows that releases of ice-trapped methane and carbon dioxide during the melting of northern hemisphere permafrost may be potentially up to 100 times greater than equivalent gas releases from the burning of fossil fuels. Moreover, the conversion of frozen water into liquid will raise sea levels; global mean sea level is predicted to rise by somewhere between 9 cm and 88 cm between the years 1990 and 2100.

Apart from the risk of low island and atoll submergence, people in the tropical Pacific island nations are also concerned about increasing carbonate solubility, or ocean acidification. As atmospheric carbon dioxide rises, oceanic water becomes more acid, which will interrupt the calcium carbonate deposition required to create invertebrate skeletons. Coelenterates, including the coral reef systems, seem to be particularly at risk.

Factors Causing Climate Change

In most regions, the earth's atmosphere maintains comfortably warm surface level temperatures of 0ºC to 50ºC. Such warm temperatures keep water molecules in a liquid state, which is essential for life as we know it. The temperature extremes outside earth's atmosphere are known to exceed the biological capacity of the lifeforms we are familiar with. For example, the airless surface of the moon rises to 100ºC on the sunlit surface and falls to −150ºC at night, with obvious implications for the various states of water. While the surface of earth is warmed by insolation, it is the long wavelength (> 0.4 micrometers) infrared energy that creates a sensation of warmth. Water vapor absorbs strongly in the band between 4 micrometers and 7 micrometers, and carbon dioxide in the band between 13 micrometers and 19 micrometers. These ranges leave a window between 7 micrometers and 13 micrometers through which more than 70 percent of insolation is reflected from the earth's surface back into space.

A "greenhouse" effect arises as a result of atmospheric absorption of a fraction of the long-wave radiation as it is reflected from the warmed surface. Among the gas components of the atmosphere responsible for the absorption of heat, it is carbon dioxide that takes up and stores by far the largest proportion; in effect, carbon dioxide is a warming blanket. Thus any increases in the atmospheric ratio of carbon dioxide are matched by an increasing greenhouse effect, raising global temperature. Such changes in the concentration of greenhouse gases alter the efficiency with which Earth cools off.

Evidence shows that carbon dioxide varied little in the several million years before the present, but in the last two centuries human activities have made dramatic changes in the once rather stable levels of greenhouse gases. Since 1960 total atmospheric carbon dioxide has increased more than 10 percent. Over the same period the average world temperature (compared to the previous three decades) increased from 15.0°C to 15.2°C. About three-quarters of the human emissions of carbon dioxide to the atmosphere during the last twenty years are the result of burning fossil fuels—the rest comes mostly from land use changes, particularly deforestation (Bosselmann et al. 2002).

The human contribution to climate change will persist for centuries after greenhouse gases are stabilized, because of their lasting effects on atmospheric composition, shifts in solar radiation effects, and the long timespan involved in any corrective reversal of the deep oceanic circulation shifts modified by changing temperatures. Global mean surface temperature increases, and rising sea levels because of the thermal expansion of the ocean, are projected to continue for the next few hundred years, even if the present levels of greenhouse gas concentration do not increase. In the Pacific Basin, perhaps the most evident and dramatic temperature and current shifts are found in the El Niño/La Niña oscillations.

Rhythms of Weather

In normal or La Niña years, the waters off the coast of South America are kept relatively cool by the upwelling of the Humboldt Current. The sun-heated surface water of the central Pacific is steered eastward by trade winds toward Australia. When the Southern Oscillation Index (SOI) is positive, the Pacific rain belt shifts to the south. The trade winds are strong, Australia receives a

longer wet season, and the northern New Zealand coast has sea breezes and warm, moist northeasterly weather patterns.

In an El Niño phase, the trade winds weaken and ponds of hot, salty, equatorial water develop in the east near Peru, creating their own pockets of moist, rising air, which condense into rain clouds. As a result, New Zealand experiences more frequent southwesterly winds. The SOI index is negative, and the Pacific rain belt shifts to the north. A ridge of high pressure settles well to the north of the country. Hence, below-average rainfall occurs in Northland and Auckland and along the east coasts from Gisborne to Canterbury, which lie in a rain shadow cast by the Western and High Country ranges. To the south, average rainfall and cooler weather persists in Southland, Westland, and Otago. Increases in the El Niño intensity as a result of global warming are predicted to lead to greater extremes of the droughts and floods that are associated regionally with El Niño events (Bosselmann et al. 2002).

The Impact on Global Communities

Climate change will exacerbate water shortages in many of the existing arid land masses—Central Asia, southern Africa, and the European and African countries bordering the Mediterranean. At the same time, other regions, such as parts of southern Asia, may receive greater precipitation. Several hundred millions of people are projected to suffer a reduction in their water supply of 10 percent or more by the year 2050 (for climate change predictions corresponding to a 1 percent annual increase in carbon dioxide emissions). The multiple effects of water scarcity and the increasing frequency and intensity of flood and drought cycles will create massive difficulties in water management from which few will escape (Bosselmann et al. 2002).

Sea-level rise predictions suggest that communities on low-lying coasts and small islands are at risk of sea water encroachment, creating severe social and economic effects. Those familiar with the effects of the 2004 tsunami on Indian Ocean coastal populations, or the hurricanes in the Gulf of Mexico in 2005, will be able to partially gauge the extent to which populations are vulnerable to sea-level rise. It is expected that sea water inundation will result in the permanent displacement of people and the loss of infrastructure. Sea-level rise, however, goes beyond the mere loss of useable land. There will be consequential losses in fishing and wildlife habitat, crucial resources for food sustainability—especially in communities with less than adequate food resources. Storm cycle events (even without considering changes to storm frequency or intensity) will generate waves and wind surges that pose additional risks for many small island communities and to some extent for coastal land strips at or below sea level.

For New Zealand, there will be gain and losses. Water is likely to become a key problem in eastern areas if a projected drying trend takes effect; equally, we are subject to IPO and El Niño oscillations, which create uncertainty. The increases in high-intensity rainfall and the higher incidence of tropical cyclones would heighten the risks to life, property, and ecosystems from flooding, storm surges, and wind damage.

However, predictions about the effects of climate change vary regionally; they do not necessarily mirror those for national or global trends. Analysis of data from the Bay of Plenty (Griffiths et al. 2003) indicates that total annual rainfall has generally decreased by 25 mm per decade over the period of recorded returns from 1910 to 2002, with fewer rainy days (1.5 fewer days per decade since 1960s). Current storm flood intensity levels, however, have grown higher. Short duration, extreme rainfall has become more common in the positive phase of the IPO compared to the negative phase. Also, during the La Niña periods seasonal extremes and rainfall intensity are above normal. The trend for air temperature shows rises up to 0.2°C per decade over the last fifty years, and the number of days exceeding 25°C has significantly increased. Temperatures are typically above normal during La Niña phases. Correspondingly, there are fewer days of frost, about one frost less per year.

In the future, mean annual rainfall is projected to decrease about 1 to 4 percent by the 2030s in the Bay of Plenty, no further drying is expected after that through the 2080s (Griffiths et al. 2003). The mean occurrence of the westerly wind component across New Zealand is expected to increase by approximately 10 percent of its current value in the next fifty years. Faster wind speeds are associated with intense convection currents, contributing to the warming and drying of the climate, and also with intense low-pressure systems, which could become more common.

These predicted outcomes collectively signal the possibility for more intense floods and, on the intensely cultivated Rangitaiki Plains, greater soil loss and increasing river sediment loading. Events of this type have downstream consequences such as sediment in-filling at the mouth of the Whakatane River and shallowing of Ohiwa Harbor.

Social Networks

Regional economic wealth and community infrastructure will to some extent determine how well populations will manage and survive these changes, but on a global scale the existing forms of infrastructure for relief and redevelopment aid are quite inadequate. Moreover, there is evidence that in larger scale or repetitive disasters, the availability of national and global aid and the nature of voluntary assistance can be determined by culture, religion, and social class of the displaced people. Communities, rather than the individual reliance on government aid and voluntary arrangements (consider the unraveling of the New Orleans crisis), may be pivotal to improving disaster preparedness. It is a community that holds and shapes the network of people whose skills and social organization can be promptly tapped and trusted in an emergency because members of a community are known to each other.

For New Zealand Maori, our *whanau* [families] are bound by kinship ties and *whakapapa* [genealogy] into *hapu* [subtribe or regional community] relationships (Mead 2003), and these provide the basis for community. Central to *hapu* organization is the *marae*, which includes a meeting house and communally owned land and buildings that offer a physical space for gatherings and decision making. The *marae* has a recognizable spiritual and symbolic strength, drawing on cultural practice and unity. The *marae*, through its assemblage of buildings on a dedicated space may, if its people so decide, provide various forms of hospitality (accommodation, sleeping, food preparation, meals), as well as the collective and social skills of the people who regularly associate with it. Heightening the usefulness of the marae in disaster preparedness is the high cultural value placed on *manaakitanga* [hosting and care giving]. It is a *hapu* responsibility to provide such hospitality without limit for those who seek to use that *marae*. *Marae* have commonly provided *manaaki* [hospitality] on a long-term basis in natural disasters, including the 1987 Edgecumbe earthquake, which caused extensive damage in the Bay of Plenty (J. Peri, personal communication, 2001). The management committees of many *marae* have well-established plans for coordinated relief and care in these situations (H. Hauwaho, personal communication, 2005).

Strategies for Addressing Climate Change

The United Nations Framework Climate Change Convention (UNFCCC) was the first international environmental agreement to be negotiated by virtually the whole international community. It seeks to integrate

Orakei Marae wharenui (meeting house) at Bastion Point in Auckland, Aotearoa/New Zealand.

an environmental approach to sustainable development with the protection of the global climate while considering the vital need of member states to pursue their own forms of economic development.

The stabilization of greenhouse gas concentrations in the atmosphere at a level that will prevent dangerous alterations in the climate system should be achieved within a time frame sufficient to allow ecosystems to adapt naturally to climate change, ensuring that food production is not threatened and that economic development can proceed in a sustainable manner.

A close inspection of the convention shows a considerable number of difficult issues that are unresolved or understated. Climate change can be attributed directly to human activity that alters the composition of global atmosphere, in addition to the natural climate variability observed over a comparable period. This places an onus on a scientific community to determine and define what constitutes a change—to distinguish human-induced from natural variability. This is a nearly impossible task.

The convention recognized that some climate change is inevitable. Its approach in measuring or recognizing effects will allow some vulnerable countries to prepare for the adverse effects of climate change. The inclusion of food production in the objectives is important for Africa in the short term. Parties should take measures and adopt policies that are precautionary, cost effective, and comprehensive but that also take into account socioeconomic contexts. They should not constitute a means of arbitrary and unjustifiable discrimination or a disguised restriction on international trade.

Unlike other regions, New Zealand's share of the three most important greenhouse gases, carbon dioxide, methane, and nitrous oxide, arises principally from the agricultural sector rather than industrial productivity. Most New Zealand methane and nitrous oxide emissions come specifically from enteric fermentation

by ruminants (such as sheep). Research indicates that there are high and low emitting animals and that feed quality can reduce ruminant emission by 10 percent—the modification of stomach bacteria seems a potential source of emission improvement. Diet manipulation and changing winter management practices offer two potential means of reducing of nitrous gases.

With respect to carbon dioxide, various attempts to reduce fossil fuel consumption through technological improvements, particularly in reducing emissions from private vehicles, transport planning, and traffic management are all ongoing issues. Of significance is our greater use of renewable energy, such as hydro, wind, and solar, as well as continuing trends toward improving building insulation, heat reflection, and better uses of entropy—that is, promoting energy efficiency. Increasing our carbon sink through afforestation and preventing deforestation are considered short-term options that help to remove carbon dioxide from the atmosphere and have the additional benefit of stabilizing land and soil.

Indigenous Knowledge Bases

Our understanding of climate change improves when scientific data can be read in conjunction with Indigenous knowledge. Valuable historical information is located within oral testimony such as *purakau* [legends], *waiata* [songs], and *whakapapa* [genealogy] that identify previous experiences of a type that will assist in the reconstruction of long-term climate trends, and to some extent in predicting future regional changes. From Tangaroa (god of the sea): *Tiaki mai I ahau, maku ano koe e tiaki* [Look after me and I will look after you].

Most *hapu* have generations of experience using environmental indicators to predict current events and changes in weather patterns. These predictions are based on consistent observations of particular events like the arrival and departure of migratory species, the calls and flock movements of birds, the onset and intensity of flowering in native plant species, and the phenomena of cloud caps and sun halos. Shifts in wind directions and wind speeds are associated with both local, short-term weather events and the onset or conclusion of seasons.

Planting, harvesting, fishing, and weaving were often regulated in accord with regional weather patterns, but the onset of all seasonal activities was largely governed by the visibility of Matariki (the Pleiades). The arrival of this cluster of stars on the eastern horizon at dawn in mid-June marks the beginning of the Maori New Year—a time of resting and preparation, with main crop planting still about three months away. When the stars of Matariki appear widely apart, a warm season is expected, but when the stars appear close together, the following growing season will be cooler. King and colleagues (2005) acknowledge the success and reliability of Maori weather and climate predictions and suggest potential benefits of creating a comprehensive model of climate understanding through the cooperative deployment of Maori knowledge and Western science.

Human Impacts on Habitat Fragmentation and Species Loss

In New Zealand more Indigenous land habitat has been converted to pastoral and horticultural farmland (about 51 percent) than the world average (37 percent) for similar agricultural practices. Our forest cover, once greater than 85 percent of the land area, has been reduced to remote mountainous areas or widely dispersed and highly fragmented lowland and coastal relics. Clear felling and extensive burning were the conventional preliminaries to pastoral farming. Habitat loss and wildlife recovery programs were not perceived as particularly important. Replacing native forests and grasslands with pastoral plant species has resulted in the contraction of territorial range for native biota and a contemporary exploitation by non-native varieties. There is an associated contraction in the populations of native earthworms, ring nematodes, land snails, and various arthropods, while in areas that have undergone land-use changes there is evidence to suggest that increases in native insect faunas have occurred.

Habitat and niche space has been reduced by the total conversion or fragmentation of existing natural habitats, including wetlands, dunelands, and tussock grasslands, to pasture. Natural habitats have secondarily been degraded by introduced but commercially significant species such as the Monterey pine, which are grown on plantations, as well as pastoral grasses and crops, introduced horticultural varieties, and invasive weeds.

Satellite images of two northern native forests, Waipoua (22,750 ha) and Omapere (5,151 ha), show that both have been fragmented into hundreds of small patches. The small size and patchiness of these forest relics have a major impact on habitat and niche space, affecting the viability of faunal home range and feeding territory. These impacts are depressing both linear and circular forms of plant succession, as well as cross pollination and seed dispersal opportunities. Similar changes affect wetlands and tussock land.

The distribution and densities of native fauna, many of which exist as ancient relics (isolated from continental evolution during the Tertiary period) have suffered huge losses because of recent human impacts. The ground-feeding, nocturnal New Zealand brown kiwi typically have a feeding and foraging range of about 100 m, but because of forest disturbances they may encounter open pasture intervals more than 300 m between foraging sites. Patch intervals have even greater significance when considered in relation to seed masting (periodic flowering, influenced by the length of days during a preceding period of physiological flower induction). In mast years it is estimated that some species (such as beech and rimu) increase seed production up to 5,000 percent.

Though the reasons for it are not fully understood, there is strong evidence that hot, dry conditions in late summer and autumn will lead to a mast year in the following spring and summer. The reliance of New Zealand flora on reptile and bird faunas for seed dispersal means that the seed mast may operate to control population sizes. In years when seed production is poor, some bird species do not attempt to breed. The warming, drying effects of climate change will drastically alter the already precipitous event of seed masting and the reliance of native species on seeds as food resources.

Impacts of Sea Level Changes on Indigenous Resources

Water, food sources, and habitat diversity are critical to Maori cultural practices. Loss of habitat stability because of climate change raises concerns about the distribution and survival of the migratory species on which Maori rely as food resources. Progressive rises in sea level will alter the character of existing estuaries. The tides may occur further upstream of current exchanges, creating additional delays to freshwater river and stream flow. There will be alterations in the dispersion of river sediments, flow channels, and estuarine biota, as wetland habitats become saline. For diadromous species, nesting sites may become lost or displaced and the passage of water, or the function of water in olfactory messaging (for example, in Kokopu egg releases), will be altered to the degree that life cycles are disrupted. *Tuna* [freshwater eel] stocks will suffer similar disruptions. Consider the implications suggested by oral *whakapapa*; the Maori use of lunar cycle timing in predicting the sea migration (pre-spawning) catadromy of *tuna* is now also used in hydro dam water release operations to prevent turbine clogging. Climate change is throwing the natural systems off the predictable cycles of the moon.

Wetlands Offer Unrealized Value

Water is the life giver and preserver—a source drawn from the mixing of the blood of Papatuanuku (Mother Earth) and the tears of Ranginui (Father Sky); therefore, water is significant in spiritual and *whakapapa* relationships. Wetlands were a recognized water source but also a living space shaped by their sponge-like quality, absorbing excess water during high rainfall and releasing it slowly during droughts. The wetlands are a major source of food, cultivation, and plant and organic materials associated with our physical culture. Maori developed an intensified form of domesticated horticulture through an interrelated complex of irrigation and drainage using wetland and associated stream networks. The complex was not simply one of drainage to secure dry pastoral land. The essential conservation of a water system, with regard for its life-bearing properties for organisms through ponds and waterway networking, was maintained within the land-use mosaic.

More importantly, wetland biodiversity offered a multitude of opportunities and resources, each one at a time within a phase of succession, seasonal rhythm, or life stage. Depending on the extent of the wetlands or their riparian character, they can provide flood and drought protection and routes for travel and communication. Waterways are home to many forms of *taniwha* that ensure physical and spiritual protection. In this way particular types or sacred areas [*wahi tapu*] of wetlands became a source of cleansing and healing. *Paru,* an organically rich, anaerobically altered, blue and black mud, came from certain wetland areas and is a traditional source for dyeing fabrics.

Maori resistance to European wetlands ownership and destruction has been continuous throughout the history of colonization—much of which was embedded in their expert knowledge and high regard for wetland resources. Pastoral farming has misinterpreted and misused the opportunity of wetland spaces. The engineering effort to discharge water (particularly floodwater) from wetlands failed to note the geomorphological principles associated with sediment transport and soil loss (as well as downstream events at river discharge points) on coastal sediment supply and its redistribution. As the climate warms, wetlands will become pivotal to water storage, filtration, and resupply. The dynamics of the floodplain will make the summertime pasture paradise containing a few cows even less desirable and economically unviable.

Changing Food Resources

Like other Indigenous peoples, those of us with access (albeit limited) to natural resources still carefully maintain, use, and enjoy their benefits. They extend or substitute our diet. Although some might simply be morsels or flavorings supplementing Western foods, we retain the knowledge of their cultivation, collection, and preparation. These may include the more unusual materials like *karengo* [red seaweed], *pikopiko* [fern fronds], *puha* [sour thistle], *karaka* berries, watercress, gourds, and rotten corn. Equally, we make wide use of each product; for example, after eating the flesh we retain the shells of *kina* [sea urchin], *tio* [oyster], *paua* [abalone] and *kutai* [mussel] for varied uses in soil enrichment, drainage and pathways, and the decorative arts. Fish heads and crayfish bodies are more of a delicacy among Maori than the other body parts (trunk and tail muscle) enjoyed by non-Indigenous people.

In an altered rainfall and temperature regime, Maori may choose to switch their principal root storage crop from *kumara* [sweet potato] plantations (warm/dry) to taro (wet/tropical) with no new social or horticultural experience required. Non-Indigenous people may not feel so comfortable about replacements for rice and potatoes—sago and tapioca come to mind. What is significant is that the retention and nurturing of traditional knowledge and expertise provides for shifts in food (or pharmacological) sources. Traditional knowledge draws on supplies outside of current commercialized crops (crops bound to a pathway of genetic selection and horticultural experience and based on the contemporary climate regime). The rapidity of climate change will almost certainly undermine commercial supplies, and as a result commercial agriculture will fail the demands of a society in which diet and food sources have become overwhelmingly conditioned by the marketplace.

Matata wetland in May 2005 (left) and in post-flood disruption in June 2005 (right) in Aotearoa/New Zealand.

Sources

Adams, J. 2005. Tracking a Homecoming: The Migration of Titi from California. *Titi Times* (Department of Zoology, University of Otago), *Kia Mau Te Titi Mo Ake Tonu Atu* research project 15:6–7.

Bosselmann, K., J. Fuller, and J. Salinger. 2002. *Climate change in New Zealand: Scientific and Legal Assessments.* New Zealand Centre for Environmental Law, Monograph Series vol. 2. Auckland: University of Auckland.

Griffiths, G., et al. 2003. The Climate of the Bay of Plenty: Past and Future? *NIWA Client Report* AKL 2003-044, National Institute of Water and Atmospheric Research, Auckland.

Hayward, B. W., H. R. Grenfell, R. Carter, and J. J. Hayward. 2004. Benthic Foraminiferal Evidence for the Neogene Palaeoceanographic History of the Southwest Pacific East of New Zealand. *Marine Geology* 205:147–184.

King, D. N. T., A. Skipper, H. Ngamane, and B. W. Tawhai. 2005. Understanding Local Weather and Climate with Maori Environmental Knowledge. National Institute of Water and Atmospheric Research (NIWA), Auckland.

Mead, H. M. 2003. *Tikanga Maori: Living by Maori Values.* Wellington: Huia Publishers.

New Zealand Geographic Board. 1990. *He Korero Purakau Mo Nga Taunahanahatanga A Nga Tupuna [Place Names of the Ancestors: A Maori Oral History Atlas].* Wellington: Government Printing Office.

Power, S., ed. 1999. Inter-decadal Modulation of the Impact of ENSO on Australia. *Climate Dynamics* 15(5):319–324.

[*] *Purakau* like the one told here are commonplace stories of explanation that come from a deeper body of knowledge held by *tohunga*, fire markers of the series of undersea active volcanoes at the western edge of the subduction zone, also known as cones and "black smokers."

IMPACTS OF GLOBAL CLIMATE CHANGE

Bradford Burnham

Master in Public Administration, The Evergreen State College (Olympia, Washington)

It is now undeniable that climate change is occurring around the world. Scientists have discovered that the average surface air temperature has risen to its warmest level in 650,000 years. The rise in temperature has already affected environments around the world and, since the temperature is predicted to continue to rise, the effects will become more widespread and intense during this century. Undoubtedly, it will soon become apparent that "combating climate change is the greatest challenge of human history" (Johansen 2003).

Earth's average surface air temperature has risen and fallen over the last 800,000 years. However, the climate change observed by scientists over the past fifty years is most likely due to human activity. We release gases into the air as we grow crops, burn wood and coal, and drive our vehicles. These gases change the way that the air around us holds heat from the sun and the ground. Some of these gases are good at trapping sunlight as it enters our atmosphere, and others are good at trapping heat that rises from the earth. These gases are called "greenhouse gases" because they act like a greenhouse and trap heat. Our activities have raised the amount or concentration of some of these gases by as much as 30 percent above the highest level for the last 800,000 years (Karl et al. 2009).

It is natural for some of the gases in the air to trap heat. This ability of the air, or atmosphere, allows us to have a livable air temperature at the surface. A balance exists between how much heat the air holds and how much it releases. However, if the balance is changed and too much of the gases that hold heat are added, the average surface temperature will rise.

The Changing Atmosphere

Our activities are changing the amounts of greenhouse gases in the atmosphere. As more greenhouse gases become more abundant, more of the sun's heat will be trapped in the atmosphere, raising the average temperature. The greenhouse gases carbon dioxide (CO_2), methane (CH_4), nitrous oxide (N_2O), and some halocarbons (CFC_{13}, CF_2C_{12}) have all increased in concentration in the atmosphere since 1750.

Carbon dioxide (CO_2), the most abundant of these gases, has increased 31 percent since 1750. This is the highest concentration of CO_2 in the past 800,000 years and probably in the past 20 million years. In fact, the rate of increase surpasses what has occurred during at least the last 20,000 years. About 80 percent of carbon dioxide emissions come from the burning of fossil fuels, and about 20 percent come from deforestation and related agricultural practices (Karl et al. 2009).

Methane (CH_4), another greenhouse gas, has increased in concentration 151 percent since 1750 and about half of the increase is due to the burning of fossil fuels and the off-gassing of landfills and agricultural practices in cattle and rice production.

The greenhouse gas nitrous oxide (N_2O) has increased by 17 percent since 1750. This is the highest concentration in 1,000 years. About 30 percent of the N_2O emissions are due to our activities, such as agriculture, cattle feed practices, and the manufacture of chemicals.

Some halocarbons (CFC_{13}, CF_2C_{12}) have decreased because of environmental regulations, but their substitutes (CHF_2CL and CF_3CH_2F), some of which are greenhouse gases, are increasing (Houghton et al. 2001).

The Changing Temperature

The average temperature of the air at ground level has increased over the last century. The global average surface temperature, which includes the entire Earth from pole to pole, has increased over the twentieth century by 0.83°C and is predicted to continue to rise over the twenty-first century an additional 1°C to 6.4°C (Karl et al. 2009). Although this increase seems small, it is a significant increase in a delicately balanced climatic system.

The nighttime low temperatures have changed more than the daytime highs. The nighttime daily minimum air temperatures over land have increased about 0.2°C per decade, and the daytime daily maximum air temperatures have increased 0.1°C per decade (Houghton et al. 2001). In addition, winter has been affected more than summer in many regions. The nighttime lows are less cold, so there are fewer below-freezing nights. This translates into a shorter winter. In the northern hemisphere, many lakes and streams freeze about a week later in early winter and thaw about ten days earlier at the end of winter (Johansen 2003).

The changes over landmasses will not be uniform. In the northern high latitudes, cold seasons will see greater increases in temperature. In fact, the northern regions of North America and northern and central Asia are predicted to have increases 40 percent greater than global mean warming (Houghton et al. 2001). All land areas are predicted to have increases in surface temperatures above the global average because the oceans can absorb more heat and moderate changes effectively.

The Climate Forecast

We can already measure the effects of climate change on regional weather events all over the world. Pronounced increases in precipitation over the past hundred years have been observed in eastern North America, southern South America, and northern Europe. On the other hand, less precipitation has fallen in the Mediterranean region, most of Africa, and southern Asia (Karl et al. 2009).

Winds may also be affected by climate change. The trade winds may change in speed, duration, and location. Wind is usually generated by differences in air temperature and air pressure from region to region. Air that is warmed rises and creates an area of low pressure because of the upward movement. Nearby cooler air is drawn into the region. The trade winds act as conveyor belts moving warm air that has risen from warm regions of the equator and mid-latitudes into the cooler higher latitudes. Scientists have already noticed a slight change in the path of the jet stream of the northern hemisphere. This wind has occasionally moved slightly higher in latitude (Johansen 2003).

The oceans have a great influence on weather in any particular region. The oceans absorb and retain a lot of heat from the sun. Global ocean heat content has risen since the late 1950s. An increase in heat may translate into more, and more intense, storms. Intense storms, such as hurricanes and typhoons, obtain much of their energy from warm ocean water. As the oceans warm, the possibility of more intense storms rises. The heat in the oceans can also affect long-term phenomena, such as the El Niño-Southern Oscillation (or ENSO), which shifts Pacific water temperatures and air surface pressures about every five years. ENSOs have been more

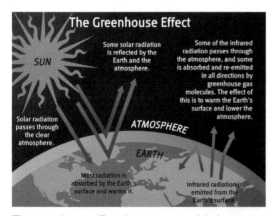

The greenhouse effect that generates global warming.

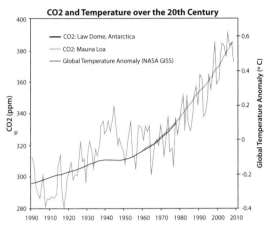

As global carbon dioxide (CO2) levels have risen from 1990 to 2010, global average temperatures have also climbed upward in a nearly parallel track.

frequent and intense since the 1970s, compared with the last hundred years (Houghton et al. 2001).

Geographic Changes

Sea level has been rising recently as the average atmospheric temperature has increased. The rises result from the melting of land-based ice sheets and the expansion of the ocean water as it heats up. The level has risen by about 20 cm over the past century. Over the past fifteen years, the rate of increase has accelerated, and sea level is now rising at a rate close to double that of the past century. The average sea level rise may be between 0.9 and 1.2 m by the end of the twenty-first century (Karl et al. 2009).

If emissions of greenhouse gases are not reduced, shorelines of the world may be dramatically redrawn due to climate change. Tens of millions of people living in low-lying areas will be affected by sea level rise. Cities such as Dhaka, Bangkok, Calcutta, and Manila, and the U.S. states of Florida and Louisiana, are particularly susceptible to rising sea levels (Rosentrater 2005).

Shifting Habitats

Global climate changes will cause a wide variety of shifts in local climate conditions, or climes, in the regions of the world. Clime changes may translate into environmental alterations great enough to reshape the conditions that many species of plants and animals need to survive. Some organisms may be able to cope with the changes and live in a slightly stressed state, while others will not be able to survive in the region any longer.

Climate change that takes place from decade to decade can affect some species more than others. Some

northeastern birds now arrive in their summer habitats an average of thirteen days earlier than in the first half of the twentieth century. Some birds that winter in South America return to the northeast an average of four days earlier (Ahmad et al. 2001). As an example of a more drastic impact, climate change may have accelerated the local extinctions of two species of checkerspot butterfly (McLaughlin et al. 2002).

Arctic environments will most likely experience the greatest climate changes. The oceans and the land have slowly been warming in recent history to the point that there has been a "possible 40% decrease in the thickness of sea ice in the Arctic during summer in recent times" (Houghton et al. 2001). The Alaska North Slope looks like Swiss cheese from a plane, with small-to-large meltwater lakes on top of the permafrost. As the permafrost thaws, "peat in the frozen subsoil can decompose, releasing greenhouse gases into the atmosphere, impacting global warming" (Bach and Beckman 2006).

The loss of sea ice affects many species. Predators of seals will also have a difficult time living with climate change. About two-thirds of the world's polar bears are projected to disappear by the middle of the twenty-first century. It is predicted that in seventy-five years, there will be no more wild polar bears in Alaska (Karl et al. 2009).

Tree lines have risen to higher elevations in many mountains due to increased average temperatures. There has been more regrowth along the tree lines, and they have slowly been moving higher up the mountains as temperature increases (Houghton et al. 2001).

Corals reefs expel their microscopic yellow-green algae when the water reaches 30°C. This phenomenon is called *coral bleaching* because the corals turn white and die after they expel the algae. The record-setting temperatures of 1997 and 1998 caused the bleaching of coral reefs in thirty-two countries (Houghton et al. 2001). Some robust corals in Australia's Great Barrier Reef were harmed as well, including one that had been dated back 700 years (Houghton et al. 2001).

Our Connection with the Changes

Studies of past climate changes have shown a connection between them and subsequent changes in society. In the Andean altiplano of South America, evidence suggests that a period of drier climate (ca. 1100–1400 CE) caused the Tiwanaku civilization (300 BCE–1100 CE) to disappear. Its agricultural production declined to the point that people had to abandon their fields (Binford and Kolata 1996).

Evidence in the Pacific islands links rapid cooling and sea level fall around 1300 CE with increased conflict and abandonment of villages. The climate changes caused water tables to fall, storms to increase, coral reefs to dry, and wetlands to alter. The changes that the peoples living on the islands experienced as a result were significant and long-lasting (Nunn 2003).

The expected rise in sea level in the twenty-first century may completely submerge some low islands and micro-atolls. Some atolls may need to be abandoned altogether. Towns on low island chains such as the Bahamas, Kiribati, the Marshall Islands, and the Maldives are only a few meters above sea level. In addition to losing land, these islands will have saltwater infiltration into their fresh water supply and septic systems. One report states that "sea-level rise and climate changes, coupled with environmental changes, have destroyed some very important and unique cultural and spiritual sites, coastal protected areas, and traditional heritage sites in the Federated States of Micronesia, Tuvalu, the Marshall Islands, Niue, and Kiribati and continue to threaten others" (Ahmad et al. 2001).

The Inuit and the Arctic

Inuvialuit hunters on Banks Island in Canada's High Arctic have strong connections with their environment. They have been noticing the climate conditions becoming more unpredictable and the landscape unfamiliar. The annual ice and multi-year ice have changed, causing problems for the wildlife and the hunters. The ice is less predictable and, therefore, less stable and safe. The freeze-up is a month later and the thaw earlier.

The change in climate has brought new phenomena, such as thunder and lightning. New species are arriving with warmer weather as well. Foreign birds such as barn swallows and robins visit now, and salmon have been caught in nearby rivers for the first time (Ashford and Castledon 2001). Ringed seals have had population size changes connected to the changes in sea ice conditions. Caribou and musk oxen have benefited from increased forage, due to increased summer rains, but they are also being negatively affected by more frequent freezing rains in autumn (Riedlinger 2001).

The Inuit of northern Canada, Alaska, Siberia, and Greenland have also already had to deal with environmental changes brought on by a warming climate. "Since the millennium they have seen their landscape, their livelihood and their very cultural identity eroded at such an alarming rate that they now look set to become the first society to fall victim to climate change in the 21st century" (Kendall 2006). The capacity to adapt is a part of Inuit life. The rate of change, however, mixed with the increased variability of seasonal conditions and extreme events, might make it difficult for them to use traditional resources effectively in coping with the predicted change (Riedlinger 2001).

Multi-year sea ice is now melting at a rate of about 10 percent per decade. At this rate, polar bears and other species dependent on stable ice platforms might become extinct before the end of the twenty-first century (Rosentrater 2005). Caleb Pungowiyi, a Yupik from Nome, Alaska, says, "Ice is a supporter of life. It brings the sea animals from the north into our area and in the fall it also becomes an extension of our land.... And it affects the animals, too. They depend on the ice for breeding, for pupping, denning, lying, and having their young. They molt on it, they migrate on it. And so ice is a very important element to us. When it starts disintegrating and disappearing faster, it affects our lives dramatically" (Moreno 2000).

Traditional knowledge and science have begun to be combined to better predict what is happening, what will happen, and how people, especially those in the northern latitudes, will cope with the changes. The former Vuntut Gwich'in chief Randall Tetlichi framed the challenges of integrating traditional and local knowledge with the work of science and policy as a need to "double understand" (Bermana and Kofinas 2004).

Conclusions

The scientific evidence shows that climate change has been accelerating in recent decades and that some of the changes have been caused by the production of greenhouse gases during the last century and a half. The current warming period is lasting longer than usual. It appears that our activities have altered the natural cycle of heating and cooling for the Earth.

Current predictions show that all landmasses will experience climate change. The mid- to high-latitude areas of the northern hemisphere and Antarctica will most likely experience the greatest change in surface air temperature. Low-lying coastal areas and island nations will most likely experience much of the geographical change as glaciers continue to melt and raise the ocean level. Coastal areas could also experience more intense storms as the oceans warm and provide more heat energy to hurricanes, cyclones, and monsoons.

The changes in climate will affect plant and animal species throughout the world. Some environments will change so dramatically that many plants and animals will not be able to move away or adapt in time. These species may disappear locally as they shift to higher latitudes or up mountain slopes, or even become globally extinct.

As we change our current habits and try to mitigate the effects of our past activities, we must also protect as many species and habitats as possible from the dramatic changes that may occur due to climate change.

Resources

Ahmad, Q. K., et al. 2001. *Climate Change 2001: Impacts, Adaptation, and Vulnerability.* A Report of Working Group II of the Intergovernmental Panel on Climate Change. Cambridge: Cambridge University Press.

Alroy, John. 2001. A Multispecies Overkill Simulation of the End-Pleistocene Megafaunal Mass Extinction. *Science* 292(5523):1893.

Ashford, Graham, and Jennifer Castleden. 2001. Inuit Observations on Climate Change. Final Report. International Institute for Sustainable Development. www.iisd.org/publications/publication.asp?pno=410.

Bach, John, and Wendy Beckman. 2006. Red Flags in the Great White North. *University of Cincinnati Research* (Summer):26–33.

Barnasky, Anthony, Elizabeth A. Hadly, and Christopher J. Bell. 2003. Mammalian Response to Global Warming on Varied Temporal Scales. *Journal of Mammalogy* 84(2):354–368.

Bermana, Matthew, and Gary Kofinas. 2004. Hunting for Models: Grounded and Rational Choice Approaches to Analyzing Climate Effects on Subsistence Hunting in an Arctic Community. *Ecological Economics* 49:31–46.

Berteaux, Dominique, Denis Reale, Andrew G. McAdam, and Stan Boutin. 2004. Keeping Pace with Fast Climate Change: Can Arctic Life Count on Evolution? *Integrative and Comparative Biology* 44:140–151.

Binford, M. B. and Alan L. Kolata. 1996. Climate variation and the Rise and Fall of an Andean Civilization. *Quaternary Research* 47:235–248.

Brooks, Robert T. 2004. Weather-Related Effects on Woodland Vernal Pool Hydrology and Hydroperiod. *Wetlands* 24(1):104–114.

Derocher, Andrew E., Nicholas J. Lunn, and Ian Stirling. 2004. Polar Bears in a Warming Climate. *Integrative and Comparative Biology* 44:163–176.

Dessai, Suraje. 2003. Heat stress and Mortality in Lisbon Part II: An Assessment of the Potential Impacts of Climate Change. *International Journal of Biometeorology* 48(1):37–44.

Esat, Tezer M., Malcolm T. McCulloch, Alexander W. Tudhope, Graham E. Mortimer, John Chappell, Bradley Pillans, Allan R. Chivas, Akio Omura. 1999. Coral Record of Equatorial Sea-Surface Temperatures during the Penultimate Deglaciation at Huon Peninsula. *Science* 283(5399):202–204.

Fenge, Terry. 2001. The Inuit and Climate Change. *Isuma: Canadian Journal of Policy Research/Revue Canadienne de Recherche sur les Politiques* 2(4):79–85.

Flinn, Edward D. 2006. GRACE Detects Loss of Antarctic Ice. *Aerospace America* 44(6):24.

Galbraith, H., R. Jones, R. Park, J. Clough, S. Herrod-Julius, B. Harrington, and G. Page. 2002. Global Climate Change and Sea Level Rise: Potential Losses of Intertidal Habitat for Shorebirds. *Waterbirds* 25(2):173–183.

Harwood, John. 2001. Marine Mammals and Their Environment in the Twenty-First Century. *Journal of Mammalogy* 82(3):630–640.

Houghton, J. T., Y. Ding, D. J. Griggs, M. Noguer, P. J. van der Linden, X. Dai, K. Maskell, and C. A. Johnson, eds. 2001. *Climate Change 2001: The Scientific Basis.* Cambridge: Cambridge University Press.

Jackson, Robert B., Stephen R. Carpenter, Clifford N. Dahm, Diane M. McKnight, Robert J. Naiman, Sandra L. Postel, and Steven W. Running. 2001. *Ecological Applications* 11(4):1027–1045.

Johansen, Bruce. 2003. *The Global Warming Desk Reference.* Westport, Conn.: Greenwood Press.

Karl, Thomas R., Jerry M. Melillo, and Thomas C. Peterson, eds. 2009. *Global Climate Change Impacts in the United States.* Cambridge: Cambridge University Press.

Kendall, Clare. 2006. Life on the Edge of a Warming World. *The Ecologist* 36(5):26.

Kennett, Douglas, and James Kennett. 2000. Competitive and Cooperative Responses to Climatic Instability in Coastal Southern California. *American Antiquity* 65(2):379–395.

Kitaysky, Alexander, and Elena Golubova. 2000. Climate Change Causes Contrasting Trends in Reproductive Performance of Planktivorous and Piscivorous Alcids. *Journal of Animal Ecology* 69:248–262.

Laval, Richard K. 2004. Impact of Global Warming and Locally Changing Climate on Tropical Cloud Forest Bats. *Journal of Mammalogy* 85(2):237–244.

Liu, Xiaodong, Liguang Sun, Zhouqing Xie, Xuebin Yin, and Yuhong Wang. 2005. A 1300-year Record of Penguin Populations at Ardley Island in the Antarctic, as Deduced from the Geochemical Data in the Ornithogenic Lake Sediments. *Arctic, Antarctic, and Alpine Research* 37(4):490–498.

Matsui Tetsuya, Tsutomu Yagihashi, Tomoki Nakaya, Hirosi Tauda, Shuichiro Yoshinaga, Hiromu

Daimaru, Nobuyuki Tanaka. 2004. Probability Distributions, Vulnerability and Sensitivity in *Fagas crenata* Forests Following Predicted Climate Change in Japan. *Journal of Vegetation Science* 15:605–614.

McFadzien, Diane, Francis Areki, Tarai Biuvakadua, and Monifa Fiu. 2004. *Climate Witness Community Toolkit.* Suva, Fiji: World Wildlife Federation, South Pacific Programme.

McLaughlin, John F., Jessica J. Hellmann, Carol L. Boggs, and Paul R. Ehrlich. 2002. Climate Change Hastens Population Extinctions. *Proceedings of the National Academy of Sciences* 99(9):6070–6074.

Meehl, Gerald A., Warren M. Washington, Benjamin D. Santer, and William D. Collins. 2006. Climate Change Projections for the Twenty-First Century and Climate Change Commitment in the CCSM3. *Journal of Climate* 19(11):2597–2616.

Moreno, Fidel. 2000. In the Arctic, Ice Is Life—and It's Disappearing. *Native Americas*, Feb. 11.

Nunn P. D. 2003. Revising Ideas about Environmental Determinism: Human-Environment Relations in the Pacific Islands. *Asia Pacific Viewpoint* 44(1):63–72.

Parker, David E. 2006. A Demonstration that Large-Scale Warming is not Urban. *Journal of Climate* 19(12):2882–2894.

Peterson, James, and Thomas Kwak. 1999. Modeling the Effects of Land Use and Climate Change on Riverine Smallmouth Bass. *Ecological Applications* 9(4):1391–1404.

Polyak, Victor J., and Yemane Asmerom. 2001. Late Holocene Climate and Cultural Changes in the Southwestern United States. *Science* 294(5540):148–151.

Riedlinger, Dyanna. 2001. Responding to Climate Change in Northern Communities: Impacts and Adaptations. *InfoNorth* (Arctic) 54(1):96–98.

Rosentrater, Lynn, ed. 2005. 2° is Too Much! Evidence and Implications of Dangerous Climate Change in the Arctic. www.panda.org/downloads/arctic/050129evidenceandimplicationshires.pdf

Shelley, Suzanne. 2006. Global Warming: On the Front Burner. *Chemical Engineering* 113(6):19–23.

Smithers, S. G., D. Hopley, and K. E. Parnell. 2006. Fringing and Nearshore Coral Reefs of the Great Barrier Reef: Episodic Holocene Development and Future Prospects. *Journal of Coastal Research* 22(1):175–187.

Solomon, S., G. K. Plattner, R. Knutti, and P. Friedlingstein. 2009. Irreversible Climate Change due to Carbon Dioxide Emissions. *Proceedings of the National Academy of Sciences* 106(6):1704–1709.

Stachowicz, John, Jeffrey Terwin, Robert Whitlatch, and Richard Osman. 2002. Linking Climate Change and Biological Invasions: Ocean Warming Facilitates Nonindigenous Species Invasions. *Proceedings of the National Academy of Sciences* 99(24):15497–15500.

Waggoner, Paul E., ed. 1990. *Climate Change and U.S. Water Resources.* New York: John Wiley and Sons.

EFFECTS OF CLIMATE CHANGE ON WOMEN'S AND CHILDREN'S HEALTH

Debra McNutt

Graduate student in Master of Public Administration Tribal Concentration program, The Evergreen State College (Olympia, Washington). This study was produced for the Community Alliance and Peacemaking Project.

"Children inherit the societies that adults build."
—*Katherine Shea & Sophie J. Balk*

Climate change is not just about global warming, but about instability in our natural world caused by the burning of carbon in oil and coal. "Rainfall and freshwater availability, average temperatures, agricultural growth zones and sea level all will change. Ecological and human health consequences can be anticipated and some are already being measured" (Shea and Balk 2007, 3). Some areas may get hotter and drier, while other areas may get cooler or wetter. But either way, the most vulnerable people in these changing conditions are women and children.

Women are extremely vulnerable to climate change and may bear an unreasonably large share of the burden of adaptation. The young, the elderly, the poor, the frail, and those who live in the top floors of apartment buildings and lack access to air conditioning, especially in large urban areas, are particularly vulnerable (Duncan 2007).

Children are likely to suffer disproportionately from the direct and indirect health consequences of a rapidly warming world … children are more sensitive than adults to harm from environmental hazards. Climate change increases these hazards by worsening air quality, stimulating more extreme weather events, creating conditions that favor increases in food-, water-, and vector-borne infections, and enhancing heat stress conditions (Shea and Balk 2007, 3–4).

Yet in the global and national discourse on climate change, the effects on women and children are strangely absent. In examining this topic, a researcher is easily struck by the lack of attention to women and children, and the assumption that climate change affects all communities equally. The international women's human rights network MADRE observes: "Worldwide, the compounded effect of poverty and gender discrimination is the single gravest threat to women's health: women have the least access to health services, nutritious food, clean water, and opportunities for rest. As overall human health declines, women face the greatest risk of illness, as well as unsustainable work burdens of caring for the sick." The organization points out how the global climate change community has failed to take these differences into account:

Most approaches to tackling the threats of climate change focus on scientific and technological aspects of the problem, ignoring its social impact. Both the Kyoto Protocol and the UN Framework Convention on Climate Change neglect to even mention gender. Yet developing a gender analysis—an understanding of the ways that men and women are differently affected by climate change and respond differently to its threats—is increasingly crucial to saving lives, saving resources, and quite possibly, saving the life of the planet (MADRE 2008).

The specific impacts of climate change even more intensely affect Indigenous women and children, in both urban and rural areas. The possible responses and solutions also differ for Native women and children, who may not be protected by more generic solutions.

Impacts of Climate Change

Heat Waves

Heat waves cause not only heat stroke, but also increased illness and death, especially in the concrete "heat islands" of cities. The effect of heat on pregnant women and their babies can increase infant mortality, but other women (such as those who cook over hot stoves) are also vulnerable. Duncan (2007) observes, "Men and women differ in their response to extreme heat. Women sweat less, have a higher metabolic rate,

and have thicker subcutaneous fat that prevents them from cooling themselves as efficiently as men. Fortunately, heat-related health impacts can be reduced through ... the use of air conditioners, increased intake of fluids, the development of community-wide heat emergency plans, and improved heat-warning systems." Air conditioning can help, but it contributes to higher electrical use that can cause more climate change.

The effects of heat waves are felt disproportionately by elders and children:

This is a concern primarily for the elderly; infants and small children are at greater risk than the average adult from heat stroke and death under extreme temperature conditions. Very small children are vulnerable because they do not have fully developed temperature regulation mechanisms and are unable to change their environments without help from adults. Older children and adolescents spend more time in vigorous activity outdoors and thus have higher exposures (Shea and Balk 2007).

Extreme Weather and Stress

In the Pacific Northwest, we are already experiencing extreme shifts in our weather, including windstorms, mudslides, and floods. Since 2006, storms have knocked out power to the region, mudslides have cut access to Skokomish, floods have closed highways near Chehalis, and waves have pushed driftwood up to a school at Quileute. Rising sea levels will make these coastal storms and erosion even worse. In a weather crisis, women protect their kids from injury and drowning, care for the hurt and ill, and provide emotional support. Duncan (2007) also comments on the social stress generated by disasters:

With increased temperatures, extreme weather events are likely to increase. Gender significantly affects the daily lives of women and men, before, during, and after an extreme event. Women who are battered, immigrants, indigenous, isolated, poor, refugee, and seniors are particularly vulnerable to such events. Police reports of domestic violence following the 1980 Mt. St. Helen's volcanic eruption increased by 46 percent.

Women and their families have to deal not only with the damage, but the stress and infections from cleaning up mold, spoiled food, and dirty or toxic water. The UN Intergovernmental Panel on Climate Change (IPCC 2007) observes that women are vulnerable in periods after disasters. "Natural disasters have been shown to result in increased domestic violence against, and posttraumatic stress disorders in, women ... Women make an important contribution to disaster reduction, often informally through participating in disaster management and acting as agents of social change. Their resilience and their networks are critical in household and community recovery."

Breathing Problems

Climate change can cause breathing problems in several ways. Warmer summers create more ozone pollution and a longer and more intense allergy season: "We expect that climate change will result in changes in the quantity, quality and distribution of pollens and other aeroallergens … asthma and allergies are likely to be worsened regionally in a warmer world" (Shea and Balk 2007, 5).

The same carbon in the atmosphere that causes climate change also causes more plants to grow, increasing pollens and allergies. Warmer summers and the spread of tree-killing insects (such as the spruce bark beetle) cause more intense wildfires. The IPCC (2007) documents that these climate changes can cause more respiratory diseases and deaths:

> In some regions, changes in temperature and precipitation are projected to increase the frequency and severity of fire events … Forest and bush fires cause burns, damage from smoke inhalation and other injuries. … Toxic gaseous and particulate air pollutants are released into the atmosphere, which can significantly contribute to acute and chronic illnesses of the respiratory system, particularly in children, including pneumonia, upper respiratory diseases, asthma and chronic obstructive pulmonary [disease] … Pollutants from forest fires can affect air quality for thousands of kilometers.

Threats to Water Quality and Quantity

Women have a special relationship to water, in birth, ceremonies, and even cooking and cleaning. Climate change is affecting the availability and reliability of fresh water, and in Northwest winters, we are seeing more rainfall and less snow in the mountains, and the retreat of the snowpack and glaciers. Streams flood more in the winter, and have reduced flow in the spring (NIARI 2006).

Flooding can carry disease, and reduced flow can make clean, fresh water scarce. Shea and Balk (2007, 6) observe, "Changes in precipitation and more extreme precipitation events are also very likely. Heavy rain is highly correlated with outbreaks of waterborne illness as surface and groundwater become contaminated by run-off and overwhelm water treatment systems. …

Infants and small children are at higher risk for complications and hospitalizations from such infections."

These infections and waterborne illnesses can have serious implications for vulnerable women and children. "Although most of these cases of waterborne disease involve mild gastrointestinal illnesses, other severe outcomes such as myocarditis are now recognized. These infections and illnesses can be chronic and even fatal in infants, the elderly, pregnant women, and people with weakened immune systems" (Patz *et al.* 2000, 371).

Disruption of Food and Natural Resources

Climate change can disrupt food supplies, and harm traditional foods. Sudden streamflow changes damage salmon habitat, ocean temperature changes kill crabs and other marine life, and warmer temperatures cause more shellfish diseases. "We are likely to witness crises in food and water supply, large-scale species extinctions, forced migrations of populations because of dramatic sea level rise, drought, and loss of natural resources" (Shea and Balk 2007, 8).

Warming also makes valued animal and plant species shift northward or up mountain slopes, outside of customary harvesting areas, as new weeds, pests, and diseases shift into the area (NIARI 2006). The increase in weeds and pests also leads to more use of herbicides and pesticides on our food. Heat waves, droughts, and floods may even affect the availability of food crops, according to the U.S. Environmental Protection Agency (2007):

> The increased potential for droughts, floods and heat waves will pose challenges for farmers. Additionally, the enduring changes in climate, water supply and soil moisture could make it less feasible to continue crop production in certain regions … increased frequency of heat stress, droughts and floods negatively affect crop yields and livestock. … Climate variability and change also modify the risks of fires, pest and pathogen outbreak.

Increase in Diseases

Warmer temperatures can encourage the growth of bacteria, causing more ailments such as salmonella: "Foodborne infections are likely to increase due in part to changes in eating behavior … and in part because food-borne pathogens grow faster in warmer weather" (Shea and Balk 2007, 6).

The northward shift of species will bring in more rodents, insects, and viruses that affect human health. Rodents, mosquitoes, and ticks serve as vectors that transport human disease, and their ranges are increasing. They can carry diseases such as West Nile Virus in mosquitoes, Lyme disease in ticks, and hantavirus in mice. (The positive growth of local gardens and backyard composting can unfortunately lead to more rodents.) According to Shea and Balk "Patterns of vector-borne illness are expected to change. Insects and rodents respond quickly to changes in temperature and moisture by migrating and, during favorable conditions, by reproducing more rapidly, often resulting in localized 'plagues.' … Warmer weather and failure of winter kills will prolong the transmission season and change the range of vectors in latitude and altitude, resulting in more illness. Amplification of West Nile Virus, for example, is associated with warmer winters and spring drought" (Shea and Balk 2007, 6).

Possible Responses

Unlike most non-Native communities, tribes still possess traditional ecological knowledge, sovereignty, and a sense of community that will be key to surviving the climate changes ahead (NIARI 2006). Native women have a special responsibility to the water, the family, and the community. Tribal youth also have the opportunity to get involved in protecting the future. Together, they can think ahead and plan for both the gradual and sudden changes of climate change, starting with particular discussions and actions:

In anticipation of more extreme weather events, we can work with local authorities to strengthen the public health infrastructure including early warning systems, and disaster preparedness and response plans. In areas where vector-borne illness is likely to increase, we can work to implement preventive strategies such as eliminating breeding grounds for rodents or mosquitoes, thus minimizing the need for the widespread use of toxic pesticides. In large cities subject to heat waves and concentrated air pollution, we can participate in programs to educate parents, teachers, child care providers and children … Each region faces different challenges (Shea and Balk 2007, 11).

Women have to be involved in this planning for it to be effective and to take into account the needs of the majority of the population who are women and children. Duncan (2007) asserts: "Women must be included in disaster prevention, mitigation, and recovery strategies. Specifically, women must be engaged in: family, household, and workplace preparation for extreme weather events; response and recovery; emergency site

organization; physical and emotional care for children; and organizing kin and friendship networks."

The importance of Indigenous women's responses to climate change is being discussed in many other countries around the world, particularly in Asia, Africa, and Latin America (Asia Pacific Forum on Women, Law and Development 2009, International Union for Conservation of Nature 2009, Tebtebba Foundation 2009). At the United Nations Permanent Forum on Indigenous Issues in 2009, the Indigenous Women's Biodiversity Network said in a statement:

Indigenous women have a vital role in the conservation and sustainable use of biological diversity, and in regards to maintaining Indigenous Peoples' traditional knowledge, cultures and languages, which we pass on from generation to generation. Indigenous women are concerned that inappropriate policies cause resource abuse, climate change, foster extractive industries, and over-harvesting all resulting in the continued loss of biodiversity. … Indigenous women have a role in the transmission of knowledge from the past to future generations. This link ensures the strengthening of our cultural values.

The Native Women's Association of Canada (2007) similarly asserts that Native women "have always played an important role in both traditional and modern societies and are integral to the wellbeing of their families and communities in times of crisis," and that they "must be encouraged and supported in the important role of educating and preparing their communities to develop general adaptation strategies in order to minimize the potential harm from impacts of climate change."

Like Native peoples in general, women possess knowledge in family health and traditional knowledge that will be critical to surviving the inevitable changes ahead.

As MADRE (2008) concludes, "Although they are the most threatened, women provide critical resources for maintaining health. Women's capacity to activate social networks for caregiving, their stewardship of medicinal plants, their expertise in traditional medicine, and—of course—the health of women themselves must be protected in order to defend women's human rights and enable communities to adapt to increased health threats associated with climate change."

Resources

Asia Pacific Forum on Women, Law and Development. 2009. Rural and Indigenous Women's Statement on Climate Change: A Submission to the Parties of the UN Framework Convention on Climate Change.

http://www.climatechangeaction.net/action/rural-and-indigenous-women%E2%80%99s-statement-climate-change

Duncan, Dr. Kirsty. 2007. "Global Climate Change and Women's Health." *Women & Environments* (Spring). http://www.redorbit.com/news/health/890479/global_climate_change_and_womens_health/index.html

Engendering Climate Change. 2012. http://genderinclimatechange.wordpress.com

Indigenous Women's Biodiversity Network. 2009. Joint Statement to UN Permanent Forum on Indigenous Issues Eighth Session (May 19). http://www.indigenousportal.com/index.php?option=com_mycontent&task=view&id=5197&Itemid=254

Intergovernmental Panel on Climate Change (IPCC). 2007. "Human Health" (Chapter 8) of Working Group II Report "Impacts, Adaptation and Vulnerability." In *IPCC Fourth Assessment Report.* http://www.ipcc.ch/ipccreports/ar4-wg2.htm

International Union for Conservation of Nature. 2009. "Indigenous women: Most vulnerable to climate change but key agents of change" (June 22). http://www.iucn.org/news_homepage/all_news_by_theme/gender_news/?3403/Indigenous-women-most-vulnerable-to-climate-change-but-key-agents-of-change

MADRE. 2008. "A Women's Rights-based Approach to Climate Change." http://www.madre.org/index/press-room-4/news/a-womens-rights-based-approach-to-climate-change-245.html

Native Women's Association of Canada. 2007. "Aboriginal Women and Climate Change: An Issue Paper." http://www.laa.gov.nl.ca/laa/naws/pdf/nwac-climate_change.pdf

Northwest Indian Applied Research Institute (NIARI). 2006. *Climate Change and Pacific Rim Indigenous Nations.* http://academic.evergreen.edu/g/grossmaz/IndigClimate2.pdf

Patz, Jonathan, et al. 2000. "The Potential Health Impacts of Climate Variability and Change for the United States: Executive Summary of the Report of the Health Sector of the U.S. National Assessment." *Environmental Health Perspectives* 108 (4). http://www.ncbi.nlm.nih.gov/pmc/articles/PMC1638004/

Shea, Katherine, and Sophie J. Balk. 2007. "Climate Change and Children's Health: What Health Professionals Need to Know and What We Can Do About It." *Health and Environment.* http://www.healthandenvironment.org/?module=uploads&func=download&fileId=418

Tebtebba Foundation. 2009. "Indigenous Women and Climate Change." In *Guide on Climate Change and Indigenous Peoples.* 2nd Edition. Baguio, Philippines. http://www.tebtebba.org/index.php/content/160-2nd-edition-of-guide-on-climate-change-and-indigenous-peoples-now-released

U.S. Environmental Protection Agency (EPA). 2007. "Agriculture and Food Supply." *Climate Change - Health and Environmental Effects.* http://www.epa.gov/climatechange/effects/agriculture.html

World Health Organization (WHO). 2003. *Climate Change and Human Health - Risks and Responses.* http://www.who.int/globalchange/climate/en/ccSCREEN.pdf

PART III. CURRENT RESPONSES

CURRENT RESPONSES

The current debates around climate change often seem depressing and overpowering. The discussion usually centers on the global scale of "global warming," and how can one person or family help protect the entire world? Changing individual habits is a start, but simply changing light bulbs is just not enough to slow the growth of fossil fuel industries that emit carbon gases into the atmosphere. Changing the habits of global industries and political institutions seems to move at a more glacial pace than the glaciers that are melting, especially when some leaders still blindly express skepticism that climate change is actually occurring.

The changes that are necessary for survival are not happening at the level of the individual nor at the level of national or global policy. The most promising responses are at the level of local communities and regional bodies, working together to mitigate or adapt to the changing climate realities. The solutions that work might not be found at the United Nations or the U.S. Congress but in stitching together a series of local responses into regional quilts that can enhance the resilience of each ecosystem. With political self-determination, traditional ecological knowledge, and a resilient sense of community, Indigenous peoples are better suited to survive climate change than the mainstream, energy-addicted, shopping-mall society.

Indigenous resilience requires turning ideas quickly into action. The Native communities that prepare the earliest for the inevitable changes will be those that make it. As the Nisqually treaty rights leader Billy Frank Jr. says, "Let's stop talking about things and let's get out there and roll up our sleeves and get busy." The Ojibwe Anishinaabe community organizer and economist Winona LaDuke wrote in *Indian Country Today* on November 4, 2009:

> We have a shot at being self-determining or we can be the victims. This is a time of tumultuous change, economic downturns, accelerating climate destabilization, and the depletion of oil supplies…. If we don't act, we will be caught in a very difficult place as indigenous peoples.
>
> We need to make decisions about the future of our communities and what that future will look like. Will we continue to rely on the outside industrial economy for our food, energy, and other basic needs or will we look to create our own local economies as a way to determine our own destiny?

Now is the time to act. We can start by assessing our current economies. Food and energy consume massive portions of our tribal economies—nearly half of the average tribal economy is spent outside the reservation on energy and food. This creates a huge economic leak. We can set a goal to re-localize tribal economies by developing energy efficiency, renewable energy, and sustainable food…. There is a great deal of work taking place in our communities to re-localize food, energy, and to build resilient and sustainable economies.

Salmon bake at the 2011 Quileute Days festival in La Push, Washington.

INDIGENOUS RESPONSES TO THE INTERNATIONAL CLIMATE CHANGE FRAMEWORK

Zoltán Grossman

Member of the Faculty in Geography/Native American and World Indigenous Peoples Studies, The Evergreen State College (Olympia, Washington)

"Humankind is capable of saving the earth if we recover the principles of solidarity, complementarity, and harmony with nature in contraposition to the reign of competition, profits, and rampant consumption of natural resources."

—Bolivian President Evo Morales, 2008

For the past decade, Indigenous non-governmental organizations (NGOs) and some Native governments have been attempting to participate in the international discussion around the climate crisis and to intervene in the international climate change regulatory framework. Since 1998, for example, they have attended the Conferences of the Parties (COPs) or the annual meetings of the state signatories of the global warming treaty, known as the United Nations Framework Convention on Climate Change (UNFCCC). Indigenous groups' goals are to urge a reduction in the greenhouse gas emissions that threaten Native lands and resources, to secure recognition of Indigenous nations as holding a "special status" in climate change negotiations, and to gain international support for their efforts to mitigate global climate change.

Indigenous NGOs and nations have also taken a number of other paths to work internationally around climate change. They have called the attention of other United Nations agencies to climate change as a pressing issue of economic, social, and cultural rights, and they have asserted the role of traditional ecological knowledge (TEK) in identifying and adapting to climate change. They have used international law to bring legal complaints to international legal forums such as the Inter-American Human Rights Commission. Indigenous nations have also asserted their sovereignty through the pursuit of "climate justice" and weighed whether to participate in carbon-trading systems or to use trust responsibility mechanisms to influence national government actions.

Indigenous organization representatives provide a briefing on their positions on the UN Framework Convention on Climate Change in Bonn, 2001. Panel (left to right): Patrina Dumaru (Fiji), Raymond de Chavez (Philippines), Alfred Ilenre (Nigeria), Sebastião Manchineri (Brazil), Hector Huertas (Panama), and Robert Gough (U.S.).

International Legal Conventions

Indigenous NGOs consistently refer to specific international legal conventions to justify their call for a stronger Indigenous role in combating and mitigating climate change. The body of international law they cite is situated at the intersection of racial minority rights, minority cultural rights, Indigenous land rights, global environmental protection, and sustainable development.

Some of the international law was established by regional bodies, such as the American Declaration of Rights and Duties of Man, in which western hemisphere governments agreed to rights protections later adopted by the Organization of American States (Inter-American Commission on Human Rights 1948). Countries such as the United States and Canada can refuse to comply with international law, but then they appear hypocritical when they denounce other countries for violating the same laws.

Most of the international law involved in Indigenous climate change advocacy was promulgated by the United Nations. For example, Indigenous groups cite the UN Convention on Racial Discrimination, which states, "Special concrete measures shall be taken in appropriate circumstances in order to secure adequate development or protection of individuals belonging to certain racial groups with the object of ensuring the full enjoyment by such individuals of human rights and fundamental freedoms" (UN Convention on Elimination of All Forms of Racial Discrimination 1963). Other UN conventions and declarations have also provided fodder for Indigenous organizations, even if the UN agencies involved do not primarily work around Indigenous issues.

International Labour Organization

The International Labour Organization (ILO) is a United Nations agency originally founded in 1919 to improve employment and workers' rights and conditions. In 1989, it adopted Convention 169, "concerning Indigenous and Tribal Peoples in Independent Countries." Much of this convention concerns employment and working conditions for Indigenous communities, but Article 4 specifically addresses environmental and cultural rights: "Special measures shall be adopted as appropriate for safeguarding the persons, institutions, property, labour, cultures, and environment of the peoples concerned. Such special measures shall not be contrary to the freely expressed wishes of the peoples concerned" (ILO 1989).

In addition, Article 7 of ILO Convention 169 relates to development activities:

> The peoples concerned shall have the right to decide their own priorities for the process of development as it affects their lives, beliefs, institutions, and spiritual well-being and the lands they occupy or otherwise use.... In addition, they shall participate in the formulation, implementation, and evaluation of plans and programs for national and regional development which may affect them directly. ... Governments shall ensure that, whenever appropriate, studies are carried out, in cooperation with the peoples concerned, to assess the social, spiritual, cultural, and environmental impact on them of planned development activities. The results of these studies shall be considered as fundamental criteria for the implementation of these activities. Governments shall take measures, in cooperation with the peoples concerned, to protect and preserve the environment of the territories they inhabit. (International Labour Organization 1989)

United Nations Environmental Summits

The United Nations Conference on Environment and Development (UNCED), popularly termed the "Earth Summit," drew world leaders to Rio de Janeiro, Brazil, in June 1992. A parallel conference was held by environmental NGOs (and some Indigenous groups) critical of the pro-corporate biases of the attending governments. At the Rio Earth Summit, world leaders signed key agreements on the global environment and sustainable economic development, some of which include references to Indigenous rights.

For example, the Rio Declaration on Environment and Development signed by world leaders at the Earth Summit included Principle 22, which states that "Indigenous people and their communities and other local communities have a vital role in environmental management and development because of their knowledge and traditional practices. States should recognize and duly support their identity, culture, and interests and enable their effective participation in the achievement of sustainable development" (Rio Declaration on Environment and Development 1992).

The Earth Summit leaders also adopted Agenda 21 as a global Program of Action on Sustainable Development to guide the United Nations and national governments. Agenda 21 includes chapter 26, specifically "Recognizing and Strengthening the Role of Indigenous Peoples and their Communities." Chapter 26 offers

> recognition that the lands of Indigenous people and their communities should be protected from activities that are environmentally unsound or that the indigenous people concerned consider to be socially and culturally inappropriate; Recognition of their values, traditional knowledge, and resource management practices with a view to promoting environmentally sound and sustainable development; Recognition that traditional and direct dependence on renewable resources and ecosystems, including sustainable harvesting, continues to be essential to the cultural, economic, and physical well-being of indigenous people and their communities. (UN Department of Economic and Social Affairs 1992)

Agenda 21 also includes chapter 11 on "Combating Deforestation," another section that Indigenous NGOs have cited to combat climate change. Chapter 11 commits governments to "undertaking supportive measures to ensure sustainable utilization of biological resources and conservation of biological diversity and the traditional forest habitats of indigenous people, for-

Bolivian President Evo Morales addresses the United Nations Permanent Forum on Indigenous Issues (UNPFII) in New York to open its special session on climate change in 2008.

est dwellers, and local communities" (UN Department of Economic and Social Affairs 1992).

Also at the 1992 Earth Summit, the United Nations issued Annex III or the "Statement of Principles for a Global Consensus on the Management, Conservation, and Sustainable Development of All Types of Forests." Principle 5(a) of this statement states that

> national forest policies should recognize and duly support the identity, culture, and the rights of indigenous people, their communities, and other communities and forest dwellers. Appropriate conditions should be promoted for these groups to enable them to have an economic stake in forest use, perform economic activities, and achieve and maintain cultural identity and social organization, as well as adequate levels of livelihood and well-being, through, inter alia, those land tenure arrangements which serve as incentives for the sustainable management of forests. (UN Conference on Environment and Development 1992)

One decade after the Earth Summit, the United Nations held the World Summit on Sustainable Development (WSSD) in Johannesburg, South Africa. The 2002 Summit assessed the progress made since the 1992 summit and so was popularly termed "Rio+10." The WSSD was also criticized by many environmental NGOs, and Indigenous NGOs held a parallel International Peoples International Summit on Sustainable Development. The Indigenous delegates issued the "Kimberley Declaration," which echoed many of the same demands as Indigenous statements at the UNFCCC and reiterated that "since 1992 the ecosystems of the earth have been

compounding in charge. We are in crisis. We are in an accelerating spiral of climate change that will not abide unsustainable greed" (Kimberley Declaration 1992). The Kimberley delegates also issued an "Indigenous People's Plan of Implementation on Sustainable Development," which specifically urged governments to ratify and strengthen the Kyoto Protocol to reduce greenhouse gas emissions but opposed UNFCCC programs for carbon sinks and carbon-trading mechanisms. The Rio+10 WSSD is being held in Rio de Janeiro in June 2012.

Convention on Biological Diversity

The Convention on Biological Diversity (CBD) was also adopted at the 1992 Rio Earth Summit but followed its own separate evolution. Indigenous representatives participated directly in the negotiations as part of national government delegations. For example, Tulalip Tribes Natural Resources Director Terry Williams, from Washington state, was a member of the U.S. State Department delegation, but in the process he had to educate his fellow delegates about Native American treaty rights. In his international climate change work, Williams uses the key precedent of the Indigenous role in the Stockholm Convention banning persistent organic pollutants (POPs)—the bioaccumulating harmful chemicals such as PCBs. Inuit organizations led the way in urging passage of the 2001 treaty, and the U.S. Environmental Protection Agency was directed to consult with U.S. tribes on the treaty process.

The final CBD included Article 8(j) on Indigenous knowledge and biological diversity, which declares that UN member states should

> respect, preserve and maintain knowledge, innovations, and practices of indigenous and local communities embodying traditional lifestyles relevant for the conservation and sustainable use of biological diversity and promote their wider application with the approval and involvement of the holders of such knowledge, innovations, and practices and encourage the equitable sharing of the benefits arising from the utilization of such knowledge, innovations, and practices. (UN Convention on Biological Diversity 1992a)

As part of the follow-up to the Convention on Biological Diversity, the CBD Secretariat developed a Working Group on Article 8(j) to recognize and deepen the role of Indigenous traditional knowledge in the implementation of UN projects (UN Convention on Biological Diversity 1992b).

Declaration on the Rights of Indigenous Peoples

Indigenous nations have taken their visions and griev-ances to international forums in Geneva since the Haudenosaunee statesman Deskaheh visited the League of Nations in 1923, and especially since Native leaders attended the United Nations Conference on Discrimi-nation Against Indigenous Peoples of the Americas in 1977. Greater world attention and understanding focused on Indigenous rights around the time of the 1992 Columbus Quincentennial and, as a result, the United Nations declared the International Decade for the World's Indigenous Peoples in 1995–2004 and established a UN Permanent Forum on Indigenous Issues (UNPFII).

As part of the international decade, intense nego-tiations began in Geneva to develop a United Nations Declaration on the Rights of Indigenous Peoples. Representatives of Indigenous NGOs and some gov-ernments attended the negotiations and often sparred with representatives of UN member states, particularly the United States, Australia, New Zealand, Canada, and Russia. After thirteen years of difficult talks, the UN General Assembly finally adopted the declaration in 2007. The United States, Canada, Australia, and New Zealand were the only member states to vote no. (Indigenous representatives quickly dubbed them the "CANZUS states.") All four governments have since reversed their stance, albeit with conditions; the last was the United States, which on December 16, 2010, stated that the declaration expresses "aspirations that this country seeks to achieve within the structure of the U.S. Constitution, laws, and international obligations, while also seeking, where appropriate, to improve our laws and policies" (U.S. Department of State 2010).

The declaration could strengthen Indigenous nations' goal of having a role in the global carbon regu-latory framework, and their appeals for governments to curb or mitigate climate change may have more force because of articles (listed below) that specifically refer to the impact of environmental and development policy on Indigenous peoples (United Nations General Assembly 2007). Article 20 states that

> Indigenous peoples have the right to main-tain and develop their political, economic and social systems or institutions, to be secure in the enjoyment of their own means of subsistence and development, and to engage freely in all their traditional and other economic activi-ties…. Indigenous peoples deprived of their means of subsistence and development are entitled to just and fair redress. …

[Article 24 adds,] Indigenous peoples have the right to their traditional medicines and to maintain their health practices, including the conservation of their vital medicinal plants, animals, and minerals.…

[Article 25 says,] Indigenous peoples have the right to maintain and strengthen their distinctive spiritual relationship with their traditionally owned or otherwise occupied and used lands, territories, waters, and coastal seas and other resources and to uphold their respon-sibilities to future generations in this regard.

[Article 28 states,] Indigenous peoples have the right to redress, by means that can include restitution or, when this is not possible, just, fair and equitable compensation, for the lands, territories and resources which they have tradi-tionally owned or otherwise occupied or used, and which have been confiscated, taken, occu-pied, used, or damaged without their free, prior, and informed consent…. Unless otherwise freely agreed upon by the peoples concerned, compensation shall take the form of lands, ter-ritories, and resources equal in quality, size, and legal status or of monetary compensation or other appropriate redress.

[And Article 29 continues,] Indigenous peoples have the right to the conservation and protection of the environment and the produc-tive capacity of their lands or territories and resources. States shall establish and implement assistance programmes for indigenous peoples for such conservation and protection, without discrimination.…

[Finally, Article 32 requires state action in pursuit of these principles:] States shall provide effective mechanisms for just and fair redress… and appropriate measures shall be taken to miti-gate adverse environmental, economic, social, cultural, or spiritual impact.

International Indigenous Efforts on Climate Change

At the 1992 Earth Summit in Rio, the participating states finalized an international environmental treaty to reduce the emission of the greenhouse gases that cause global warming. The United Nations Framework Convention on Climate Change (UNFCCC) has since been signed by 189 nations, including the United States. The UNFCCC provides the main arena for global climate change politics. But some Indigenous peoples have also been pursuing other strategies outside the United Nations framework, by using human rights avenues, building partnerships with city governments, and adopting renewable energy technologies to pull themselves away from fossil fuels. All these international strategies build not only from international legal doctrines, but also from the cultural and spiritual values of Indigenous peoples themselves.

United Nations Framework Convention on Climate Change

The initial UNFCCC treaty included no mandatory limits on greenhouse gases and no enforcement mechanisms, but it established a framework for "protocols" or updates to reach emission targets in the protocols. Every year, the signatory countries attend a Conference of the Parties (COP) to develop and approve protocols. In 1997, nations attending COP-3 in Kyoto, Japan, approved a protocol that established mandatory targets would reduce emissions by 6 to 9 percent below 1990 levels by 2012. The Bush administration refused to ratify this Kyoto Protocol, which was approved by 163 signatory countries (that emit 65 percent of all greenhouse gases). The Kyoto Protocol, which has become better known than the UNFCCC itself, entered into effect in 2005.

Indigenous representatives have attended the proceedings of every COP since 1998 and met together as a caucus, sometimes in conjunction with local (non-Native) governments. Most of the Indigenous delegates represent NGOs, but some have also represented a few Indigenous governments (such as the Buffalo River Dene Nation in Canada). Before or during the COPs, the Native representatives have met in an International Indigenous Forum on Climate Change, issued declarations or caucus statements that outlined their interest in climate change, and made demands of the UNFCCC Secretariat and the treaty's signatory countries (table 4B.1). These demands would be even stronger if they were made by Indigenous government officials, as part of a government-to-government relationship with a signatory state.

The annual declarations establish an inherent Indigenous interest in the effects of climate change and call for a reduction in greenhouse gas emissions for the sake of the survival of Indigenous cultures and the planet. Indigenous nations (like many of the smaller "Annex II" countries that signed the Kyoto Protocol) emit very few greenhouse gases themselves and are therefore at the mercy of the "Annex I" countries that emit the lion's share of the harmful gases. (Indigenous nations are counted by the United Nations as merely part of these larger countries.)

The declarations all point to the need for the UNFCCC to recognize the "special status" of Indigenous nations in the international regulatory processes addressing climate change, and they insist that Indigenous peoples should have their own seat at the table. Indigenous nations possess TEK that offers early warnings of impending climatic changes (and traditional practices that can reduce emissions), as well as the ability to "test drive" possible models for community mitigation of and adaptation to these changes. Indigenous nations are neither NGOs nor full UN member states, so they need their own intermediate political niche in order to participate fully in the regulatory process. A special status could involve a formal observer status at the UNFCCC, the creation of a "contact group" on climate change and Indigenous peoples, or folding Indigenous governments into the category of Indigenous Parliamentarians (who are formally represented in the UNPFII).

The declarations specifically demand the creation of an open-ended intersessional working group on Indigenous peoples and climate change, to study and propose effective solutions to respond to the emergency situations caused by climate change's effects on Indigenous peoples. In addition, many of the declarations demand that Indigenous concerns become a permanent agenda item for each COP and that the UNFCCC Secretariat in Bonn create a permanent division for Indigenous concerns. In 2003, the UNPFII demanded that the UNFCCC include a Working Group on Indigenous Peoples and Climate Change—a demand that was reinforced at the 2008 session of the UNPFII after the passage of the UN Declaration on the Rights of Indigenous Peoples.

The Indigenous declarations point out that Native peoples possess practices and knowledge for minimizing the emission of greenhouse gases and are currently undertaking scientific and technical initiatives based on their traditional practices, which generate knowledge about production systems that have a minimal greenhouse effect. Native people also volunteer their communities for renewable energy projects that use solar, wind, and other alternative power sources, both to provide models for energy conversion and to strengthen the sovereignty and economic sustainability of their communities.

The declarations also point out that Indigenous peoples have funding needs if they are to participate fully and equally in the climate change research and regulatory processes. These needs include financial sup-

Table 1

Indigenous Statements to Conferences of the Parties (COPs) of the United Nations Framework Convention on Climate Change (UNFCCC)

Year	COP	City	International Indigenous Forum on Climate Change statement or other Indigenous declaration on UNFCCC
1998	4	Buenos Aires	Albuquerque Declaration
1999	5	Bonn	
2000	6	The Hague	1st IIFCC (Lyon), 2nd IIFCC (Hague), Quito Declaration
2001	6-B	Bonn	3rd IIFCC (Bonn Declaration)
2001	7	Marrakech	Indigenous Peoples and Local Communities Caucus
2002	8	New Delhi	Indigenous Caucus Statement on Climate Change
2003	9	Milan	6th IIFCC (Milan Declaration)
2004	10	Buenos Aires	Buenos Aires Declaration
2005	11	Montreal	Tiohtiá:ke (Montreal) Declaration, Arctic Indigenous Statement/Youth Statement
2006	12	Nairobi	Statement of Indigenous Peoples to UN Commission on Sustainable Development
2007	13	Bali	International Forum of Indigenous Peoples on Climate Change
2008	14	Poznan	International Indigenous Peoples Forum on Climate Change
2009	15	Copenhagen	Anchorage Declaration of Indigenous Peoples Global Summit on Climate Change; IIPFCC; Mystic Lake Declaration
2010	16	Cancun	International Indigenous Peoples Forum on Climate Change
2011	17	Durban	International Forum of Indigenous Peoples on Climate Change
2012	18	Doha	International Forum of Indigenous Peoples on Climate Change

port for attending COPs and other UNFCCC meetings and funding for research on the local impacts of climate change and particularly for the mitigation of harmful effects on their communities. They ask that any Climate Impact Assessments of their regions involve full and equal Indigenous participation (including Indigenous knowledge systems and observations) and generate scientific data and knowledge that can be used directly by the affected communities. The statements generally do not ask for solely financial compensation for harmful effects, given that the gases' specific origins are difficult to track in any legal sense and that money is no compensation for the loss of ancient cultural systems and natural resources.

Conferences of the Parties of the UNFCCC

Indigenous Declarations to the UNFCCC

The first Indigenous statement intended to intervene directly in the United Nations climate change treaty process, the 1998 Albuquerque Declaration, stated:

There is a direct relationship between the denial of Indigenous Peoples' land and water rights, along with the appropriation without consent of Indigenous Peoples' natural resources, and the causes of global climate change today.… The four elements of fire, water, earth, and air sustain all life. These elements of life are being destroyed and misused by the modern world. Fire gives life and understanding, but is being disrespected by technology of the industrialized world that allows it to take life such as the fire in the coal-fired powered plants, the toxic waste incinerators, the fossil-fuel combustion engine, and other polluting technologies that add to greenhouse gases. Coal extraction from sacred earth is being used to fuel the greenhouse gases that are causing global climate warming. (Albuquerque Declaration 1998)

The 2000 Lyon Declaration stated:

Despite the recognition of our role in preventing global warming, when it comes time to sign international conventions … once again, our right to participate in national and international discussions that directly affect our Peoples and territories is denied. Our active opposition to oil exploration, logging, and mining helps prevent the accelerated deterioration of the climate. Nonetheless, our territories have been handed over to national and multinational corpora-

tions who exploit our natural resources in an indiscriminate and unsustainable fashion. Any decision or action … must include our full and effective participation. (International Indigenous Peoples Forum on Climate Change 2000a)

The 2000 Hague Declaration of Indigenous Peoples and Climate Change asserted:

We are profoundly concerned that current discussions … do not recognise our right to adequate participation. These policies and mechanisms exclude us as participants, deny our contributions, and marginalize our peoples. These policies and mechanisms will permit developed countries to avoid their responsibility to reduce emissions at source, promote the expansion of global capital, and deepen our marginalization…. Concepts, practices, and measures, such as plantations, carbon sinks and tradeable emissions, will result in projects which adversely impact upon our natural, sensitive, and fragile ecosystems, contaminating our soils, forests, and waters. In the past, even well intentioned development policies and projects have resulted in disastrous social and ecological consequences. (International Indigenous Peoples Forum on Climate Change 2000b).

The 2001 Bonn Declaration added:

The discussions under the UNFCCC and the Kyoto Protocol have totally excluded the indigenous peoples to the extent that neither recognizes the right of indigenous peoples to full and effective participation and to contribute to discussions and debates. This contrasts with other international processes which assure our participation and contribution within discussions…. We openly oppose the measures to mitigate climate change under discussion that are based essentially on a mercantilist and utilitarian vision of the forests, seas, territories, and resources of Indigenous Peoples, which are being exclusively valued for their capacity to absorb CO_2 and produce oxygen, and which negate our traditional cultural practices and spiritual values. (International Indigenous Peoples Forum on Climate Change 2001)

At COP-7 in Marrakech, Morocco, the Indigenous Peoples and Local Communities Caucus stated:

No development mechanism can be clean, from our point of view, if it does not guarantee the rights of Indigenous Peoples including the right to free, prior, informed consent of indigenous and local communities and the respect of our cultures, practices, sciences, and knowledge. Nonetheless, we resolve to continue contributing with our knowledge of nature conservation and management to prevent and mitigate the effects of climate change…. To correct this inconsistency, we need an adequate space and special status in the structure of the UNFCCC. (Indigenous Peoples and Local Communities Caucus 2001)

At COP-8 in New Delhi, India, in 2002, the Indigenous Peoples' Caucus Statement on Climate Change asserted:

We, Indigenous Peoples, live in sensitive zones where effects of climate change are most devastating. Traditional ways of life are disproportionately affected by climate change particularly in polar and arid zones, forest, wetland, river and coastal areas…. The Kyoto Protocol is not sufficient to reverse, mitigate, or stop the catastrophes that threaten our Mother Planet Earth. (Indigenous Peoples' Caucus Statement on Climate Change 2002).

At COP-9 in Milan, Italy, in 2003, the Sixth International Indigenous Peoples Forum stated:

Our special relationship with Mother Earth is sacred and must be honored, protected, and loved. We further declare our holistic vision, which strongly binds biological diversity, cultural and spiritual identity and unites people with its ancestral territories. Our ancestral territories, spiritual, social, biological, and cultural resources are the fundamental basis for our existence, health, and livelihoods but are threatened and destroyed by climate change and its consequences…. The United Nations has clearly recognized our rights to participate in the UN processes…. We call upon the UNFCCC to recognize that through the protection and promotion of Indigenous Peoples' rights and through recognizing and integrating our dynamic and holistic visions, we are securing not only our future, but the future of humanity and social and environmental justice for all. (International Forum of Indigenous Peoples on Climate Change 2003)

The frustration of Indigenous representatives became evident at COP-10 in Buenos Aires, Argentina, in 2004:

We are hearing from the states the same old arguments being discussed on how to alleviate and mitigate the climate disasters that affect all humanity. These arguments do not address the mounting costs of adapting to climate changes within our Indigenous communities, exemplified by the Indigenous peoples of the Arctic region whose lands are literally melting before their eyes…. We consider this planet our Mother Earth where all humanity is born and nurtured. It is time that we looked to each other and that we listen to each other, recognizing and valuing the cultural and human qualities within each of us…. [We ask] why our previous requests to…provide a mechanism for us to actively participate in the UNFCCC were not listened to, are we not part of this planet? (Buenos Aires Declaration 2004)

COP-11 was held in Montreal, Canada, in 2005, the first year after the implementation of the Kyoto Protocol. Indigenous representatives issued the Tiohtiá:ke Declaration, named after a Mohawk word for Montreal:

We reaffirm our inherent rights over our territories, lands, and resources. Our cosmovision strongly binds biological diversity, cultural and spiritual identity and unites our peoples with our ancestral territories. This is the fundamental basis for our existence, health and livelihoods which are being disproportionately threatened and destroyed by climate change and its consequences. Indigenous Peoples require a human rights–based approach in addressing climate change…. Establish a process that works towards the full phase-out of fossil fuels, with a just transition to sustainable jobs, energy, and environment. We are against the expansion of and new exploration for the extraction of oil, natural gas, and coal within and near Indigenous lands, especially in pristine and sensitive areas…. We once again remind you that one is only as healthy as the air we breathe, the water that we quench our thirst with each day, and the earth in which we plant our seeds to have the various products of sustenance for the duration of our journey here on Mother Earth (International Indigenous Peoples Forum on Climate Change 2005).

At COP-13 in Bali, Indonesia, the International Forum of Indigenous Peoples on Climate Change focused on the United Nations' Reduced Emissions from Deforestation and Degradation (REDD) program:

REDD will not benefit Indigenous Peoples, but in fact, it will result in more violations of Indigenous Peoples' Rights. It will increase the violation of our Human Rights, our rights to our lands, territories, and resources, steal our land, cause forced evictions, prevent access and threaten indigenous agriculture practices, destroy biodiversity and culture diversity, and cause social conflicts. Under REDD, States and Carbon Traders will take more control over our forests (International Forum of Indigenous Peoples on Climate Change 2007).

At COP-14 in Poznan, Poland, the International Indigenous Peoples' Forum on Climate Change focused its statement on

the human rights violations caused by the Clean Development Mechanisms (CDM) [establishing a global carbon trading system] and other carbon trading and carbon offset regimes.... Any further expansion of the CDM is an excuse to avoid real emissions reductions. The CDM and the carbon market are instruments that commodify the atmosphere, promote privatization, and concentrate resources in the hands of a few, taking away the rights of many to live with dignity. CDM are not a mechanism for mitigating climate change. It is not just "carbon" or pollution that is being traded, but people's lives. (International Indigenous Peoples' Forum on Climate Change 2008)

Both the carbon-trading and carbon offset programs treat carbon as a tradable commodity, and allow Annex I (developed) countries to buy and sell carbon credits on a global market. Carbon-trading mechanisms allow corporations that emit carbon to avoid fines and penalties by offsetting their carbon production, in the form of buying carbon credits from operations that reduce the amount of carbon emissions.

In April 2008, the UNPFII met at UN headquarters in New York. For the first time, the central theme of this seventh session was "Climate Change, Bio-Cultural Diversity, and Livelihoods: The Stewardship Role of Indigenous Peoples and New Challenges." Indigenous

opposition to the CDM and REDD was a major theme of the NGO delegates and Indigenous Parliamentarians, who spoke to the UNPFII, a panel of experts chaired by Philippine Igorot activist Victoria Tauli-Corpuz. The final session report initially overlooked the strong opposition to carbon offset systems and even praised several World Bank CDM projects, including a wind farm in northern Colombia (Tauli-Corpuz and Aggaluk 2008). The wind project provides electricity for the expansion of the gigantic El Cerrejón coal mine and is part of a larger industrial complex that has been opposed by Wayúu Indigenous communities, leading to assassinations of project opponents (Emanuelson 2004). In response to this report, Indigenous representatives made strong statements against carbon trading and held a brief floor demonstration (United Nations Economic and Social Council 2008). The UNPFII was convinced to change the final text to acknowledge that most representatives oppose CDM projects (UNPFII 2008).

Evo Morales, the first elected Indigenous leader of Bolivia, had delivered the keynote speech to the UNPFII session, the first head of state ever to address the body. He delivered "Ten Commandments to Save the Planet," including his sixth commandment: "Respect for the mother Earth. Learn from the historic teachings of native and indigenous peoples with regard to the respect for the mother Earth. A collective social consciousness must be developed among all sectors of society, recognizing that the Earth is our mother" (Morales 2008a).

In a later climate change statement to the UNFCCC, President Morales urged Indigenous peoples and the world community to "Save the Planet from Capitalism":

"Climate change" has placed all humankind before great choice: to continue in the ways of capitalism and death, or to start down the path of harmony with nature and respect for life.... It is fundamental to guarantee the participation of our people as active stakeholders at a national, regional, and worldwide level, especially taking into account those sectors most affected, such as indigenous peoples who have always promoted the defense of Mother Earth. (Morales 2008b)

Partly in preparation for COP-15 in Copenhagen, Denmark, Indigenous peoples from around the world met in Anchorage, Alaska, for the Indigenous Peoples Global Forum on Climate Change. The Anchorage Declaration added a note of urgency and hope to the Indigenous demands (See Section I)

Bolivian indigenous parliamentary delegates to the UNPFII special session on climate at UN Headquarters in 2008.

In preparation for COP-15 in Copenhagen, the International Indigenous Peoples Forum on Climate Change released a thirteen-point statement that included language on adaptation strategies:

We have intrinsic contributions towards addressing the climate crisis, and renewing the relationships between humans and nature. For generations, we have managed ecosystems nurturing its integrity and complexity in sustainable and culturally diverse ways....Traditional knowledge, innovations, and adaptation practices embody local adaptative management to the changing environment, and complement scientific research, observations, and monitoring. (International Indigenous Peoples Forum on Climate Change 2009)

In response of the failure of the Copenhagen conference (engineered by the United States, Denmark, and China), Bolivian President Evo Morales called the World People's Conference on Climate Change and the Rights of Mother Earth in April 2010 in the city of Cochabamba (known as a center of Indigenous resistance to water privatization). The People's Agreement of Cochabamba went beyond opposition to the UNFCCC policy by stating,

It is imperative that we forge a new system that restores harmony with nature and among human beings. And in order for there to be balance with nature, there must first be equity among human beings. We propose to the peoples of the world the recovery, revalorization, and strengthening of the knowledge, wisdom,

and ancestral practices of Indigenous Peoples, which are affirmed in the thought and practices of "Living Well," recognizing Mother Earth as a living being with which we have an indivisible, interdependent, complementary, and spiritual relationship. To face climate change, we must recognize Mother Earth as the source of life and forge a new system. (World's Peoples Conference 2010)

Indigenous peoples had a more sizable presence at COP-16 in Cancún, Mexico, including a series of Canadian First Nations protests against expanded fossil fuel extraction in the Alberta Tar Sands. The Indigenous representatives decisively rejected carbon market mechanisms that the COP adopted without protections for indigenous lands.

The Cancun Indigenous delegates stated that

it is crucial that indigenous peoples' demands are realized in the climate negotiations in Cancun and beyond. Indigenous peoples did not cause climate change. Their low-carbon lifeways, traditional knowledge and practices, protection and sustainable use of their forests and resources present alternative solutions to the current climate crisis" (Indigenous Climate Portal 2010).

Before COP-17 in Durban, South Africa, the joint declaration of Indigenous organizations addressed issues of adaptation and mitigation, demanding that any programs:

Guarantee respect, protection and promotion of indigenous peoples' traditional knowledge and sustainable livelihoods, including the cultural and spiritual aspects. Public policies and funds should prioritize full recognition of indigenous peoples' territory.... All mitigation and adaptation evaluation, recovery and development actions should incorporate indigenous peoples' knowledge and technologies, subject to their free, prior and informed consent and also guarantee the full participation of indigenous experts....The States should ensure that Indigenous peoples have the right of mobility and are not forced to relocate far from their traditional territories and lands and that the rights of peoples in voluntary isolation are respected" (International Indigenous Peoples' Forum on Climate Change 2011).

Should Indigenous Nations Be Involved in the UN Process?

Most Indigenous NGO representatives to the UNPFII oppose the Kyoto Protocol, and question whether the UNFCCC has been weakened to such an extent that it actually undermines Indigenous interests. They cite the UN Intergovernmental Panel on Climate Change (IPCC) report, which states that greenhouse gas emissions need to be immediately reduced by 60 percent in order to stabilize global temperatures—far above the Kyoto targets. As stated earlier, the representatives also oppose the implementation of carbon sinks and carbon-trading mechanisms in the CDM of the Kyoto Protocol and the World Bank's Prototype Carbon Fund.

A growing number of environmental groups point out that these carbon commodity mechanisms allow developed countries such as the United States to continue polluting the atmosphere with greenhouse gases, if they offset the pollution in other ways. For example, the CDM identifies large forests (in countries such as Canada or Brazil) as "carbon sinks" that absorb carbon dioxide, thereby allowing those countries to emit more gases. Indigenous declarations to the UNFCCC have pointed out that so-called "conservation" measures in these "carbon sink" forests and fields exclude Indigenous harvesting practices, and that fast-growing pine or eucalyptus plantations have displaced Indigenous peoples in the name of protecting the atmosphere (Lohmann 2006).

The REDD forestry plan "focuses largely on the programs of NGOs and nation-states, overlooking and in some cases prohibiting indigenous forest stewardship…. Even indigenous leaders who favor the potentially lucrative program have complained that REDD plans have targeted their land without their authentic consent" (Martinez 2010).

Indigenous critics of carbon trading also fear that the CDM goal to "clean up" the emissions of burning fossil fuels may lead to the mining of more oil, coal, and natural gas on Native lands.

Indigenous activists and governments have been in the forefront of opposition to proposed pipelines taking oil from the Alberta tar sands to the international market (such as the Enbridge pipeline across British Columbia and Keystone pipeline across the Great Plains), and the proposed "Heavy Haul" of extractive equipment from Columbia River ports to Alberta (Indigenous Environmental Network 2012).

Since nuclear power is increasingly promoted as an alternative to fossil fuels (even though it emits high levels of greenhouse gases in the enrichment stage), plans for new uranium mines are accelerating. Some southwestern U.S. reservations have already seen detrimental effects from fossil fuel and uranium extraction on their air, surface waters, water tables, and human health. A commitment by resource-rich tribes not to participate in this fossil fuel extraction would send a strong message to the energy industry, yet the mining revenues would need to be replaced to sustain tribal economic development. The Council of Energy Resources Tribes (CERT), which represents these resource-rich tribes, has initiated renewable energy projects but has not yet discussed reducing fossil fuel extraction in order to secure offset compensation, as is being discussed in Ecuador.

In contrast, northern Australian Aboriginal communities have reduced greenhouse gas emissions by instituting traditional fire abatement practices. As an essay in *Cultural Survival* notes, "The scheme, called the Arnhem Land Fire Abatement Project, is aimed not only at reducing the severity of wildfires, but also at providing a substantial stream of income for Aboriginal peoples through international carbon-emissions trading programs" (Cherrington 2006). The Nez Perce Tribe has replanted trees on bare Idaho rangeland as a way to acquire income from the Chicago Carbon Exchange. It is restoring the cleared forests as a "carbon crop"—part of a "tribal portfolio" created by the National Carbon Offset Coalition. Yet the tribe has encountered numerous barriers and has yet to see a profit from the venture (Tribal Climate Change Forum 2009). A few tribal governments have symbolically ratified the Kyoto Protocol but "comply" by pledging to use more renewable energies instead of trading carbon (Little Traverse Bay Bands of Odawa Indians 2005).

At the 2004 World Summit on Sustainable Development in South Africa, numerous environmental and Indigenous organizations signed onto the Durban Declaration, which rejected carbon trading as a mechanism to reduce global warming. It stated,

> The carbon market creates transferable rights to dump carbon in the air, oceans, soil, and vegetation far in excess of the capacity of these systems to hold it. Billions of dollars worth of these rights are to be awarded free of charge to the biggest corporate emitters of greenhouse gases…. Costs of future reductions in fossil fuel use are likely to fall disproportionately on the public sector, communities, Indigenous peoples, and individual taxpayers. (Durban Declaration on Climate Change 2004)

Tribes can certainly view carbon credits as a source of income but not as a long-term strategy to curb global warming. The Kyoto Protocol expires in 2013 and is being replaced by a new regulatory system. Global pressure is now being exerted to replace carbon trading with a system of taxes or "dumping fees" on carbon emissions, or to fine contributors to global warming. Indigenous governments could begin to add to these voices for real change.

Indigenous Pacific island states, such as Tuvalu and Kiribati, stand to lose the most from climate change, and in fact they have already begun to lose islands to rising sea levels. Since they emit only a tiny fraction of the world's atmospheric carbon and have few forests that qualify as carbon sinks, they tend to favor carbon credit systems. Although they are fully sovereign and are member states of the United Nations, they have nonetheless had little say in global climate change forums. As former colonies that only recently gained independence, they are subject to neocolonialism, with their economic survival still dependent on the United States, Australia, New Zealand, and other former colonial powers.

The experience of Pacific island states is an object lesson for Indigenous nations that are not fully sovereign and are counted as part of states in "developed regions" that are large carbon emitters. If independent UN member states do not have powers to protect their environment and cultures, how can semi-sovereign Indigenous nations? Yet Pacific island states do have a positive track record of cooperation in the face of a common environmental threat. In 1968, France began conducting nuclear weapons tests in its colony of French Polynesia, contaminating some fishing zones with radiation. In the 1970s and 1980s, the tests were met with united objections by member states of the South Pacific Forum and protests by Greenpeace and other environmental groups. In 1985, French intelligence agents bombed a Greenpeace vessel in New Zealand, killing one. The strong protests of South Pacific states, including New Zealand, played a role in pressuring nuclear powers to develop and sign the 1996 Comprehensive Test Ban Treaty.

This environmental mobilization of South Pacific island peoples is being repeated as they face the new threat of global climate change. Yet as small countries, they face an uphill struggle against global warming and have been critical of many aspects of the UNFCCC process. As Tuvalu leaders told the UNFCCC: "Providing us with capacity building, adaptation, and other imaginative measures to mitigate climate change while refusing to institute domestic policy and political mea-

sures that will genuinely reduce global emissions is like treating us like the pig you fatten for slaughter at your eldest son's 21st birthday party" (South Pacific Regional Environment Program 1999).

Human Rights Strategies

Besides the UNFCCC process, Indigenous organizations and nations have also used other approaches to internationalize their climate change demands and expand the scope of their local, community-based concerns to the global level. One of these approaches is to use the argument of human rights, for which there is a much more developed body of international law than for Indigenous rights or cross-border rights of environmental protection. The downside of using a human rights argument is that it focuses attention on the individual Indigenous person, rather than the collective sovereignty inherent in the nation. It also tends to focus on compensation or financial settlement rather than land returns or cultural revitalization.

Yet human rights approaches have also drawn attention to the violations of Indigenous rights, most notably the recognized right of cultural and religious expression of Native peoples. For example, in 1998 the United Nations Commission on Human Rights dispatched its Special Rapporteur on Religious Intolerance (for the first time) to investigate the United States for human rights violations. After visiting Native American communities and interviewing tribal members, he issued a statement and report that condemned the United States for desecrating Native American sacred sites and relocating Native peoples away from ceremonial areas in their homelands (Amor 1998). Indigenous NGOs have asked for a Special Rapporteur who could be charged with investigating the impacts of climate change on Indigenous peoples, including their rights to traditional foods and other natural resources.

The most publicized use of human rights conventions has been by the Inuit Circumpolar Council (ICC), which represents Inuit communities in Alaska, Canada, Greenland, and Russia. In 2005, the ICC submitted a petition to the Inter-American Commission on Human Rights (IACHR, based in the Organization of American States in Washington, D.C.), seeking relief from violations of the human rights of Inuit resulting from global warming caused by greenhouse gas emissions from the United States. At COP-11 in Montreal, ICC Chair Sheila Watt-Cloutier explained that the petition was intended to pressure the United States to reduce greenhouse gas emissions and not to secure financial compensation:

This petition is not about money, it is about encouraging the United States of America to join the world community to agree to deep cuts in greenhouse gas emissions needed to protect the Arctic environment and Inuit culture and, ultimately, the world. We submit this petition not in a spirit of confrontation—that is not the Inuit way—but as a means of inviting and promoting dialogue with the United States of America within the context of the climate change convention. (Inuit Circumpolar Conference 2005)

The ICC Chair further explained that the petition

will seek a declaration in international law that the erosion and potential destruction of the Inuit way of life brought about by climate change resulting from emission of greenhouse gases amounts to a violation of the fundamental human rights of Inuit. It will draw on the compelling combination of official science and traditional knowledge within the Arctic Climate Impact Assessment to focus political attention on the Arctic and Inuit dimensions to this global issue. (Watt-Cloutier 2004)

She cited the previous success of the ICC in using scientific assessments of persistent organic pollutants (POPs) in the Arctic to pressure states to agree to the Stockholm Convention, which banned the harmful chemicals. The POPs treaty specifically cited Arctic peoples in its preamble, so it serves as precedent for small Indigenous peoples having a central role in global environmental policy.

ICC attorneys Martin Goldberg and Martin Wagner submitted the petition to the IACHR. They cited the 1948 American Declaration of the Rights and Duties of Man as the basis for the complaint:

Many rights contained in the American Declaration, including the rights to life and personal security; to residence and movement; to inviolability of the home; to the benefits of culture; and to work and to fair remuneration, could serve as the basis of a complaint.... A healthy environment is fundamental to the enjoyment of nearly all of the most fundamental human rights. (Goldberg and Wagner 2004)

In December 2006, the IACHR refused to consider the petition, but it did hold a hearing on climate change at the ICC's request in March 2007.

Energy Partnership Strategies

Another innovative Indigenous approach to combating climate change has stepped outside of international legal regimes entirely by developing direct relationships between Indigenous peoples affected by climate change and local governments in cities that generate greenhouse gases. The Climate Alliance of European Cities with Indigenous Rain Forest Peoples is an association of European Union cities linked in partnership with Indigenous peoples living in the Amazon Basin. The 1,300 member municipalities in the Climate Alliance (representing about 50 million citizens) have resolved to reduce their emissions of the greenhouse gas carbon dioxide by 10 percent every five years, thus cutting their 1990 levels of greenhouse gas emissions in half (Environment News Service 2006).

Through the Climate Alliance, the European cities and municipalities are partnering with the Coordinating Body for the Indigenous Organizations of the Amazon Basin (COICA), a network of more than four hundred Indigenous peoples, and the International Alliance of the Indigenous-Tribal Peoples of the Tropical Forests (IAIP). The European cities "partner" with the Indigenous peoples by taking action to support Indigenous demands to protect the rain forest from the effects of climate change. The Manifesto of European Cities on an Alliance with Amazonian Indian Peoples supports "the preservation of the tropical rain forest, the basis of [the Indigenous people's] very existence, through the demarcation and sustainable use of the Amazonian territories. Their defense of the forests and rivers is a contribution to sustaining the earth's atmosphere for future generations as the basic precondition for human existence" (Climate Alliance 1990).

Conclusions

Native peoples face a global "triple whammy" from fossil fuel development. First, their lands are developed without their consent and their natural resources and health damaged—whether in the rainforests of Ecuador or Colombia, the Niger Delta of Nigeria, the Arctic National Wildlife Refuge, or the Tar Sands of Alberta. Second, they are experiencing the earliest and most severe damage from climate change, generated mainly by the burning of fossil fuels. Third, they are being removed or restricted from their traditional lands by monocrop tree plantations that are intended to "offset" continued carbon emissions. Amazingly, some Native peoples around the world (such as those in Ecuador and Colombia) are actually affected by all three phenomena: the extraction of the fossil fuels that cause climate change, the harmful climate change itself, and then by displacement in the name of mitigating the effects of climate change!

Most of the international strategies that Indigenous peoples have pursued to combat harmful climate change have been in arenas dominated by the same settler states that have colonized Native lands. The countries that emit the greatest quantity of greenhouse gases have been those that have been the most resistant to the recognition of Indigenous sovereignty in the international legal system—particularly the United States. Although inserting an Indigenous voice into the United Nations process is critical to raising awareness and applying moral pressure, it would do little practically to curb climate change. Obtaining "special status" for Indigenous peoples within the UNFCCC process would at least offer Native representatives a place at the table, even if they are being dealt a stacked deck.

Yet Indigenous peoples are not dealing entirely with a rigged game. The direct involvement of recognized Indigenous governments in the UN processes could gain more results than have been gained by NGOs, by framing Native concerns in a government-to-government context (and in the United States, in terms of the federal trust responsibility). The Biodiversity and POPs treaties both involved direct Indigenous input, which may have been critical to their success. The Inuit petition to the IACHR could serve as a precedent for more international legal challenges to emissions of greenhouse gases, which in turn can (as in the POPs treaty process) exert real pressure on the U.S. government.

The passage of the United Nations Declaration on the Rights of Indigenous Peoples offers opportunities to argue for Indigenous rights as a strategy for responding to the fossil fuels crisis and climate change. If Indigenous peoples' voices are heard and their rights of self-determination are recognized, it will be good not only for Indigenous nations, but also for the survival of the rest of humanity.

Resources

Albuquerque Declaration. 1998. "Circles of Wisdom" Native Peoples/Native Homelands Climate Change Summit, November 1. www.nativevillage.org/Inspiration-/Albuquerque%20Convention.htm

Amor, Abdelfattah. 1998. Report submitted by Special Rapporteur on Religious Intolerance, in accordance with the United Nations Commission on Human Rights Resolution 1998/18. United Nations Doc. # E/CN.4/1999/58/Add.1, December 9. www.nativeweb.org//pages/legal/intolerance.html

Anchorage Declaration. See Indigenous Peoples Global Forum on Climate Change

"Arctic Indigenous Peoples Unveil Statement on Climate Change." 2005. Arctic Day News. (June 12). www.arcticpeoples.org/Newsletter/Documents/ArcticDayNews.doc Blackfire. 2007. Overwhelming. From [Silence] is a Weapon. Flagstaff, AZ, Tacoho Records. http://www.blackfire.net; video: http://www.youtube.com/watch?v=onWcHYTjiKg

Buenos Aires Declaration. 2004. Declaration of the Indigenous Peoples Attending COP 10, UNFCCC, Argentina, December 6–17. http://www.tebtebba.org/index.php/all-resources/category/80-ipfccc-meetings-2000-2004

Cherrington, Mark. 2006. Aboriginal Practices Play a Role in Reducing Global Warming. Cultural Survival Voices, March 1. http://www.culturalsurvival.org/publications/voices/mark-cherrington/aboriginal-practices-play-role-reducing-global-warming

Climate Alliance. 2009. http://www.klimabuendnis.org

Declaration on Climate Change from Youth of the Arctic. 2005. www.taiga.net/ayn/declaration.html

Durban Declaration on Climate Change. 2004. World Summit on Sustainable Development, October 10. http://www.durbanclimatejustice.org/durban-declaration/english.html

Emanuelson, Dick. 2004. Por eso asesinan y desplazan a los indígenas Wayúu. Agencia Prensa Rural, May 29. http://www.prensarural.org/emanuelsson20040529.htm

Environment News Service. 2006. European Cities Pledge to Slash Greenhouse Gas Emissions, May 9. http://www.ens-newswire.com/ens/may2006/2006-05-09-06.asp

Goldberg, Daniel M., and Martin Wagner. 2004. Petitioning for Adverse Impacts of Global Warming in the Inter-American Human Rights System. In Climate Change Five Years After Kyoto, ed. Vela I. Glover, 191–210. Enfield, NH: Science Publishers. http://www.ciel.org/Publications/Petitioning_GlobalWarming_IAHR.pdf

Honor the Earth. 2005. Climate Change and Native Communities. http://www.honorearth.org/climate-change-and-native-communities

Indigenous Caucus Statement on Climate Change. 2002. Conference of the Parties 8, UNFCCC, New Delhi, India, October 23–November 1. http://www.tebtebba.org/index.php/all-resources/category/80-ipfccc-meetings-2000-2004

Indigenous Climate Portal. 2010. Press Statement: Indigenous Peoples' Ambitions for Cancun,

December 3. http://www.indigenousclimate.org/

Indigenous Environmental Network. 2012. Pipeline and Heavy Haul Campaign. http://www.ienearth.org/ien-pipeline.html

Indigenous Peoples Global Forum on Climate Change. 2009. Anchorage Declaration, April 24.

Indigenous Peoples and Local Communities Caucus. 2001. Conference of the Parties 7, UNFCCC, Marrakech, Morocco, October 29–November 10. http://www.treatycouncil.org/new_page_5231311.htm

Inter-American Commission on Human Rights. 1948. American Declaration of Rights and Duties of Man. 9th International Conference of American States, Bogotá, Colombia, March–May. http://www1.umn.edu/humanrts/oasinstr/zoas2dec.htm

International Forum of Indigenous Peoples on Climate Change. 2000a. Lyon Declaration of the First International Forum, France, September 4–6. http://www.treatycouncil.org/new_page_5211.htm

———. 2000b. The Hague Declaration of the Second International Forum, COP 6, UNFCCC, Netherlands, November 11–12. http://www.tebtebba.org/index.php/all-resources/category/80-ipfccc-meetings-2000-2004

———. 2001. Bonn Declaration of the Third International Forum, COP 6-B, UNFCCC, Germany, July 14–15. http://www.wrm.org.uy/actors/CCC/IPBonn.html

———. 2003. Milan Declaration of the Sixth International Forum, COP 9, UNFCCC, Italy, November 29–30. http://www.ilc.unsw.edu.au/sites/ilc.unsw.edu.au/files/mdocs/milan%20declaration.pdf

———. 2005. The Tiohtiá:ke Declaration: Statement to the State Parties of the COP 11/MOP 1, UNFCCC, Montreal (Tiohtiá:ke), Canada, November 28–December 9. http://www.forestpeoples.org/topics/un-framework-convention-climate-change-unfccc/news/2011/05/tiohtiake-declaration

———. 2007. Statement on "Reduced Emissions from Deforestation and Forest Degradation" (REDD), Bali, Indonesia, November 1. http://www.indigenousclimate.org/index.php?option=com_content&view=article&id=59&Itemid=60&lang=en

International Indigenous Peoples' Forum on Climate Change. 2007. Closing Statement in Bali, December 12. http://www.redd-monitor.org/2009/01/09/redd-text-is-insufficient-and-offensive-closing-statement-of-the-international-indigenous-peoples%E2%80%99-forum-on-climate-change/

———. 2009. Policy Paper on Climate Change. http://www.indigenousportal.com/Climate-Change/IIPFCC-Policy-Paper-on-Climate-Change-September-27-2009.html

International Indigenous Peoples' Forum on Climate Change. 2011. Declaration of the Indigenous Peoples of the World to the UNFCCC COP 17, Durban, South Africa (Dec. 2) http://www.forestpeoples.org/topics/un-framework-convention-climate-change-unfccc/publication/2011/declaration-indigenous-peoples

International Labour Organization (ILO). 1989. Convention (No. 169) concerning Indigenous and Tribal Peoples in Independent Countries, June 27. http://www.ilo.org/ilolex/cgi-lex/convde.pl?C169

Inuit Circumpolar Conference (ICC). 2005. Inuit Petition Inter-American Commission on Human Rights to Oppose Climate Change Caused by the United States, December 7. http://www.inuitcircumpolar.com/index.php?ID=316&Lang=En

Kimberley Declaration. 2002. International Indigenous Peoples Summit on Sustainable Development, Khoi-San Territory, South Africa, August 20–23. http://www.tebtebba.org/index.php/all-resources/category/17-rio-10-world-summit-on-sustainable-development?download=476:the-kimberley-declaration

LaDuke, Winona. 1999. *All Our Relations: Native Struggles for Land and Life.* Boston: South End Press.

Little Traverse Bay Bands of Odawa Indians. 2005. Resolution on Climate Change and Kyoto Protocol. Harbor Springs, MI: Little Traverse Bay Bands of Odawa Indians.

Lohmann, Larry. 2006. *Carbon Trading: A Critical Conversation on Climate Change, Privatisation and Power.* Durban Group for Climate Justice and Dag Hammarskjöld Foundation. London: The Corner House. http://www.thecornerhouse.org.uk/summary.shtml?x=544225

Martinez, Dennis. 2010. The Missing Delegate at Cancún: Indigenous Peoples. *National Geographic NewsWatch* (Dec. 8) http://newswatch.nationalgeographic.com/2010/12/08/the-invisible-delegate-at-cancun-indigenous-peoples/

Morales Ayma, Evo. 2008a. Evo Morales' 10 Commandments to Save the Planet, April 28. http://www.cadtm.org/Evo-Morales-10-Commandments-to

———. 2008b. Save the Planet from Capitalism: Letter from President Evo Morales to UNFCCC about Climate Change and the International Crisis,

November 28. http://www.ienearth.org/news/evo_morales_letter.html

Quito Declaration on Climate Change Negotiations. 2000. Ecuador (May 4–6). http://www.tebtebba.org/index.php/all-resources/category/80-ipfccc-meetings-2000-2004

Rio Declaration on Environment and Development. 1992. United Nations Conference on Environment and Development, Rio de Janeiro, June 3–14. http://www.brazil.org.uk/environment/riodeclaration.html

South Pacific Regional Environment Program. 1999. UNFCCC Fifth Conference of Parties Climate Convention Update: Pacific Delegations, November 1.

Statements of Indigenous Peoples to UN Commission on Sustainable Development. 2006. http://www.un.org/esa/sustdev/mgroups/about_mgroups/amg_indigenous_main.htm

Tauli-Corpuz, Victoria, and Aqqaluk Lynge. 2008. Impact of Climate Change Mitigation Measures on Indigenous Peoples and on their Territories and Lands. United Nations Permanent Forum on Indigenous Issues, United Nations Economic and Social Council E/C.19/2008/10, March 19. http://www.un.org/esa/socdev/unpfii/documents/E_C19_2008_10.pdf.

Tribal Climate Change Forum. 2009. University of Oregon, Many Nations Longhouse, October 16. Oral presentation.

United Nations Conference on Environment and Development. 1992. Report of the UNCED Annex III: Non-Legally-Binding Authoritative Statement of Principles for a Global Consensus on the Management, Conservation, and Sustainable Development of All Types of Forests. http://www.un.org/documents/ga/conf151/aconf15126-3annex3.htm

United Nations Convention on Biological Diversity. 1992a. http://www.biodiv.org/convention/articles.asp?a=cbd-08

———. 1992b. Working Group on Article 8(j): Traditional Knowledge and the Convention on Biological Diversity. http://www.biodiv.org/programmes/socio-eco/traditional/default.aspx

United Nations Convention on Elimination of All Forms of Racial Discrimination. 1963. Proclaimed by General Assembly Resolution 1904 (XVIII) of 20 November 1963. http://www2.ohchr.org/english/law/cerd.htm

United Nations Department of Economic and Social Affairs, Division for Sustainable Development. 1992. Agenda 21: Program of Action on Sustainable Development. http://www.un.org/esa/dsd/agenda21/?utm_source=OldRedirect&utm_medium=redirect&utm_content=dsd&utm_campaign=OldRedirect

United Nations Division for Sustainable Development. 2008. Major Groups: Indigenous Peoples. http://www.un.org/esa/sustdev/mgroups/about_mgroups/amg_indigenous_main.htm

United Nations Economic and Social Council (ECOSOC). 2008a. Indigenous Peoples must be Included in Global Negotiations Aimed at Combating Climate Change, Say Speakers in Permanent Forum. United Nations Department of Public Information HR/4946, April 22. http://www.un.org/News/Press/docs/2008/hr4946.doc.htm

United Nations Economic and Social Council (ECOSOC). 2008b. Permanent Forum on Indigenous Issues discusses ways to more effectively promote countries' implementation of Declaration on Rights. United Nations Department of Public Information HR/4950 (April 29). http://www.un.org/News/Press/docs/2008/hr4950.doc.htm

United Nations Economic and Social Council (ECOSOC). 2008c. Indigenous peoples face growing crisis as climate change, unchecked economic growth, unfavorable domestic laws force them from lands, Forum told. United Nations Department of Public Information HR/4951 (April 30). http:// www.un.org/News/Press/docs/2008/hr4951.doc. htm

United Nations General Assembly. 2007. United Nations Declaration on Rights of Indigenous Peoples, September 13. http://www.un.org/esa/socdev/unpfii/documents/DRIPS_en.pdf

United Nations Human Rights Council. 2006. Implementation of General Assembly Resolution 60/251 of 15 March 2006 entitled Human Rights Council: Draft United Nations Declaration on the Rights of Indigenous Peoples, June 23. http://www2.ohchr.org/english/bodies/hrcouncil/2session/documents.htm

United Nations Permanent Forum on Indigenous Issues (UNPFII). 2008a. 7th Session: Climate Change, Bio-Cultural Diversity and Livelihoods: The Stewardship Role of Indigenous Peoples and New Challenges. http://social.un.org/index/IndigenousPeoples/UNPFIISessions/Seventh.aspx

United Nations Permanent Forum on Indigenous

Issues (UNPFII). 2008b. Guide on Climate Change and Indigenous Peoples. http://www.tebtebba.org/index. php/content/160-2nd-edition-of-guide-on-climate-change-and-indigenous-peoples-now-released

United States Department of State. 2010. Announcement of U.S. Support for the United Nations Declaration on the Rights of Indigenous Peoples, December 16. http://www.state. gov/documents/organization/153223.pdf

Watt-Cloutier, Sheila. 2004. Climate Change and Human Rights—Human Rights Dialogue: Environmental Rights. *Carnegie Council on Ethics and International Affairs* 2(1) http://www.carnegiecouncil.org/resources/publications/dialogue/2_11/section_1/4445.html

World People's Conference on Climate Change and the Rights of Mother Earth. 2010. People's Agreement of Cochabamba. Cochabamba, Bolivia, April 22. http://pwccc.wordpress.com/support/

ON OUR OWN: ADAPTING TO CLIMATE CHANGE

Finding an Internal and an Intergovernmental Framework for an Adaptation Strategy

Rudolph C. Rÿser

Dr. Rudolph C. Rÿser is chair of the Center for World Indigenous Studies (CWIS) in Olympia, Washington.

Editors' note: This essay derives from an interview with Dr. Rÿser conducted by Zoltán Grossman on October 5, 2009; the interview was transcribed by Courtney Hayden and revised by Dr. Rÿser.

If one wants to find the green parts of the world, look only where the Indigenous people live and there's a reason for that. There is a strong motive to duplicate that, which means relying more heavily on Indigenous people.

The climate change issue is fundamentally an issue of Indigenous peoples' sovereignty, cutting across virtually every topic of importance to a society. Without exercising authority to define risks and vulnerabilities across a wide range of interrelated parts in a society (as any tribal community might want to do), Indigenous nations cannot establish themselves as regulators or set standards that respond to the adverse affects of cli-

mate change—as they must. This ends up being a very significant problem for Indigenous people worldwide. This is true since they are faced with the threats and the realities of human-induced climate change. Indigenous peoples are not being asked nor are they vigorously offering themselves to act in the capacity as governing authorities, as regulators and standard setters, but it is apparent that if they do not, they risk marginalization at best and exploitation to their detriment at worst. Adaptation and responses to the adverse affects of climate requires firm leadership, sustained responses, and steady negotiations to ensure the tribal social, economic, political, and cultural survival—in other words, Indigenous governments acting as sovereign powers.

Tribal peoples must reach into their cultural toolbox to draw out resources that will enable them to adapt to climate change challenges internally. At the same time they must meet the challenge of negotiating with neighboring peoples and institutions to prevent encroachments on their sovereign powers.

In the face of a growing interest to participate in the global and regional climate change dialogue, representatives from Indigenous nations or organizations attending international conferences demand to be heard. They call on states' government officials to hear them and most particularly hear that they possess traditional knowledge that must be a part of the dialogue. Indigenous representatives have a problem when they are asked to share that knowledge—to explain what that traditional knowledge is and how it can enrich the debate about responses to changing climate. Too often, proponents of traditional knowledge fall silent about the actual content of their traditional knowledge, leaving the debate to conventional scientists and states' government political leaders. Instead of falling silent, Indigenous representatives should be prepared to step forward with constructive analysis and proposals.

Traditional knowledge is a resource held within all Indigenous communities, yet for many reasons we have often not been able to explore and apply this knowledge to the issue of climate change. This may occur for several reasons: 1) this knowledge may be held secret or protected; or conversely, it may be lost or in the process of being forgotten. 2) similarly, because traditional knowledge has not been valued by conventional science or has been relegated to a secondary or adjunctive model, many people feel hesitant to proffer information that will be rejected, and finally, 3) since traditional knowledge is often locally specific it has not been shared or tested across communities. Now is the time to overcome all of these obstacles and to assert the primacy of traditional knowledge in solving many of our environmental problems. But first we must acknowledge and resolve the historical and community traumas that may preclude its application because of adherence to the myth of the primacy of conventional science.

In the U.S. Pacific Northwest we are very interested in traditional forest management practices, but it is also the case that we end up with a lot of conventional scientific methods used to manage forests—methods that may not be effective in preventing carbon emissions or in increasing the capacity of that forest to absorb carbon emissions. This view is not to suggest that conventional science is wrong. It is really to say that conventional science and Native sciences rooted in traditional knowledge must be applied together where possible.

The Menominee nation, located in Wisconsin, applies a sustainable forest management model that relies in part on traditional thinking. They harvest selectively. Menominee foresters harvest trees that are dead or dying or to clear areas to allow for the stronger trees to grow. Even though this method is more expensive, it has produced a hugely productive natural forest alive with diversity. The methods used now ensure a forest that appears from space as a large dark green rectangle in Wisconsin when the snows come, applying a blanket of white over the remainder of the state. The Menominee maintain a vital forest while earning revenues at the same time. Traditional knowledge has much to offer.

Other tribes that are forest dependent need to cut trees to make money, but when they cut trees, of course, they reduce the capacity of the forest to absorb carbon. But at the same time they are eliminating carbon and expelling it into the broader environment by cutting the trees. Timber-dependent tribes must confront this difficult conundrum. What does traditional knowledge offer here? The Menominee forest management system may be a good answer.

Tribes face financial obstacles when economic interest is a primary motive that stands against cultural interest and ultimately the environmental interest. The Clinton administration advocated in the 1990s a policy that says everyone can have "economy and environment at the same time," without clearly explaining how you do that. Each tribe is faced with virtually the same question when it comes to forest management: How do you make the money required by members while ensuring the low carbon footprint necessary for environmental balance?

Thus are defined two major aspects of the internal tribal dialogue: one is cultural relationships, the rela-

tionships that the culture permits people to have with the environment (food, medicines, fresh water, shelter). To the extent that there is a codependence between people and the environment ensuring life, we must ask: how do we preserve, promote, and maintain that relationship as environmental circumstances change? Secondly, where do we get the financial resources to respond to change in a way that is sustainable for the tribe? Naturally, the inclination is to talk about things like cap and trade or state taxation of carbon emitters and to provide money off those receipts to those who don't produce carbon and greenhouse gases. Yet this also poses difficult challenges as it leads to increased dependence by the tribal community on the production of things that are carbon emitters and requires more capital investment even as they [tribal members] become more dependent on currency.

Despite all our current and projected efforts, the ocean is rising and will continue to do so for some time. Indigenous nations must act in collaboration with others and on their own to reduce the adverse effects of climate change, while working to develop strategies based in traditional knowledge and conventional science to *adapt*. Adaptation and collaboration are the major strategic actions that we have identified as viable approaches at the Center for World Indigenous Studies while working on behalf of the Quinault Nation and other nations in Africa and Canada in the international dialogue on climate change.

Adaptation

Responding to the adverse effects of climate change is essentially a matter of Indigenous peoples' adaptation. Adaptation strategies and policies are matters of local as well as international concern. The local reality is that Indigenous peoples (unlike other populations who are dependent on industrialized cities) have a biocultural relationship that is either dormant or active within one or more ecological zones. If the relationship is dormant or even damaged, it must be reactivated. What does this mean? It means that the culture of a people interacting with the biological and mineral environment is essential to the continuity of human life. Humans, as it is increasingly apparent, are a part of nature, not, as the Bishop of Hippo long ago argued, "separate from" nature and exercising the power over nature. Ample evidence exists in the growing literature that human beings have long actively engaged in a symbiotic relationship with the natural environment—giving and receiving the benefits of nature's generosity. When human beings or any other life form takes more than nature's capacity to reproduce, then humans or that life form suffers while the natural world licks its wounds. Hazel Wolfe, that wonderfully vigorous advocate of environmental protection and human cooperation, once observed with that special twinkle in her nearly 100-year-old eyes, "Earth is to humans as a dog is to fleas. Humans are an irritant when they act badly and like fleas on a dog, the humans are expendable; the earth and its environments, like the dog, will go on."

Concerted and accelerated collaboration for adaptation is not new. Humans have long had to adapt to changes in the environment either because of human migration or as a result of sharp or evolving changes in the environment. Long ago, Indigenous peoples in Africa, Asia, Europe, and the Americas engaged in what we might now call "terraforming," the act of intentional modification of the earth's surfaces, caring for the flora and fauna in the "natural garden." The Passamaquoddy, Wampanoag, and Massachusetts nations, along with many of their neighbors, transformed the northeastern coasts of Massachusetts, Maine, and Rhode Island by carefully and systematically selecting plants, animals, and lands for sustainability. Were they natural environmentalists? No, they were opportunists who recognized that knowledge gained from observing nature can be applied to nature in a cooperative fashion, benefiting humans as well as the environment.

The upshot was, well before the formation of the United States of America, a highly productive food, medicine, shelter, clothing, and health environment for the peoples while maintaining a balance in the environment. Notably, when the people along the coast of what is now Maine and Massachusetts died from introduced diseases from Northern Europe, the natural garden they created returned to the wild—demonstrating that the productive natural garden was dependent on human beings. Their longtime residence along the coast demonstrated the great benefits humans received from their "natural garden." Similarly, Indigenous peoples in what are now the Ohio and the Mississippi valleys, the California coast and the Southwest, Haida Gwaii and the Pacific Northwest, all engaged in terraforming—cooperatively engineering natural changes in the environment that enhanced and balanced human activity with the natural processes of the areas. This type of opportunism must once again contribute to restoring balance in the natural environment.

In the Pacific Northwest where certain habitats were out of balance, people intervened (as is happening now on some reservations) to restore such habitats—increasing fish, plants, and various animals to an area. These traditional knowledge techniques included slash and burn, river and creek redirection, and adaptation of tools that encouraged desirable plants (consider the Quamash digging stick). Animals were encouraged by the clearing of meadows of brush to increase deer, elk, or moose grazing. All this occurred amidst adherence to systematic cultural rules for wild food and medicine harvesting. The technique of slash and burn ensured a strengthening of the soil while returning most of the wood fiber carbons to the soil. This increased the "living soil" quality and ensured increased storage of carbon while providing lands for new plants and animals.

The Quinault Indian nation recently completed the first phase of a long-term project to restore ecosystem functions in the Upper Quinault River through the installation of engineered log jams, in cooperation with the U.S. National Park Service, Forest Service, local property owners, and others. The project was designed to stabilize flows and channel structures from extreme flows, provide spawning and rearing habitat for salmon, and protect roads and property from excessive erosion. The Quinault restoration effort will require several years and millions of dollars to complete. The Quinault nation assigned this long-term, expensive project a high priority to protect their Blueback (a unique run of sockeye). This special salmon has sustained the Quinault nation's culture and economy for millennia. The terraforming project reversed the continued degradation of habitat from development and water flows that have become increasingly extreme in recent years. The nation adapted the earth to restore it.

Food security, emergency services, and a range of other social and economic vulnerabilities threaten Indigenous peoples, and thus they give rise to the need for adaptation strategies. Adaptation now must mean reclaiming these and other cultural practices to rehabilitate on a larger scale whole ecosystems that have been damaged by sometimes more than a hundred years of destructive, industrial-scale exploitation by newcomers who assumed wrongly the resources were unlimited and free for the taking. Not only are plants and animals limited, there is a substantial price that must be paid, as is now quite evident.

These adaptation measures can reduce the adverse affects of climate change by increasing carbon sequestration in soils. Soils that are alive and vital can sequester three times more carbon than can plants and water systems, seas and streams. Managing ecosystems and re-establishing human/earth symbiosis through terraforming and selective plant management can provide a healthy and productive way of life once again for tribal peoples.

Tribal Canoe Journey on Seattle's Lake Washington in 2006, hosted by the Muckleshoot Tribe.

Adaptation and Collaboration

Indian nations are not alone. Other Indigenous nations, counties, states, the federal union, and the international community all challenge the tribal governments and their communities. Competing interests surround Indian nations. They are compelled to negotiate within their territories among their own people and between territories—with neighboring tribal peoples and other jurisdictions. Negotiations among Indigenous peoples of the UN Framework Convention on Climate Change (UNFCCC) treaty involve a serious discussion of "adaptation" from the tribal perspective. Members of each Indian community must engage the difficult task of carrying out an internal dialogue. How will each community respond to climate change? While those discussions don't now always deal with the details of specific measures one takes to adapt, they do need to focus on the framework for Indigenous peoples' collaborative involvement in the process of adaptation. Once a framework for the discussion is developed, it becomes possible to discuss the details for meeting the adaptation demands.

The International Indigenous Peoples Forum on Climate Change is an ad hoc body of Indigenous organizations and Indigenous nations that has worked since 2002 at the international level in climate change negotiations based on the UNFCCC treaty. Adaptation has slowly become an increasingly important topic in the international debate.

In a jointly developed statement, adaptation was addressed by the IIPFCC this way:

Parties shall recognize customary methods of adaptation employed by Indigenous peoples and local communities; and further acknowl-edge the benefits to Indigenous peoples guided and informed by the best available science and traditional knowledge, innovations, and practices as obligatory for community adaptation, disaster planning, and response. Indigenous peoples' law, regulations, plans, and customary standards shall be recognized as authoritative and determinative as to adaptation risks, values, and benefits within the Indigenous peoples' territorial jurisdiction. Full and effective participation of Indigenous peoples subject to their free, prior, and informed consent—at all stages of the adaptation process, including governance and disbursement of adaptation finance, planning, implementation, monitoring, and reporting consistent with the United Nations Declaration on the Rights of Indigenous Peoples. (IIPFCC non-paper 8—negotiating text, 26 November 2009)

The focus of this critique is on Indigenous peoples acting in the capacity of governing authorities. This is an essential element in the development of an adaptation strategy.

Even if we have an international treaty and we all agree to do something, it will ultimately come down to what do we do in our own backyard. Do we produce more carbon in our little backyard, or do we take action that promotes the sequestration of carbon? Do we use chemicals that continue to pollute the waters, or do we not use chemicals to pollute the waters? Do we establish procedures where we can specifically identify a single source pollutant, or do we have to look around and establish a completely new system to find out multiple-source pollution? Can we apply this to each one of these eco-niches? Ultimately if we can, it could be far more effective than anything else.

Collaboration and the recognition of the essential benefits of subsistence and symbiotic Earth/human relations must precede a treaty. We must recognize the practical circumstance: if we don't do something, our house will fill up with water and trees will fall on us. One would hope tribes could succeed by collaborating with neighboring jurisdictions. Yet the problem is that most neighboring jurisdictions (counties, states, etcetera) don't want to recognize that the tribal population has either the authority, right, or interest to act and collaborate. Tribal officials must work to change this political environment, and that is where dealing with the state, the federal government, and the international community becomes essential.

Between 1964 and 1984, many Indian leaders developed a real understanding of the importance of intergovernmental relationships. As Quinault leader Joe DeLaCruz famously said, "We aren't going away and the state is not going away, so we better figure out a way to deal with the state and vice versa." That principle has held sway ever since. The impetus, though, for developing effective intergovernmental mechanisms simply hasn't fully developed. The consequence of that is that we have a lot of language that says, "We ought to be following a policy," but we don't do the hard work of creating the tools to implement the policy. That is what we have to be doing now, because the practical reality is that failure to do so creates enormous problems with climate change. Let's say Tribe A decided to develop a set of regulations and cultural standards that assert, "This is how we are going to deal with this particular problem and these are the dos and don'ts." The state has not had that conversation with you, but it is separately developing its rules and regulations—they could be *simpático* or they could be in conflict. Absent an intergovernmental framework for working out the differences between tribal and state rules and regulations on climate change, both governments face growing jurisdictional conflicts. Intertribal conflicts over regulations emerge as a possibility as well.

When tribal leaders negotiated the Centennial Accord with the Governor's Office of the state of Washington in 1989, we failed to create a framework for its onward operation, we just laid out the principles of co-management of natural resources. Now, that was interesting and a valuable first step, but here we are many years later, and there is still no framework for working out fundamental intergovernmental conflicts over jurisdiction. As it turns out, there's equally no framework for tribal governments dealing with the United States either. We discussed developing a tripartite intergovernmental mechanism that involved tribes, the federal government, and the states when tribal governments sponsored a yearlong study by the Inter-tribal Study Group on Tribal–State Relations (Joe DeLaCruz, president of Quinault, and Russell Jim, councilman from Yakama, co-chaired). What that proposal would have initially required is the underlying tribal governmental structure that we now have with the self-governance mechanism negotiated through self-government compacts in 1990. So, it's now more possible to do a tripartite intergovernmental mechanism than it was in 1980 when the study group first developed the idea. I have a lot of optimism, *but* there isn't an awful lot of memory about how any of this works. Because we don't have the political leadership who has that historical memory, it's becoming incumbent upon some of us who do remember to try to remind people or let people know that this

initial work has happened and the framework is there to create this mechanism. Northwest tribal governments have led on the formulation of new tribal-state-federal policy in many ways, in large measure because of the visionary leadership we had, including people like Joe DeLaCruz (Quinault), Lucy Covington and Mel Tonasket (Colville Confederated Tribes), Bob Jim, Roger Jim, and Russell Jim (Yakama Nation), Cal Peters (Squaxin Island), Sam Cagey (Lummi), Tandy Wilbur (Swinomish), and Joe Garry (Spokane).

A similar framework for intergovernmental relations has become essential at the international level as well. There is currently no such intergovernmental mechanism. Such a mechanism can facilitate negotiations and mediation between tribal governments and states' governments over climate change policy or any other policy.

Changes since the Boldt Decision

In the Northwest, we had a whole host of agreements between tribes in the late sixties and the seventies. The tribes frequently met en masse and discussed public policy and common threats and how they were going to deal with them. During that time, up into the eighties, we had political leaders who understood that the key issues were protection of our land base, development of our tribal government, and preservation of our culture. The fourth issue was treaty rights. Every issue that came to the table was about how we achieved those four things.

But as we got through the federal court's Boldt Decision recognizing Washington tribes' treaty rights in 1974, we were increasingly asked to have technical people address various technical problems associated with fisheries management. The people of vision—the political leaders—stepped back. This led to more people who had managerial and technical knowledge at the table. Meetings were no longer about these four subjects, they were about things like, "How does a liver fall out of a fish and how do we prevent that?" or "Is a hatchery better than wild fish?" and those kinds of questions. Biologists and engineers were talking, but most political leaders had no knowledge about what any of this really meant. It's not that they were ignorant; it just wasn't their area of expertise. Because these discussions and outcomes were never clearly linked back to the four major subjects, treaty rights, culture, strengthening tribal government, and affirming the land base as a matter of the tribal vision, it resulted in a schism between traditional knowledge, science, and political action that we are trying to mend.

As time went along, we ended up with a new generation of elected officials who are quite distant from those early mandates. Tribal vision as the defining force was set aside and replaced with efforts to mirror the behaviors of the United States. If the U.S. had certain kinds of scientists, tribes had to have the same. Often, since the U.S. paid for much of what tribal communities began to do technically, the capabilities became focused on duplicating U.S. capabilities. It created a greater distance between political leadership and the population with whom they were supposed to be identified. A language barrier evolved between technical or official language and what people knew as the vision. The population, for a hundred years, understood treaty rights. They understood cultural development and preservation of culture. They understood land rights. These were ideas that people had become accustomed to thinking about. They understood, increasingly, the tribal government ideas of sovereignty and self-government, even though these ideas were often shrouded in official language. This language was obscure and unrelated to ordinary experience, and it excluded people. The efforts of earlier political leaders were about inclusion and not specialization that excluded the participation of whole parts of the Indian population.

After the Boldt Decision was finalized, we began to create a hybrid understanding of the relationship between European science and Native science. Nobody called the practical/everyday/integrated approach to things "Native science," but that's what it is. And it did have an influence: Many of the political/cultural leaders would say, "The wild fish are the essential part of our understanding of good fish," and a biologist would say, "Why would that be true?" Then they would come up with a biological explanation of why whatever the leader said was true. And then they could go to court, which is the motivation for doing this in the first place, and argue that you must have wild fish because of the biological argument. And we say, okay that's fine, but what that represented was an attempt at integrating Native science and Western science, so they could be used simultaneously.

The tribes in the Northwest began combining conventional science and Native science not only on fish, but also in the Hanford nuclear waste cleanup efforts, involving the Yakama nation, Umatilla, and Nez Perce, and the hydroelectric discussions about dams involving the Colville Confederated Tribes and the Lower Elwha Klallam. It isn't as if there has been a total separation—there just hasn't been a total integration of Native and Western science. The development of the climate change challenge and the need for an intergovernmen-

tal framework combined to make it necessary to integrate the two. That is the nature of the discussions the Quinault Indian nation has had with the United States on climate change. I expect it will take many more years before there is a full understanding and appreciation of how that intergovernmental process works. The Quinault government has a great deal of responsibility to demonstrate how it works. If we can show how the two sciences working together function, then it becomes a case example of what the United States and other jurisdictions should apply to [their] adaptation needs.

International Climate Change Discussions

For the past several years at the international level, there has been a functional impasse between Indigenous peoples and the UN member states' governments. The states' governments have essentially placated Indigenous peoples in a sustained attempt at relieving a political pressure valve [without] actually conced[ing] to Indigenous peoples' demands. The relationship between Indigenous peoples and states' governments became stagnant. The Quinault government took a proactive approach to change the dynamics by offering themselves as a governing authority instead of the usual approaches used by non-governmental organizations. To test out some potential solutions, we began discussions with selected states' government representatives directly, instead of meeting through UN organs and representatives. We discovered there was a considerable interest in an aggressive action on the part of Indigenous peoples to put recommendations and proposals on the table, acting as governing authorities with responsibilities similar to states' governments. The response was very different from what had been going on for many years. Indigenous peoples acting in the role of non-governmental organizations would approach UN member states' delegates and say, "Well, what are you going to do for me today?" And of course, the states would say, "Talk to your own state because they represent you." Indigenous nations had classified themselves as non-governmental entities functioning within the context of "civil society." States' governments simply responded in a normal manner to representatives from within their states.

What we and the Quinault government discovered was that states' government officials would deal with Indigenous peoples if they saw them as governing authorities and acting within a particular jurisdiction. An Indian government with jurisdictional responsibilities and accountability to constituencies [was] understandable. Once an Indian government presented itself as an equal, the member states' governments began to say, "Yes, of course we should be able to talk. Because you have regulations and we have regulations, and you have rules and we have rules, and you make laws and we make laws, and we don't want to create problems for ourselves…. We ought to find a way to work together."

The Quinault government proposed the creation of the International Intergovernmental Contact Group on Climate Change, identified as the "Five States, Five Nations" solution. Basically, what the proposal provided was an integrated approach to addressing climate change and a focus for Indigenous peoples and state governments to deal with the proposals from the Indigenous table. The proposal was carried directly to individual states' governments. The position taken by virtually all Indigenous peoples' actors before this proposal was to present themselves as a civil society interest. As civil society participants in international meetings, Indigenous peoples or their organizations and communities took the position that they may advise on treaty language, but they cannot have a role in decision making to settle the outcome.

The UN system is obligated to listen to civil society, and representatives of non-governmental organizations do get an opportunity to speak or submit a paper. But that doesn't guarantee that anything gets qualified as a part of the final decision. And as Indigenous peoples, there is no way to leverage influence to decide what is done. First, Indigenous communities don't have enough people. The Indigenous population relative to the size of other populations is nil. One and seven tenths percent of the total U.S. population is made up of more than 560 tribal communities; and either individually or collectively, these communities have no representatives in the Congress of the United States, no political tool other than the ability to lobby. So, if tribal communities want climate change legislation, they can offer a viewpoint, but they will have a tough time competing with the coal companies.

What we found with the Quinault leadership is that when Indian nations assert their governmental role and they are prepared and willing to act as governing authorities (to not only impose but enforce their rules), then the other government representative on the other side of the table says, "I recognize what that is: that's the kind of thing we do."

International Rulemaking

In spring 2009, Indigenous delegations came together in Anchorage, Alaska, and at the end of several days of deliberation participants issued a declaration. Contained in their declaration are a number of measures that were formulated into legal proposals that require ratification and approval of Indigenous peoples back home. That's what we ought to be doing if we are going to face up to the role of Indigenous peoples as parties to international rulemaking. Waiting to deliver a message to a panel of experts at the United Nations generates at least thirty years of possible discussion, and maybe two sentences about something or other in a UN convention somewhere. We don't have time like that.

Indigenous peoples have the ability that the UN system doesn't have, if they would just take advantage of it. They don't have a lot of bureaucracy, so they can act more quickly on their own and establish rules, even if they can't get the UN member states' governments to agree to them now. We have to be aggressively advocating for ourselves. We have to aggressively promote, develop, and execute solutions. We can't ask somebody else who created the problem to come and solve our problem.

What we need to know from tribes is what *you* can do about the problems. And if you have a solution—tell us about it. If you have a proposal for steps to be taken—lay them out. We can work together to try to find a way to do that. Indigenous peoples are not homogeneous, and we are going to have different points of view; that should be accepted. The only reason we talk about having a unified position now is that member state governments demand it—that's the only reason. Offering a coherent policy or plan—even different policies and plans—can nevertheless produce important progress. Indeed, proffering policies and plans suitable for different ecosystems is essential for each nation.

Asserting Local Solutions

How do we succeed amidst all the opposition, given that states, organizations, and corporations do not wish to accept the presence of Indigenous nations in the international dialogue? We set the schedule, we define the question and redefine it when necessary, and then we offer the solutions and set about addressing them. We have had these successes in self-governance, child foster care laws, housing—because the tribes pushed and created a little wave. They proactively set the agenda and said what *must* be done. They didn't say "We've got a housing problem, what do you think I ought to do?" No, they said, "Here is the solution to the housing market" and pushed it.

The same thing has been happening as we push forward on climate change. We are saying, "These are the things that have to be done. The ecosystem is really the focus." We can have a profound effect on climate change—far more significant than treaties, or, frankly, states' government legislation. All of the solutions are really at the ground level. Yes, you will have pipes spilling pollutants, but if you have pockets in the world that are actually getting cleaner and working better, tribal communities have the ability to survive. Once we can survive, then we can begin to deal with everybody else.

There should be thousands of agreements, and you cannot deal with Indian Country as one country. It is more than 560 countries in the United States and even more. So we must deal with each one, and while it is the case that bureaucracy loves to have limited numbers, we are going to have to overcome and go past that. That means bypass the bureaucracy to be able to address the practical reality that we have all these tribes, all these different ecosystems that need to be addressed, and they must be dealt with by the merits of each one.

Tribal communities are already making important and immediate changes. The Hoh on the Pacific west coast discovered they had to move their whole village to avoid the overwhelming floods that had been building for a hundred years [see Papiez, Part II]. The Hoh government began that process in 2008. The Quinault observed that sixty yards of their beach has eroded and the water is now sixty yards closer. That doesn't mean fifty years from now, it means we have less than five or ten years, and so the whole village of Taholah has to be moved or new adaptation measures have to be developed. The first step is to establish the principles upon which a tribal community is going to operate. We may want to prioritize emergency services, hazard relief, the construction of buildings, public health, and food security. For example, how do we address the fact that berries are not there anymore and the deer aren't coming down close enough to catch them? These are the kinds of questions that tribal communities will need to ask. First a preliminary assessment is required and then the commitment to conducting a life way risk assessment, which is an entirely locally focused review of all the different vulnerabilities. Only after taking these steps can a community begin to identify ways to respond to vulnerabilities.

Native Science and the Failure of Carbon Trading

The European Union had quite a number of years' experience attempting to commodify carbon, and they found that it didn't really work when they used a cap-and-trade system. A lot of that had to do with the fact that they gave away a lot of permits and a lot of companies made a lot of money off of those free permits. This has led to the conclusion that regarding the commodifica-

tion of carbon and greenhouse gases, a straightforward taxation system is going to be necessary.

The identification of various forests for carbon sequestration as a part of the formulation of permits systems also has very serious problems, because there are no consistent methods of measurement. The local rule of Indigenous peoples is ultimately going to have to be the solution, which is to say they define what is available.

This leads back to applying Native science to these kinds of problems. Conventional sciences have something to offer and we can agree to that, but we must have reciprocity, and the states' governments must agree to accept the conclusions of Native science. We know that even the Western sciences aren't generally accepted. There has to be an agreement on the integration on these two bodies of knowledge so we can make some judgments. There is also a tendency to ignore the fact that Native sciences are not absolute, which is to say they recognize variances that take place. The problem with a lot of Western science is that it's supposed to be absolute and actually it is conditional. Once the scientist made the truth, then it's supposed to be the truth. Of course, we've discovered that statement isn't altogether true. We need to accept the natural variances in how we measure things—it alters how we define the value of carbon in the forests or in the soil or in the ocean or wherever it is—we allow for changes to take place over time, we allow for the nonfinancial value of things, and that's where Native science allows you to step in. You can say that things are life-supporting in ways that have nothing to do with the medium of exchange and push for the definition of life values. I think "life values" is one of the things that Indigenous peoples can place on the negotiating table.

Native Advantages

There was a belief for a while that each tribe could act autonomously (with all of its resources) to achieve whatever it wanted, but on some issues like climate change that crosscut so many different areas of human concern, it is impossible to do that. Individual tribal communities can affect their own ecosystem and make internal decisions that have benefits, but how are they going to deal with somebody who is spewing smoke out forty-eight miles away and off your territory? They have to coordinate their responses with other nations and apply the intergovernmental process as well. Tribes have experience with the intergovernmental process and they don't fear it. We used to fear it, but we don't anymore.

Native societies have advantages by definition, not only here in the Northwest but everywhere. They have the benefit of broader resources, not only in terms of financial and institutional resources, but they also have technical personnel with enough experience. They can make quicker decisions (if they choose to do so) and recognize that they themselves could take the initiative and make decisions that would actually have effects. When they do make those decisions, they have ripple political effects on all other jurisdictions around them. Understanding that is crucial, and I think the tribes in the Northwest have demonstrated their understanding over the years. When they have taken the initiative, they have developed political leverage, proactively defined the agenda, and they have identified a process by which they will achieve a solution—and they proposed a solution that can be negotiated. The intergovernmental framework needed has yet to be developed, and when it is developed, it becomes possible for Indian nations to act as equal partners in the international dialogue to develop adaptation strategies and effect responses to climate change.

SWINOMISH CLIMATE CHANGE INITIATIVE

Executive Summary of the Swinomish Climate Change Initiative Draft Impact Assessment Technical Report (October 2009)

In recognition of a growing body of scientific evidence and in response to certain specific local events, the Swinomish Indian Senate issued a proclamation in 2007 directing action to study the possible effects of climate change on the Swinomish Indian Reservation community, lands, and resources and determine appropriate responses. Following this proclamation, the Tribe initiated a two-year project in late 2008 to assess how climate change may affect the Swinomish Indian Reservation and to develop strategies to address potential impacts.

The outcome of this project is the production of three key reports: this Impact Assessment Technical Report, a preliminary Adaptation Strategy Report, and a Community Action Plan with recommendations for future adaptation options and strategies. This technical report comprises the first milestone of the project. It represents the work of a multidisciplinary team led by staff of the Swinomish Office of Planning and Community Development, in partnership with the University of Washington Climate Impacts Group (CIG), and with further scientific assistance from Skagit River

System Cooperative (SRSC). The report describes the scientific data and potential climate change scenarios, assesses possible local impacts, and identifies specific areas of potential risk and vulnerability to climate change effects.

Key Findings

The fourth assessment of climate change released in 2007 by the Intergovernmental Panel on Climate Change (IPCC) represents a consensus among scientists worldwide that rising greenhouse gas concentrations are unequivocally attributable to human activities. In an effort to understand the response of climate to increased greenhouse gas emissions in the future, the IPCC coordinated global climate change modeling experiments carried out by the international research community under various emissions scenarios. The models under all scenarios point to a warmer future climate, with the rate of warming correlated to the rate of greenhouse gas emissions. The key to preparing for the potential impacts of climate change is to understand the extent of possible effects on our regional biological, physical, and human systems under future scenarios of varying ranges. To assess the range of impacts and the potential risks posed to human and natural systems, this report considers several regional climate change scenarios, from low to high impacts, to accommodate the inherent uncertainty in the climate change models and future greenhouse gas emissions scenarios. This approach is consistent with similar analyses and efforts by other governmental entities, such as King County, Washington, and the California Coastal Commission.

Based on assessment of current documented models and scenarios, the principal areas and resources within the Swinomish Indian Reservation vulnerable to climate change impacts are shorelines, beaches, low-lying terrain, and forests, along with the assets within those areas. Impacts to some of these vulnerable areas are potentially high within twenty to fifty years, increasing through the end of the century and beyond. Other areas and resources may have moderate impacts during this time frame. Significant among these potential impacts are the following:

➤ Over 1,100 acres of Swinomish Reservation lands, or approximately 15 percent of Reservation uplands, are potentially at risk of inundation from increasing sea level rise, including the only agricultural lands within the Reservation, the Tribe's primary economic development lands, and sensitive shoreline areas.

➤ Approximately 160 residential structures are potentially at risk of inundation from sea level rise and/or tidal surge, with a total estimated value of over $83 million.

- ► Approximately eighteen nonresidential or commercial structures are potentially at risk of inundation from sea level rise and/or tidal surge, with a total estimated value of almost $19 million.

- ► Approximately 2,218 acres of uplands and over 1,500 properties are in a high-risk zone for potential wildfire based on projected increase in temperatures; total value of structures and properties within this zone is estimated to be more than $518 million. Most other areas within the Reservation are at least at moderate risk of wildfire.

- ► Vital transportation links and access routes to the Reservation are at risk of inundation, with the potential to isolate the Reservation from the mainland during increasingly high tidal events.

- ► Beach seining sites and shellfish beds along the west shore of the Reservation, areas of traditional tribal harvest, are at significant risk of permanent inundation and potential loss. Important "keystone" species such as shellfish and salmon are at risk of higher levels of contamination from algal blooms and other diseases that may be exacerbated by increased temperature and other changes.

- ► The Reservation population as a whole, particularly those who are ill or elderly, are potentially at risk of a variety of heat-related illnesses during isolated or extended high heat episodes as average temperatures increase, and tribal members in particular may be at risk of increased incidence of respiratory ailments such as asthma from potential increase in synergistic impacts of pollutants.

- ► Sensitive cultural sites within low-lying areas may face permanent inundation, and traditional native species may be lost as they are forced to migrate or adapt to hotter, drier climatic conditions.

The risk of inundation of shorelines and low-lying areas is expected to increase over the long term with gradual sea level rise and projections of more frequent and intense storm/tidal surges. Global projections of sea level rise indicate a range from lower estimates of 18–59 cm (~1½ to 2 feet) by the end of the century (IPCC, 2007) to higher estimates of up to 55–125 cm (~2 to 4+ feet) within the same time frame (Rahmstorf, 2007; Pacific Institute, 2009). Regional estimates of sea level rise depend on the local effects of wind patterns, atmospheric pressure, and vertical land movement caused by tectonic activity. Considering these local conditions, Mote et al. (2008) estimates regional sea level rise for

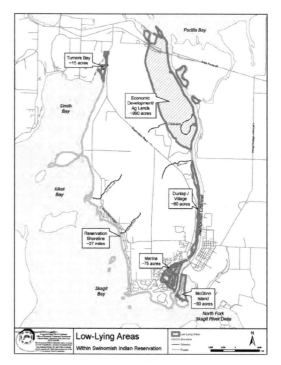

Low-lying areas within the Swinomish Indian Reservation affected by coastal flooding.

the Puget Sound to span from very low estimates of 16 cm (6 in) to very high estimates of 128 cm (50 in) by the end of the twenty-first century. Structures, roads, utilities, and other assets within nearshore or low-lying areas will be increasingly impacted by sea level rise and tidal surge events to the extent that adaptation measures will not be available or able to forestall, protect against, or prevent such impacts.

Based on projections of potential sea level rise and tidal surge, risk zones were mapped for the Reservation, and an inventory of properties and improvements within these risk zones identified almost 200 properties potentially at risk, including residential structures and non-residential facilities. Any revenues generated from leasing or other commercial activity on these properties would also be at risk of loss. Low-lying agricultural and shellfish areas could ultimately be lost entirely, and primary economic development land could be significantly impacted. Certain impacts also carry potentially significant secondary consequences as well, such as inundated access routes causing isolation of the Reservation from the mainland or inundation of low-lying development zones, affecting or preventing implementation of critical economic development projects. Such secondary consequences have the potential to extend impacts to the entire Reservation population and stress the ability of Tribal and other governmental entities to respond.

Forested areas and resources are projected to experience different but equally significant impacts. Annual mean temperature is projected to increase in the Northwest by up to 3–4°F by 2040 and perhaps as much as 7–8°F by the end of the century (DOE, 2006; CIG, 2009). Gradually increasing average and summer temperatures will decrease moisture content in soils and vegetation and increase the potential for devastating wildfire throughout forested areas of the Reservation, but with potentially greatest impact in the urban/forest interface. Other impacts on forest resources include increasing drought stress with rising temperatures and an associated proliferation of drought-tolerant species, such as fir, and decline in drought-susceptible species, such as western red cedar. Changes in the species composition of large trees could also be accompanied by a shift in understory species and ground vegetation, potentially resulting in the loss or migration of certain native plants. Additionally, higher temperatures are projected to create a more suitable environment for the spread of forest pests and diseases, such as bark beetles and various fungi that would previously have been suppressed by colder winters.

Other significant or notable impacts include effects on public health, marine resources, and cultural resources. Increasing temperatures are projected to impact human health in numerous ways; heat stress will have a variety of impacts on the general population, as already seen in some parts of the globe, and rising temperatures will create a more suitable environment for pathogens and their vectors that is not normally prevalent in a colder regime. Likewise, rising ocean acidity and shifts in tidal zones will put additional stress on nearshore marine resources such as shellfish and viable habitats. Cultural resources may be impacted both positively and negatively by tidal inundation. Gradual sea level rise will increasingly submerge nearshore or low-lying buried artifacts and sites, both protecting them and making investigation more difficult, while strong storm surges may uncover some sites or artifacts, rendering them vulnerable to weathering and tampering. Cultural use areas may be impacted by either inundation in nearshore or low-lying areas or wildfire in forested areas, rendering them unusable in either case for some extended period of time.

Report Disposition

This report is intended to lay a solid foundation for assessment of appropriate response strategies, and the ultimate intent of this and following reports is to provide a well-documented path to actions taken based on such strategy recommendations. Concurrent with this report, an advisory group has been working toward a broad planning and policy framework for preparedness response strategies and an assessment of critical issues and impacted disciplines. The ultimate goal is to develop an action plan for the future by combining the results of this technical report with the advisory group's recommendations. The action plan will include a range of strategy recommendations for adaptation to potential climate change impacts. It will also describe areas of recommended or necessary coordination with local jurisdictions where common interests exist between the Tribe and other jurisdictions, and it will examine capacity and funding requirements for implementation. The final report will be shared as a model to assist other tribal governments and jurisdictions in planning for adaptation. It is expected that this project will be but the first in a continuing and long term series of steps to determine how best to cope with and adapt to some of the most significant issues to face the Tribe in its long history. Final publication of the action plan is scheduled for September 2010.

Genesis of the Project

The geographic characteristics and coastal location of the Swinomish Indian Reservation place community assets, vital infrastructure, natural resources, sensitive cultural areas, low-lying economic development areas, and community health at risk from projected wide-ranging and long-term impacts of climate change. A climate change report issued by the state of Washington in 2006 identified the lower Skagit River area as a high risk for sea level rise. The destructive potential of sea level rise was demonstrated in February 2006 when a strong storm surge pushed water levels several feet above normal, resulting in some flooding and damage to property on the Reservation and in La Conner. This was followed in November 2006 by a strong winter storm that downed trees and power lines across the Reservation, isolating the Reservation community for three days and prompting plans for evacuation of residents to the local Tribal gymnasium. These events heightened awareness of climate impacts in general and the lack of preparedness within the community, and they helped provide a catalyst for action to determine appropriate responses to climate impacts.

In recognition of this, the Swinomish Indian Senate issued a proclamation in October of 2007 directing action to assess potential climate change challenges and develop appropriate responses. Following this proclamation, the Tribe won funding through the U.S. Department of Health and Human Services' Administration for Native Americans (ANA) to support a major new $400,000 Climate Change Initiative. While the Tribe acknowledges the need for action toward mitigating the causes of climate change, the approach of this project has been consciously directed toward adaptation actions to counter the anticipated effects of climate change on the Reservation community. This is based in large part on the recognition that while society at large must play a part in mitigation, it primarily falls to each community and local government to determine needed action toward adaptation.

Climate Change Adaptation Strategy Toolbox

Prepared by the Swinomish Indian Tribal Community for the Swinomish Climate Change Initiative.

Suggested strategy evaluation objectives

1. Comprehensiveness: addresses range of anticipated impacts and risk
2. Effectiveness: provides a long-term solution, not a band-aid fix
3. Dynamic/adaptive approach: responds to changing facts, circumstances
4. Fiscal impact and feasibility: carefully evaluates financial requirements, commitments, terms
5. Incentives: encourages use of cooperative, non-regulatory, and programmatic measures
6. Community goals: aligns strategies and options with desires and needs of the community

Sample list of potential adaptation tools

1. Incentives/programmatic/non-regulatory:
   - Land acquisition
   - Open space buffers
   - "Local Impact Districts" to fund adaptation (or flood/wildfire control districts)
   - Incentives for relocation/avoidance
   - Incentives for emerging technologies to benefit adaptation
   - Penalties for high-risk activities (disincentive)
   - "Distressed community" assistance/grants/loans
   - Volunteer/community action groups on climate issues

2. Regulatory/code controls:
   - Zoning restrictions and buffers, rolling setbacks
   - Tightened building code standards in risk zones
   - Low-impact development standards
   - Climate impact screening in development review
   - Climate impact screening in environmental review (critical/sensitive areas)
   - Streamlined permitting for adaptation activities
   - Risk response/management plan and requirements (risk zones)
   - Disclosure requirements for real estate transactions

3. Practical/engineering solutions:
   - Loss-prevention design, engineering standards (new construction)
   - Retrofitting (existing structures)
   - Armoring in place
   - Fortification (seawalls, diking, etc.)
   - Preventive maintenance (e.g., culverts, storm system, repainting, replanting)
   - Relocation (horizontally, vertically)
   - Replacement
   - Abandonment

4. Risk prevention planning:
   - Amend Hazard Mitigation/Emergency Management plans to address climate issues
   - Amend/revise resource plans to address climate impact issues (public water supplies, forestry resources, etc.)
   - Amend/revise Comprehensive Plan for climate change issues
   - Implement climate and risk assessment planning across disciplines (transportation planning, capital facilities, open space, etc.)
   - Identify/establish clearinghouse for climate change data/information

5. Emergency preparedness:
   - Emergency planning for extreme weather events (flood, heat, fire, drought, storm)
   - Provision of emergency shelters and supplies
   - Enhancement and training of emergency response teams
   - Community education/outreach on emergency planning
   - Citizen action plans
   - Event exercises (tabletop and live simulations)

6. Other:
   - Litigation (where only option available, or option of last resort?)
   - Legislation (currently primarily focused on mitigation rather than adaptation)

Principles and objectives for effective implementation

► Public education/outreach: communication, information, training, build support

► Relevancy: relate to facts, current issues, real world situations

► Political realities: address constraints and institutional barriers realistically

► Incremental approach: to cope with governmental inertia, funding requirements, etc.

► Regional approach/partnerships: problems larger than jurisdictions, cooperative efforts to promote effectiveness, capacity for response; or where necessary

Executive Summary of the Swinomish Climate Change Initiative Climate Adaptation Action Plan (October 2010)

In the fall of 2008 the Swinomish Indian Tribal Community started work on a landmark two-year Climate Change Initiative to study the impacts of climate change on the resources, assets, and community of the Swinomish Indian Reservation and to develop recommendations on actions to adapt to projected impacts. This followed issuance of a proclamation by the Tribal Senate in 2007 directing action to study and assess climate change impacts on the Reservation. Under the guidance and coordination of the Swinomish Office of Planning and Community Development, the first year of the project was devoted to assessment of projected impacts, as presented in an Impact Assessment Technical Report issued in the fall of 2009 [see the first part of this chapter]. The second year of the project was focused on evaluation of strategies and options for recommended actions to counter identified impacts, which resulted in preparation and release of this report. The ultimate goal of the project was to help ensure an enduring and climate-resilient community that can meet the challenges of anticipated impacts in the years to come. The information in this report, therefore, completes the Tribe's first critical assessment of climate change issues and actions, as a basis for the next steps in what is expected to be an ongoing effort to address a daunting array of complex issues.

The Tribe was assisted during the two years of this project by the University of Washington Climate Impacts Group as science advisors, who provided expert assistance with analysis and interpretation of climate data and models. Given a mix of inter-jurisdictional issues involved, the Tribe also solicited the assistance of a strategy advisory group comprised of representatives of Skagit County, the town of LaConner, and

the Shelter Bay Community. In addition, project staff worked with a tribal community interest group, led by a communications/outreach facilitator, to communicate information on particular significant potential impacts to tribal traditions and practices and to solicit feedback on concerns and issues. Working with these partners and groups, project staff evaluated a broad range of potential strategy options for targeting various climate impacts and developed a comprehensive list of recommendations for actions to address specified impacts.

Strategy Evaluation

In determining appropriate adaptation strategies, project staff worked with participants to survey a wide range of potential strategy options and develop a process for evaluation and prioritization of targeted strategies. To address projected impacts identified in the technical report, such as inundation of shoreline areas and resources, wildfire risk, and health impacts from higher temperatures, the project team assembled a comprehensive adaptation strategy toolbox [see above]. Types of options in the toolbox included nonregulatory tools such as buffers and incentives, regulatory controls such as shoreline restrictions and setbacks, options to allow shoreward migration of beaches and habitat, practical engineering techniques such as bank protection or raising/hardening structures where desired or appropriate, and improved risk prevention planning. To establish a rational process for evaluating strategy options, the team worked with project advisors to develop a set of evaluation objectives against which to do initial screening of options.

The basic set of evaluation objectives included:

► Comprehensiveness: Does the proposed strategy address the range of anticipated impacts and risk for the affected asset/resource, or is it limited in application?

► Long-term sustainability: Does the proposed strategy promote a sustainable long-term solution, rather than a short-term "band-aid" fix?

► Dynamic/adaptive approach: Does application of the proposed strategy allow for responding to changing facts and circumstances, or is it rigid and inflexible?

► Fiscal impact and feasibility: What is the degree of fiscal impact of the proposed strategy based on estimated financial requirements, commitments, and terms?

► Non-regulatory approaches: To what extent does the proposed strategy encourage the use of non-regulatory approaches, such as cooperative, programmatic, or incentive measures?

► Community goals: Does the proposed strategy align with desires and needs of the Reservation community as expressed through Tribal planning documents and other sources?

Once evaluation objectives were established, potential adaptation tools were screened and targeted to specific projected climate change impacts; the resulting correlation of potential tools and impacts were then assembled in a matrix into which additional considerations were factored, including:

► Existing governmental authority and capacity for implementation of given strategies;

► The level and/or type of authority or capacity needed for implementation;

► Internal and external partners needed for implementation; and

► Approximate time frame anticipated for potential implementation, as stated in increments of one to three years, three to ten years, and greater than ten years.

This completed strategy evaluation matrix formed the basis for identification and prioritization of major action recommendations.

Action Recommendations

Based on the above-mentioned evaluation matrix, an exhaustive list of targeted impacts and strategies was assessed to identify priority issues and preferred options. The primary methodology used for prioritizing impacts was based on a correlation of vulnerability and risk for the given impacts, following from impact assessment performed as discussed in the previous technical report. While strategy options were scoped and targeted for all identified impacts, for the purposes of this report the prioritization was applied to narrow the list of impacts and actions to those deemed to be most significant, as a means of focusing discussion and subsequent decisions on implementation. In addition, the list of major impacts and associated actions were reorganized under four basic categories that reflect the orientation and organization of community life on the Swinomish Indian Reservation; these four categories are coastal resources, upland resources, physical health, and community infrastructure and services. A fifth overarching category, cultural traditions and community health, has threads to all categories, given the ties and significance of cultural and community health to a great number of the issues, and as such is the subject of special focus.

After exhaustive analysis and assessment as described above, the major potential impacts and associated types of recommended actions can be summarized as follows:

Coastal Resources:
► Impact: Inundation from sea level rise and storm surge; includes impacts on shoreline areas, structures, habitat, and natural resources within those areas.
  – Actions: Shoreline controls (risk zones, setbacks, rolling easements, restrictions); physical controls (bulkhead removal for shoreward migration, armoring for shoreline protection, raising/ hardening structures, raising/extending dikes); habitat enhancement (fill removal, sediment input); land acquisition.
► Impact: Decreased habitat viability due to changing water quality parameters.
  – Actions: Aquaculture operations.

Upland Resources:
► Impact: Increased wildfire risk.
  – Actions: Adopt/maintain Firewise standards/ buffers; enhanced training/support for wildfire response; improve forest management policies/ practices.

Physical Health:
► Impact: Heat related illness.
  – Actions: Education/emergency preparation (cooling center)/weather warnings; housing design/ retrofit for improved cooling efficiency.
► Impact: Increased risk of respiratory disease.
  – Actions: Improved monitoring/reporting; maintain/strengthen health services.
► Impact: Toxic seafood contamination.
  – Actions: Strengthen traditional roles for food safety; aquaculture operations.

Community Infrastructure and Services:
► Impact: Inundation of low-lying roads and bridge approaches.
  – Actions: Build/raise dikes; raise road level; relocate route; abandon route.
► Impact: Road closure from storm/tidal surge event and/or wildfire.
  – Actions: Alternate route plans; restrict road construction in risk zones.
► Impact: Reduced potable water supplies due to decreased sources.
  – Actions: Water management plan for droughts; incentives/disincentives for water usage; wastewater treatment of grey water; water-efficient appliances/fixtures; voluntary/mandated water

restrictions; water conservation/education; additional water storage.

► Impact: Contamination of drinking water supplies from flooding.
  – Actions: Stockpile/maintain emergency water supplies; identify/protect vulnerable facilities; well testing/disinfection program; increase stormwater management capacity.
► Impact: Service disruption of communication and energy systems.
  – Actions: Develop alternate energy and communications systems.

Implementation

Making an action plan useful relies on effective implementation. Project participants identified a number of issues considered important for successful implementation or pertinent to specific proposed strategies or actions, including key principles for implementation of adaptation actions, as follows:

► Flexibility in approaches: Because of the number and complexity of many climate change issues, there may be few common solutions to the same basic impacts as they affect different areas; adaptive response may be required for changing circumstances.
► Public education/outreach: Communication, information, and training on identified issues are vital to building support within the community for action.
► Relevancy: Relating to facts, current issues, and real world situations will help to make issues and actions more relevant to the local community.
► Political realities: Address political constraints and institutional barriers realistically. Issues of organizational capacity and mainstreaming adaptation actions must be addressed.
► Incremental approach: Phasing and scaling of actions may help to cope with issues such as governmental inertia and challenging funding requirements.
► Regional approach/partnerships: Some issues are larger than individual jurisdictions; cooperative efforts may be useful or necessary to promote effectiveness or to increase capacity for response.

Finally, several implementation project concepts have been formulated based on review of priority issues and actions in each of the major categories described above. These potential project concepts address a representational cross-section of the more critical issues identified, and they range from relatively inexpensive planning activities to more costly and technically complex adaptation projects, as follows:

► Coastal protection implementation;
► Coastal resource research;
► Dike maintenance authority and program;
► Regional access preservation and coordination;
► Wildfire risk management and mitigation; and
► Local emergency planning.

Cultural Setting of Swinomish

Important information about the Swinomish culture is recognized in an ongoing health study that includes neighboring Coast Salish tribes. This study is holistic, including spiritual and community factors as well as physical health. The first phase is reported in Table II on page 140. Results of the health study will help guide the climate change adaptation effort. Future phases will identify specific actions.

The first phase sets the stage by clarifying what is meant by "community health." The study identified five indicators of community health as listed in the following table. Quotes from Coast Salish tribal members about the connections between natural resources and health provide explanatory illustrations.

Community Health in Native Communities

In many Native American communities, Swinomish included, health is defined on a community level, consisting of inseparable strands of human health, ecological health, and cultural health woven together, all equally important. Within this definition, many of the dimensions of good health as defined by the Swinomish are difficult to quantify, such as participation in spiritual ceremonies, intergenerational education opportunities, and traditional harvesting practices, yet they may be negatively impacted or even destroyed when resources are scarce or disappear (Arquette et al. 2002; Harris and Harper 1997, 2000, 2001; Wolfley 1998).

Community cohesion means to actively participate in one or more roles within your community network. Two of the main duties in regards to natural resources are harvesting and preparation, which are imbued with thoughtful significance and intention. Particular methods have been honed over countless generations. Community members each have a role in the process, and each role carries with it an identity and the pride of being a needed part of the entire process. For example, the cooks' role is looked upon with reverence equal to that of the harvesters' role. These roles are often learned and passed down through the generations such that some may be known as the best salmon cooks, while others may be known as expert clam diggers or hunters. Community members know each other's roles and who

Table II.

Suggested Top Five Tribal Health Factors and Associated Health Indicators (Salish Sea natural resources, including seafood, seaweeds, shells, etc.)

Five Health Factors	*Fifteen Health Indicators with Definitions for Each*
Community Cohesion	*Participation and cooperation*: the community members depend on each other; strong support network (e.g., everyone supports the maintenance, harvest, and distribution of resources)
	Roles (e.g., harvest, prepare, preserve natural resources): each member of the community has a role that is respected
	Familiarity: food roles are known and trusted; therefore, it is assumed food is safe
Food Security	*Availability*: natural resources are abundant and healthy
	Access: all resource use areas (i.e., "usual and accustomed" areas) are allowed to be harvested, with an emphasis on local resources for subsistence consumers.
	Sharing: ensuring that everyone in the community receives natural resources from the Salish Sea, especially elders
Ceremonial Use	*Gatherings and ceremonies*: particular community assemblies that require natural resources from the Salish Sea
	Giving thanks: thanking Nature or the Spirit for providing the natural resources when harvesting and preparing them; done with prayers and thoughtful intentions
	Feeding the Spirit: using natural resources from the Salish Sea to satisfy a spiritual "hunger" (e.g., consuming traditional foods)
Knowledge Transmission	*The Teachings*: knowledge, values, and beliefs about tribal health in connection with the Salish Sea
	Elders: the knowledge keepers; they have and are able to pass on the knowledge
	Youth: the future; they receive and respect the knowledge
Self-Determination	*Healing*: ability to choose lifestyle desired for what is considered "good health" (e.g., traditional medicines, language programs)
	Development: community enrichment opportunities directed by and for the community
	Restoration: environmental or habitat restoration projects that are community driven

taught them the roles. That familiarity instills trust that the food handlers have performed their jobs accordingly and the food is "healthy" and "safe" to consume. As one tribal community member explained, "There is an importance that you know where it [a natural resource] is caught and you know that it is part of our community and it's been part of our history for that long period of time if it's caught by local community members."

Food security depends on 1) availability of natural resources (i.e., that there are sufficient stocks to be harvested), 2) access (i.e., that harvesters are allowed access to those stocks), and 3) sharing. Particularly for elders, who have passed on their food harvest and preparation knowledge but can no longer harvest or cook for themselves, they rely heavily on the food sharing networks. Food security is defined by the United States Department of Agriculture as "access by all members at all times to enough food for an active, healthy life. Food security includes at a minimum (1) the ready availability of nutritionally adequate and safe foods, and (2) assured ability to acquire acceptable foods in socially acceptable ways" (2007).

Shellfish are a high-protein, low-fat, nutrient-rich food source (Dong 2001) that comprises a large portion of the Swinomish traditional diet. Shellfish (crab, shrimp, and clams) are in the top five types of seafood eaten at Swinomish, second only to salmon, and clams provide a stable food source. A saying shared by many Coast Salish tribes is "when the tide is out, the table is set."

The natural resources of the Salish Sea are also a significant part of the subsistence economy—part of a long history of Native peoples' supplementing economic income with subsistence foods to ensure food security. Subsistence relies on functioning kinship or community relationships sustained by internal support networks and honored and reinforced through ceremonies and gatherings (Usher et al. 2003).

As is often repeated in Coast Salish communities, one young fisher stated, "Yes, I always believe in sharing my catch because it's always been taught to me to do this, and I always try and give mostly to elders, the ones that can't get out on the water and get their own."

Ceremonial use is more than the ceremonies and gatherings themselves. It also means the importance of giving thanks to the spirits of the natural resources when harvesting and preparing them, and the necessity to feed the spirit of oneself by consuming natural resource foods or feeding the spirit of a relative who has passed away by offering natural resources. Ceremonies, also referred to as gatherings, involve natural resources such as salmon, duck, and clams and are viewed as an

At the 2007 Coast Salish Gathering, Cowichan Chief Harvey Alphonse from B.C. and Swinomish Chairman Brian Cladoosby from Washington sign an accord on deer hide to restore and protect the Salish Sea.

important part of the food-sharing network. Ceremonies provide the environment in which healing can take place (as health is both a physical and spiritual state [Garrett 1999]). Community members look forward to ceremonies for the natural resources and the company as well as the spiritual significance. Ceremonies are the best way to reinforce ties with other community members and members of other tribal communities, and they are especially important to elders, many of whom only have access to natural resources at these events throughout the year.

When asked about the importance of having natural resources at gatherings and ceremonies, the majority of interviewees said that the *events would be changed or impossible without natural resources such as seafood.* For example, the First Salmon ceremony occurs at the beginning of the fishing season. The ceremony thanks the marine natural resources for returning and allowing the people to harvest them and also asks for protection and guidance for the fishers to safely obtain plentiful catches. The proper way to harvest and prepare a natural resource is to pray and give thanks to the Spirit for offering itself to the community. In this way, the natural resource is empowered with the prayers, which in turn nourishes both the body and the soul.

The knowledge transmission health indicator encompasses the teachings about how to gather, prepare, preserve, distribute, and employ natural resources, passed down from the elders to the youth. One example of knowledge transmission that also ties in all of the aforementioned key health indictors of community cohesion, ceremonies, and food security is the role of food preparation, as recounted by a Coast Salish community member:

I fostered two teenagers.... They were getting into a little bit of trouble. They wore the backwards baseball hats that read "Native Pride," you know, "I'm Indian, I'm Indian." I said, "Well, you really want to show your people how to be Indian, here I'll show you." I just happened to be filleting fish; I had a lot of fish. And we filleted them and I made them help me.... I said, "This is how you want to be Indian is you provide food for your people. It's not standing on the corner with a Native Pride hat acting tough. That's not Indian." The effect on them was just passing that on … the importance, and emphasizes that we have to gather these foods; we have to provide these foods in the wintertime when we're putting food on the table for the smokehouse.

Self-determination is a key health indicator that incorporates healing, restoration, and development, all enacted by and at a community/ local level. Self-determination means the freedom to decide how to create and sustain "good health." This final health indicator is unique in that, unlike the four previous indicators, asserting self-determination was not necessary until externally imposed trauma occurred. "Externally imposed trauma is defined as 'events that overwhelm a community's capacities to function in stable and generative ways'" (Korn 2002). Community trauma results from externally imposed habitat destruction, economic dislocation, food security interruption, social order disruption, and physical relocation.

"Educational colonization, religious conversion, natural resource piracy, distortion of decision-making, and externally imposed priority-making are all together and individually factors that can give rise to community trauma" (Korn and Dyser 2008). Forced assimilation through boarding schools and government laws outlawing Native people from practicing their traditional ways are examples of trauma that negatively impact self-determination (cf. Adams 1995; Collins 1997; Hoxie 1984).

Self-determination is the ability to exercise sovereign rights. The first component, healing, is the availability of and access to traditional medicines, language programs, and other culturally integral community health. Development, the second component, is the ability for a community to determine and enact their own, chosen community and economic enrichment activities in their homelands (this includes both reservation areas and traditional fishing, hunting, and gathering areas). The third component of self-determination is restoration, the ability for a community to determine and enact

their own chosen environmental or habitat restoration programs. A germane example of this third component is the current dam removal and river restoration project occurring on the Elwha River.

The remaining two steps in this health study are to summarize the tribal knowledge base of threats to and status of the Salish Sea and develop criteria and priorities for actions that address the threats and tradeoffs to tribal human health. Once the tribal knowledge base of the threats to and status of the Salish Sea has been established, a closer look at threats specifically from climate change may be undertaken and priorities altered based on potential findings.

Cultural Resilience: New Term, Old Meaning

What is cultural resilience and why is it important? "Cultural resilience is a relatively new term, but it is a concept that predates the so called 'discovery' of our people," wrote Iris HeavyRunner (Blackfeet), when she was coordinator of a tribal college faculty development project at the University of Minnesota.

> The elders teach us that our children are gifts from the Creator and it is the family, community, school, and tribe's responsibility to nurture, protect, and guide them. We have long recognized how important it is for children to have people in their lives who nurture their spirit, stand by them, encourage and support them. This traditional process is what contemporary researchers, educators, and social service providers are now calling fostering resilience. Thus, resilience is not new to our people; it is a concept that has been taught for centuries. The word is new; the meaning is old. (HeavyRunner and Morris 1997)

Community Wellness and Circles of Support

Resilience is important because certain impacts of climate change may lead to *grief* and *despair*, e.g., from decline of shellfish, salmon, land animals such as elk, the loss of traditional gathering and hunting places, and impacts to traditional plants. Resilience also is key to a five-phase model of wraparound services to help native youth succeed in education (Gillory). The Climate Change Education and Awareness Group (CCEAG) seeks to integrate such a model into its youth leadership work tied to climate change.

The book *A Gathering of Wisdoms* (second edition), produced by the Swinomish Tribal Behavioral Health Program, speaks to the importance of supporting tribal mental health and community wellness; it describes how to develop a behavioral health program in a tribal setting. It provides guidance that could be a resource for addressing community health issues of grief and despair and in building resilience.

Tribal Traditions and Effective Adaptation Planning

This chapter describes ongoing work to communicate climate change issues and to recruit people to join in adaptation measures. It describes the importance of tribal traditions and the challenges of developing effective adaptation measures that respond to those traditions. It also discusses the need to have an ethical response that respects and preserves the sensitive nature of traditional knowledge and specifies ongoing work to connect elders with youth for intergenerational sharing of spiritual and other traditional environmental knowledge. Specific actions that will help ensure the connection between tribal traditions and effective adaptation planning are outlined below.

Integrate Indigenous knowledge into ongoing planning and programs. There are a number of efforts underway regionally that are exploring ways to incorporate Indigenous knowledge into planning efforts to address climate change issues, and these vary considerably in their approach. A selection of such efforts includes the following:

► The Tulalip Tribes are pursuing a codification approach to institutionalizing traditional knowledge by creating an ethical construct that functions as an indirect representation of more sensitive knowledge concepts, in the interest of respecting and protecting such core knowledge.

► Creation of tribal review boards as a vehicle for formal screening and approval of traditional knowledge and sources. This would allow for and create a pathway for application of traditional knowledge, while providing a direct means of protecting such knowledge from misappropriation and misuse.

► Application of the community health indicators referenced in this chapter as a tool for measuring success and effectiveness of implemented programs and actions. This could be done both at the outset of proposed implementation and as part of ongoing monitoring of actions.

Explore treaty implications of adaptation planning. Because climate change can jeopardize the exercise of treaty-protected rights and thus impact tribal sovereignty, a range of actions needs to be explored to address such possibility. For example, "in-lieu" claims can assert the precedent of hydropower impacts on

fisheries to help support efforts to replace impacted resources. Likewise, agreements for cooperative, defensive, or info-sharing purposes can be negotiated. Such a strategy might be useful, for example, to bolster the joint efforts of local tribes to preserve or enhance impacted resources.

During a Climate Change Summit of the Coast Salish Gathering in the spring of 2010, participants discussed many aspects of treaty resources and climate change impacts. A recurring focus of discussion concerned the subject of "first foods," meaning traditional or cultural foods that tribes have used for centuries. Participants heard about and discussed impacts already affecting the collection and use of first foods, as well as emerging strategies and programs to address these issues. Following the summit, a position paper was developed to summarize an adaptation path that includes a closer look at first foods and treaty rights (Donatuto and O'Neill 2010).

Food takes care of the people and the people take care of the food: Umatilla case study. As presented at the Coast Salish Climate Change Summit, a pertinent tribal case study and approach that addresses both incorporation of traditional knowledge and the treaty implications of climate change is that developed by the Confederated Tribes of Umatilla (Coast Salish Climate Change Summit 2010) .

Eric Quaempts, Natural Resources Director, created the Umatillas' approach to first foods after many interviews with tribal members. Their management program assigns natural resource management branches to the appropriate food. Quaempts noted the order of food to the table and its importance. Water is the first food to the table and the most important; Water Resources and Fisheries are responsible for the water. The Cultural Resources Program is responsible for all the first foods such as salmon, deer, roots, and berries, along with the other natural resource departments that manage specific foods. The reciprocity of first foods is that they are managed to return to the people. It is a food-associated culture. It means tribal members have:

► Access to them, as provided by their treaty rights;
► The teaching of first foods and learning about them;
► Harvest and how and when to do that;
► The preparation that is passed on;
► Consumption;
► Celebration, sharing, and caring.

Teara Farrow Ferman, Cultural Resources Protection Program Manager for the Umatilla, states: "We manage our foods to have the foods into the future and preserve our culture." Historically, the men harvested and presented the salmon and the deer. The women collected and presented the roots and berries. Today, because climate change is already affecting the availability of foods, some Umatilla tribal members teach their sons *and* daughters to collect both foods. "We don't know what foods will be available [to tribal members] in the future or who they will marry," said Farrow. "We want them to be able to identify these foods and know how to prepare them. It's central to who we are." Farrow emphasized the importance of collecting oral histories and archaeological data and archiving the information. "What this information has shown us is that we have adapted to changes in our environment; where we once used dip nets, we now have to use gill nets, but this hasn't changed the importance of the food," said Farrow. "The community still values the first foods order even though species have changed." If physical and ecological processes change, however, our foods will change. Umatilla tribal members are already seeing roots and berries become less available in areas they traditionally gather from, so they are having to look beyond those areas, further away. If climate change continues to have an effect, we may have to go back to our old ways of trade, trading food resources with family members of other tribes (Coast Salish Climate Change Summit 2010).

References

Adams, D. W. 1995. *Education for Extinction: American Indians and the Boarding School Experience, 1875–1928.* Lawrence: University Press of Kansas.

Arquette, M., M. Cole, K. Cook, B. LaFrance, M. Peters, J. Ransom, E. Sargent, V. Smoke, and A. Stairs. 2002. Holistic Risk-Based Environmental Decision Making: A Native Perspective. *Environmental Health Perspectives* 110(2):259–264.

Coast Salish Climate Change Summit. 2010. http://blogs.nwifc.org/climatechangesummit/tag/confederated-tribes-of-the-umatilla/

Collins, C. C. 1997. Through the Lens of Assimilation: Edwin L. Chalcraft and Chemawa Indian School. *Oregon Historical Quarterly* 98(14):390–425.

Donatuto, Jamie, and Catherine A. O'Neill. 2010. Protecting First Foods in the Face of Climate Change, Summary and Call to Action. A paper presented at the Coast Salish Gathering Climate Change Summit, July.

Dong, F. M. 2001. The Nutritional Value of Shellfish. Washington SeaGrant Program, University of Washington, Seattle. http://www.wsg.washington.edu/communications/online/nutritional.pdf (last accessed May 2008).

Garrett, M. T. 1999. Understanding the "Medicine" of Native American Traditional Views: An Integrative Review. *Counseling and Values* 43:85–98.

Gillory, Justin. Generational Resilience, Strategies for Reproducing Success in Native College Students.

Harris, S., and B. L. Harper 2001. Lifestyles, Diets, and Native American Exposure Factors to Possible Lead Exposures and Toxicity. *Environmental Research* 86:140–148.

HeavyRunner, Iris, and J. B. Morris 1997. Traditional Native Culture and Resilience. *Research and Practice* 5(1). Available from the Center for Applied Research and Educational Improvement, College of Education and Human Development, University of Minnesota.

Hoxie, F. E. 1984. *A Final Promise: The Campaign to Assimilate the Indians.* Lincoln: University of Nebraska Press.

IPCC Working Group I. 2007. *Climate Change 2007: The Physical Science Basis.* Contribution of Working Group I to the Fourth Assessment Report of the Intergovernmental Panel on Climate Change. S. Solomon, D. Qin, M. Manning, Z. Chen, M. Marquis, K. B. Avery, M. Tignor, and H. L. Miller, eds. Cambridge: Cambridge University Press.

IPCC Working Group II. 2007. *Climate Change 2007: Impacts, Adaptation and Vulnerability.* Contribution of Working Group II to the Fourth Assessment Report of the Intergovernmental Panel on Climate Change. M. L. Parry, O. F. Canziani, L. P. Palutikof, P. J. van der Linden, and C. E. Hanson, eds. Cambridge: Cambridge University Press.

Korn, L. 2002. Community Trauma and Development. *Fourth World Journal* 5(1):1–9.

———, and R. Dyser. 2008. *Preventing and Treating Diabetes Naturally: The Native Way.* Olympia, WA: DayKeeper Press.

Rahmstorf, S. 2007. A Semi-Empirical Approach to Projecting Future Sea Level Rise. *Science* 315:368–370.

Swinomish Climate Change Initiative. 2009–. http://www.swinomish-nsn.gov/climate_change/project/reports.html

U.S. Department of Agriculture. 2007. Food Security in the United States: Measuring Household Food Security. http://www.ers.usda.gov/Briefing/FoodSecurity/measurement.htm (last accessed December 2007).

Usher, P., G. Duhaime, and E. Searles. 2003. The Household as an Economic Unit in Arctic

Aboriginal Communities and its Measurement by means of a Comprehensive Survey. *Social Indicators Research* 61:175–202.

Wolfley, J. 1998. Ecological Risk Assessment and Management: Their Failure to Value Indigenous Traditional Ecological Knowledge and Protect Tribal Homelands. *American Indian Culture and Research Journal* 22(2):151–169.

PULLING TOGETHER: HONORABLE COMMUNITY ENGAGEMENT

Shelly Vendiola

Swinomish Climate Change Initiative (SCCI) Climate Change Education and Awareness Group

This is a tribute to two of my teachers—Nilak Butler, an Inuit warrior woman, brave freedom fighter, friend, and enduring spirit to many, who taught me to have a voice; and to subiyay (Bruce Miller), a Skokomish teacher, healer, and cultural preservationist who taught me about the ancient teaching of the Tree People. These are my anchors.

Introduction

As the climate continues to change largely due to man-made global warming, we are witness to many impacts that are obvious and evident on a global level and are facing the biggest challenge ever—the climate crisis. One great spiritual leader, Thomas Banyacya, a revered Hopi elder and medicine man, spoke about "life out of balance" as he predicted the extreme changes that would occur when the wires (the electric grid) would cross the earth and energy consumption would be out of control and beyond the Earth's maximum carrying capacity. That time is now.

For several generations Billy Frank, Jr., a Nisqually elder and international leader, has led the struggle for the survival of the salmon and the protection of Pacific Northwest tribal sovereignty and treaty rights. These rights are outlined in one of the treaties, the Point Elliott Treaty of 1855, which covers the Swinomish Indian tribal community:

> The right of taking fish at usual and accustomed grounds and stations is further secured to said Indians in common with all citizens of the Territory, and of erecting temporary houses for the purpose of curing, together with the privilege of hunting and gathering roots and berries on open and unclaimed lands. Provided, however, that they shall not take shell-fish from any beds staked or cultivated by citizens. (Point Elliott Treaty of 1855:Article 5)

Billy's activism and diplomacy has been instrumental in bringing the plight of the salmon into the national arena, and he was arrested on numerous occasions for courageously exercising his treaty right to fish. His activism and that of others eventually led to litigation by the United States and numerous tribes seeking a declaration of their treaty rights and an injunction against interference in the exercise of those rights by the state of Washington. Judge George Boldt issued a landmark ruling in 1974 that reaffirmed the right of these "Treaty Tribes" to fish at their usual and accustomed fishing areas. His decision was limited to off-reservation areas, as it was understood that the Indians had maintained the right to fish on-reservation. Thus, this decision provided judicial support for Indian people's continuing access to their traditional foods, including the seafood and salmon. As such, Article 5 of the Point Elliott Treaty is of particular interest in the context of climate change, the ecological degradation it causes, and its impact on tribal access to food.

In the Pacific Northwest tribes have referred to themselves as farmers of the sea. The salmon is a traditional food for the Coast Salish tribes in the Pacific Northwest, and it is threatened by climate change. In 2009, the U.S. Fish and Wildlife Service reported:

> Salmon runs through the Northwest under the jurisdiction of the National Marine Fisheries Service (NMFS) are threatened due to climate change.... Adding to the complexity of threats facing salmon and steelhead are such new dangers as human-induced climate change. Recovering imperiled Pacific salmon and steelhead is complicated by the patchwork of federal, tribal, state, county, city, and private land ownership and regulatory authorities across the salmon and steelhead landscape. (U.S. Fish and Wildlife Service 2009)

In this time of great change because of the warming temperatures over time, the question of sovereignty raises a question of whether tribal communities should be compensated for rights that are being lost due to the loss of land, cultural use places, traditional foods, and viable economies, particularly those who are geographically located on low-lying coastal areas, where storm surges and hurricanes are becoming more frequent and sea level rise is worsening.

Tribal Sovereignty and Treaty Rights

Sovereignty here in the United States will support Indigenous peoples globally. Looking through an Indigenous lens, sovereignty includes natural and traditional law. The laws of the Coast Salish peoples are found in their language, ceremonies, songs, dances, and other cultural practices. Sovereignty also includes our status as citizens of a tribe located in a specific place and linked to a tribal government. Ideally, these tribal governments enjoy a "government-to-government" relationship with other governing bodies, whether federal, state, county, or municipal.

During the 1800s, a time of forced relocations and violent assimilation policies, many tribal leaders signed treaties with the U.S. government. The fourteen tribes along the Salish Sea signed the Point Elliot Treaty of 1855 in this period. Tribes signing these treaties intended to retain their rights to access to land and cultural resources for generations to come. Our leaders who went before us may have felt signing the treaties was the best possible option during a time of genocide and grave distress for Native people, who were being violently forced from their original homelands and made to live, in some cases, among hostile tribal groups. They signed the treaties to survive and avoid complete extermination and starvation. We are still healing from this violence of our past—from what is commonly known as "historical trauma."

Before the violent relocations, when Indigenous peoples were living under the natural laws that govern this place we know as the Salish Sea, they followed a much healthier and sustainable way of life. This is a time we remember during the Annual Treaty Day gathering at the smokehouse or the Salmon Homecoming Ceremony (which is practiced by most tribes along the sea and river systems in the region).

Natural law is a way of living in the world that is governed by nature and tied to our Native language, creation stories, songs, and ceremonies. Indigenous peoples from this place continue to conduct their lives according to an oral tradition comprised of stories and teachings that have been passed on from generation to generation in turn since the beginning of time. However, as we continue to adapt to current situations, new stories and knowledge are gained. It is because of this process and community longevity that Indigenous science is a precise and ancient guide to adaptation. As such, Indigenous science is a way of life attached to place and is constantly evolving.

The warnings have been issued and the evidence is clear: we face the greatest challenge ever put before humankind—global warming, a time of climate crisis and "life out of balance." Carrie Dann is a Western Shoshone elder and international leader on tribal sovereignty and human rights issues. The Western Shoshone tribal territory includes the Nevada Nuclear Test Site, an area considered the most bombed place on the planet. During the 2009 Indigenous Peoples Global Summit on Climate Change, Carrie stated, "Mother Earth has a fever. As a result of our human assault on her, she is sick and we must do what we can for the next generations to come."

Place-based People

Swinomish means "people by the water," and the Swinomish reservation is surrounded by water. The tribe is geographically located on the Salish Sea in the Pacific Northwest, approximately 70 miles north of Seattle. Throughout the Coast Salish territories, where numerous Indian villages are located, people have a saying that captures the traditional practice of shellfish harvesting: "When the tide is out, the table is set!" The peoples[1] who make up the present-day Swinomish Indian tribal community once had numerous thriving fishing villages; however, over time many of their cultural resources have been lost because of overharvesting, contamination, and development by non-Indians, as well as an interruption of cultural knowledge as a result of the process of colonization over time. Our cultural stories are tied to the water and land. A Samish creation story tells of the Maiden of Deception Pass:

> It was a time when the people were starving and without food. One day the village chief's daughter was walking along the shoreline and was spotted by the king of the sea, who fell in love with her beauty and grace. In exchange for her hand in marriage, the king promised the chief that he would leave an abundance of salmon so the tribe would never go hungry again.

This story is an example of the Coast Salish worldview, which connects us to a place. As farmers of the sea, the Swinomish are place-based people tied to the water and land. From the mountains to the Salish Sea, our people continue to hunt, gather, and prepare traditional foods today. Natural law teaches us to give reverence, have gratitude, and never waste those resources. The natural law is to take only what is needed and to leave the rest, and to offer thanks to Creator for pro-

[1] The Swinomish Indian tribal community is a political successor to several treaty-time bands and groups, including aboriginal Swinomish, Kikiallus, aboriginal Samish, and Lower Skagit.

viding our traditional foods. Today people can be seen obeying these laws during many community gatherings, including our annual First Salmon Ceremony. The teachings revealed in the Coast Salish people's 13 Moons Calendar (see Mahle & Tom, Part I), as it applies to the Salish Sea, are another illustration of the cultural practices of fishing, hunting, and gathering for our people. Naturally, because food was and is so important to the way of life of traditional peoples, many moon names reflected transitions—what was growing during a particular time of year, or what people were doing to secure food for the community. Many tribes (such as the Wsanec or Saanich) continue to reclaim and revitalize this ancient knowledge.

One of our honored Skokomish leaders was subiyay (Bruce Miller), who often spoke of the ancient story of the Tree People. This story says that the trees have much to teach us and describes their diversity and symbiotic nature with all life. Under the forest floor there is an intricate and vast root system that keeps the forest strong. The richness and symbolism of the worldview in this story cannot be captured in a summary; the teaching and worldview can only be sustained when they are shared in the old way—face-to-face, in the presence of a traditional or cultural teacher repeatedly over time—and when they are practiced throughout one's life.

While it is difficult to translate this story into text, it captures an important message for building alliances, communal strength, and diversity and appreciating the roles that each member has in the web of the whole community. Together we are stronger. As subiyay said,

> Trees were to develop the most powerful method of teaching, and the teaching doesn't utilize language, or words, because it's teaching by example. Which is the strongest form of teaching that can be established.
>
> Animals teach, by example, their young ones. Things that teach by example have the unique absence of lying. Language gives us the ability to become liars and develop false sets of values. Trees were placed on earth as an example of what true harmony and contribution is to the world. Their example was: they existed side by side from the beginning of time with no criticism of one another. (subiyay 2000)

It is vital that such teachings are not lost or forgotten. These teachings are our *Indigenous science* and are crucial to the survival of future generations because they offer guidance and instruction in how one should live and thrive in what has become a very complex world—"life out of balance."

Threats

Our Swinomish tribe is working hard to protect, restore, and revitalize what remains of our traditional way of life. As with most cultural resources under threat, the fish and shellfish loss we are experiencing is mostly due to overharvesting by non-Indians, habitat loss and degradation, and Indians' loss of access to their usual and accustomed fishing sites. For example, on the western side of March Point (just northwest of the Swinomish reservation on Padilla Bay), there is a stretch of beach approximately one mile long where my father and relatives used to gather clams. In 1873 President Ulysses S. Grant excluded March Point from the reservation when he issued an executive order defining its northern boundary. That land had been and continues to be claimed and settled by non-Indian people.

Then in the 1950s, following the closure of the many timber mills and canneries in Anacortes, the city northwest of the reservation, Shell Oil Company and Texaco Anacortes Refinery (also operated by Shell) opened operations on March Point. Later, Tesoro bought Shell Oil. Today tribal members are required to get permission from the oil companies that occupy this traditional place to gather clams and other seafood. This requirement, as well as the knowledge that the seafood is contaminated by refinery operations, discourages tribal members from exercising their treaty rights. Another deterrent is the March Point (or Whitmarsh) Landfill, which was operated from 1950 to 1973. It initially served as an unregulated dump and later as a county disposal area. The Washington Department of Ecology is now trying to clean up the area:

> As 2009 begins, work continues on projects that will clean up contaminated land and in-water sites in Skagit County's Fidalgo and Padilla bays…. The Washington Department of Ecology (Ecology) identified the bays as high-priority, "early-action" cleanup areas under the Puget Sound Initiative…. Along Padilla Bay, Ecology is working with Texaco Inc., Shell Oil Co., Skagit County, and the Washington Department of Natural Resources (DNR) to outline a cleanup of the old March Point Landfill…. Site samples show a variety of contaminants, including metals, polycyclic aromatic hydrocarbons (PAHs), dioxins and furans. (Washington Department of Ecology 2009)

The Swinomish Water Quality Department analyzed sediment and shellfish for heavy metals, polychlorinated biphenyls (PCBs), and polybrominated diphenyl ethers (PBDEs) in 2002 and 2004. Unfortunately, due in large part to limited access to beaches surrounding the Swinomish reservation, tribal members are squeezed onto curtailed harvesting areas, such as Lone Tree Point, a traditional gathering place for Swinomish. This too has resulted in reduced resources and harvests for tribal members. As a result, in recent years people had to buy the clams from Taylor Shellfish Farms, an outside source with operations at multiple locations in the Salish Sea, for an annual traditional Swinomish community clambake hosted by the tribe's Water Quality Department.

Encroachment: Fishing Wars

According to the Yelm History Project,

> *U.S. vs. Washington* (1974) became more popularly known as the "Boldt Decision" when a ruling was issued by Federal District Court Judge George Boldt. Judge Boldt took the view that *there must be real communication and cooperation* between Indian and non-Indian groups. Boldt then stated that he agreed with the interpretation of Barbara Lane, the anthropologist called by the tribes and the federal government to help explain the origins of the treaties and the traditional Indian fishing rights and practices. Dr. Lane had pointed out that the tribes in their original treaty negotiations had most likely believed that "usual and accustomed" grounds meant fishing as it always had been conducted. Judge Boldt realized that a major problem existed in the language used in the Medicine Creek Treaty of 1855. When the treaty was translated into a language that the Indians could understand (before they signed it), it was translated into a language called Chinook Jargon. This language is not really an Indian language at all…. It was not very useful for the precise legal language used in writing laws. (Yelm History Project 2010)

What is evident is that in the thirty years since the Boldt Decision, which recognize that tribes in the state of Washington have the right to 50 percent of the fish and shellfish harvest, tribal fisherfolk are unable to sustain a living and fishing has been severely diminished. This situation is in large part due to the loss of habitat and healthy ecosystems for fish and wildlife, as well as ever-changing climate conditions.

Toxic Contamination

Professor Rebecca Tsosie (2009) has said, "Sovereignty extends beyond the reservation…. It's the people's rela-

tionship to the land and water! Does it make economic sense to continue unsustainable economic models or create new renewable and sustainable economic models?" Scientists at both the Swinomish Water Quality and Air Quality departments have provided evidence of toxic contamination of shellfish in gathering areas in and around the Swinomish reservation. Tony Basabe, a Swinomish air quality scientist, has documented the poor air quality in the area because the tribe is downwind of the Tesoro and Texaco oil refineries. The tribe has also been affected by ocean acidification, which is caused by rising carbon dioxide emissions. The National Oceanic and Atmospheric Administration (NOAA) stated in a May 2008 report that:

> The oceans have absorbed about 50% of the carbon dioxide (CO_2) released from the burning of fossil fuels, resulting in chemical reactions that lower ocean pH. This has caused an increase in hydrogen ions (acidity) of about 30% since the start of the industrial age through a process known as "ocean acidification." A growing number of studies have demonstrated *adverse impacts* on marine organisms, including:
> The rate at which reef-building corals produce their skeletons decreases.
> The ability of marine algae and free-swimming zooplankton to maintain protective shells is reduced.
> The survival of larval marine species, including commercial fish and shellfish, is reduced. (NOAA 2008)

Further, the Intergovernmental Panel on Climate Change (IPCC) wrote in its *Climate Change 2007 Synthesis Report*:

> While the effects of observed ocean acidification on the marine biosphere are as yet undocumented, the progressive acidification of oceans is expected to have negative impacts on marine shell-forming organisms (e.g., corals) and their dependent species…. The resilience of many ecosystems is likely to be exceeded in this century by an unprecedented combination of climate change, associated disturbances (e.g., flooding, drought, wildfire, insects, ocean acidification) and other global change drivers (e.g., land use change, pollution, fragmentation of natural systems, overexploitation of resources). (IPCC 2007)

Challenges

Bottom-line thinking can sabotage and diminish a concerted community-based approach to organizing to prepare for climate change. Tribal communities are building capacity for economic development while simultaneously addressing a multitude of systemic challenges, including poverty, substance abuse, domestic violence, unemployment, soaring rates of high school dropouts, environmental disasters and degradation, and vast climate crisis impacts. Many tribal programs are understaffed and underfunded, and therefore cannot provide the professional development training needed so that staff members can run programs effectively.

This paradigm and the challenges that Indigenous people face leave communities disenfranchised and continually functioning in a "crisis management" mode. This situation can become costly over time as staff turnover mounts and educated tribal members choose to work away from their tribe because of a lack of opportunities for employment. It effectively leaves tribes vulnerable to hiring and engaging non-tribal members who may be unfamiliar with tribal culture and therefore lack cultural competency and sensitivity, tribal program experience, and tend to employ "top down" approaches to management. Because of these problems, when a prevention model of programming and community engagement is introduced, it is often competing for time, attention, and funding with other programs that may seem more urgent.

Moreover, many tribal governments in the United States are set up as federally recognized Indian Reorganization Act (IRA) systems that emulate the U.S. governmental structure and favor corporate management in leadership and business. As such, this model is hierarchical and based upon a business model with complex rules and regulations for delivering programs. Consequently, it lacks processes for *honorable community engagement*. When tribal programs are set up according to this model, community members often are fearful and distrustful, thus creating disconnects between departments and programs.

Climate Change

The Swinomish reservation is surrounded by water. In 2007 the Swinomish Tribal Senate passed a proclamation to address climate change impacts following a storm surge in 2006 that caused flooding in the towns of La Conner (adjacent to the reservation) and Shelter Bay (within the reservation boundaries), and on leased land on the western side of the reservation. A 2006 Washington Department of Ecology report found that Swinomish reservation lies in the second highest-risk area for sea level rise in the state.

As a result of this threat, tribes are mobilizing and our elders are speaking about the natural world and our roles and responsibilities in paying reverence to Mother Earth. The Affiliated Tribes of Northwest Indians (ATNI) is the regional body where resolutions affecting issues important to tribes are brought forward. Several climate change resolutions have been passed by this body in recent years. During the ATNI 2008 winter conference, the Community Alliance and Peacemaking Project (CAPP) submitted Resolution #08-30, "Support for the Northwest Indigenous Alliance Initiative on Community Wellness and Climate Change," which was subsequently adopted. The resolution calls for raising awareness and educating tribal youth about climate change impacts and paths for adaptation. The Swinomish tribe became a natural partner because of its Swinomish Climate Change Initiative (see the previous chapter) and establishment of a tribal community-interest group.

Education

Our elders encourage our young people to complete their school education so they are better prepared to think, succeed in their chosen career or vocation, and contribute to their tribal communities. Denny Hurtado (Skokomish), Washington State Indian Education Director, works tirelessly promoting a tribal sovereignty curriculum for grades K–12. He is seeking support and collaboration from the twenty-nine tribes in the state of Washington, Affiliated Tribes of Northwest Indians, Northwest Indian College, and public school districts' Indian education programs to promote the use of this course of study, which can be easily integrated into the public school curriculum throughout the state, so citizens can better understand the origins of Native rights and the continuity of tribal cultures.

A coalition of colleges and universities, including The Evergreen State College, Washington State University, and Northwest Indian College, hosted an education conference for the purpose of exploring the challenges and many ways of engaging the resources of our institutions to serve Native American students: The Pathways conference was "built around a new report on what more than forty Washington colleges and universities are doing to serve Native students. The program features best practices, model programs, and policies and practices to promote student access and student success." (Pathways for Native Students 2010)

During Justin Gillory's workshop, "Generational Resilience: Strategies for Reproducing Success in Native

College Students," he shared a five-phase resiliency model of wraparound services to enable our Native young people to succeed in education. This model is similar to the one our CAPP community-interest group strives to follow as we strengthen relationships among youth, elders, and tribal departments and programs. Our model is likened to a collective leadership model.

The Lummi CEDAR Project, in collaboration with the Kellogg Foundation, developed a collective leadership model called Organized Generations. Provided there is an interest, a partnership between CAPP and the Swinomish Youth and Education Programs could be established to develop a similar model for enabling youth to participate in leadership development as a cohort. The intention is to have a strong cadre of young leaders who will collaborate with the tribal environmental educators, community development planners, select tribal departments, the community at large, and the schools to continually hone and modify the climate change adaptation plans as needed. Implementing this program would require support from tribal leadership, appropriate funding, and the advocacy of concerned community members. Continual adaptation planning that addresses the increasing changes and impacts is an important way of maintaining our way of life in the face of this climate crisis.

Swinomish Climate Change Initiative

The Swinomish Office of Planning and Community Development has coordinated and managed the Swinomish Climate Change Initiative, and a key staff team consisting of the senior planner, associate planner, and environmental management coordinator provide primary direct services to the two-year project. A technical team consisting of the director and grants coordinator and staff from the Environmental Science, Environmental Policy, and Legal departments and the Skagit River Systems Cooperative (fisheries) assisted with the initial technical review of project work. An administrative team (consisting of a GIS specialist and an assistant) helped with the technical analysis, report preparation, project mailings and publication, and related administrative tasks.

The staff team was assisted by professional staff of the University of Washington Climate Impacts Group (CIG), who possess substantial and regionally recognized knowledge and expertise in the area of climate change science. CIG staff plays a crucial role in reviewing scientific data, reports, and project documents and advising on appropriateness, validity, and accuracy in the use of scientific data and information in the project.

The first step was to provide a scientific analysis of

the impacts, and the Swinomish Technical Assessment Impact Report was produced in 2009. Second, the tribe wrote a report in 2010 to explain the tribe's initial plans for adaptation to ongoing changes. The adaptation report is indeed a "living document" that will be revised and updated as necessary over time. It is evident that a strategy for making ongoing assessments and updates must be institutionalized and clearly described within all future reports and during each phase of the process.

An Honorable Community Engagement Strategy

Eight months into the tribe's assessment of climate change impacts, CAPP was asked to establish and support an effective outreach and community engagement process. CAPP is a Native non-profit organization specializing in alliance building, facilitation, peacemaking, and youth leadership development. Terms were negotiated and my services were contracted through CAPP to serve as the communications facilitator to the Swinomish Climate Change Initiative.

This work entailed establishing an honorable engagement process for the Swinomish community by raising awareness about climate crisis impacts on the tribal community, and it opened a pathway for community influence on policy and decisions about how the tribe will adapt, prepare, and deal with the accelerating effects of global warming.

CAPP's first effort was to form a Swinomish tribal community-interest group known as the Climate Change Education and Awareness Group (CCEAG). The group was established to assist with honorable community engagement planning and public awareness, and to continually provide input to the climate change initiative planning, policy development, and review. In this manner, the advisory group provided leadership, advice, and direction to the communications facilitator and the community at large.

The CCEAG efforts were inspired by the book *The Gathering of Wisdoms*, which was produced by the Swinomish Indian Tribal Community Mental Health Program and is in its second edition as of 2002. This book expresses the wisdom of tribal mental health and community wellness; it describes how to develop a behavioral health program in a tribal setting and how to enhance community resiliency through circles of support. In my opinion, *The Gathering of Wisdoms* illustrates an Indigenous worldview and model of a tribal system that encompasses human health, the natural world, cultural traditions, and systems or circles of support. CCEAG followed the book's example in developing community engagement, although on a smaller scale and with limited funding. At the 2009 retreat,

Caroline Edwards, the Swinomish CCEAG Youth Liaison, said of climate change, "We are becoming more aware through community education and witnessing the differences and changes. I feel Mother Earth more here in this place than anywhere else." This statement suggests that CCEAG is having an effect.

The tribal community-interest group served as a subcommittee of the tribe's Climate Change Initiative and included select tribal members who represented each of the respective families within the tribal community and who were linked to various tribal committees, groups, and sectors of the community—such as the Swinomish Youth Recreation and Prevention program, education, cultural groups, social services, health, elders, law enforcement, fisheries, tribal council leaders, and other key sectors or programs. The participation of tribal members was voluntary; however, in respect and recognition of cultural mores, non-staff tribal elders were gifted an honorarium in the amount of $40 for attending CCEAG meetings and events.

Each CCEAG member was encouraged to provide insights as to how the climate change initiative would intersect with their tribal practices, program, or sector. The intent was to create a pathway for community members to engage in one or more aspects of the initiative. For example, public health is one sector that will be affected by increases in greenhouse gases, industrial pollutants, and milder, longer growing seasons. Such changes could lead to an increase in respiratory disease, including higher asthma rates, and heat-related illnesses such as heat stroke and cardiac arrest. Sea level rise would cause flooding and disturb cultural and natural resources such as shellfish habitat.

Possible outcomes of CCEAG consultations include the following:

➤ Public awareness about climate change and ecological degradation of cultural and natural resources, marine life, and forests in the Swinomish community and surrounding communities.
➤ Community and youth participation in planning for adapting to changes.
➤ Engagement of youth in creating public education materials, running sustainable community projects, and using technology to create reports, stories, and films in various digital media.
➤ Opportunities for students to be educated about environmental planning specific to climate change adaptation.
➤ Collaboration and coordination between the tribal community and government departments in mitigating environmental disasters (i.e., flooding, fires,

diseases, public safety, forests, plants, marine life, and so forth).
➤ Collaboration and strengthened relationships between Swinomish and neighboring communities in Skagit County.
➤ Opportunities to influence policy on climate change and ecological degradation impacts, adaptation, and mitigation.

The CCEAG met regularly each month and participated in community-wide events to raise awareness about climate change, share information, and get the word out about future efforts for community education and empowerment. The group planned a series of community meetings and interviews that brought folks together to talk about the changes they've seen, hopes and concerns. *The voice of the community will help inform and guide plans* for adapting to changes that will occur.

After the technical report was completed in December 2009, the Communications Facilitator developed a fact sheet that summarized the impacts and translated the language of Western science in which the report was written into language that was meaningful and more understandable for tribal members and community members without scientific training. The fact sheet was then shared with Swinomish community members during many community gatherings.

"A Way Forward" Retreat

During a CCEAG retreat members were asked to brainstorm on the question, "What is the value of CCEAG to the Swinomish community?"

In summary, we found that there is a dire need to continue to do the following:

➤ Raise awareness within our community,
➤ Keep the community updated on the facts,
➤ Seek out intertribal opportunities for raising awareness and making partnerships, and
➤ Strengthen the relationships between youth and elders to share cultural teachings.

We must recapture the values and wisdom of the old ones (elders) and understand our deep connection to this place. We need to connect our youth to opportunities to learn Native science and Western science so they become well informed and can participate meaningfully and appropriately in the decision-making process. We need to keep our youth encouraged and invite them to help us to document how things have changed over time. This community work should be tied to the school curriculum or course incentives. Finally, and most importantly, we need to continue to help develop strategies for addressing the impacts of climate change and ecological degradation on fish, shellfish, forest, drinking water, and so forth. Any economic development and construction must be green and sustainable so as to lessen our carbon footprint, save money, and go back to the old ways of living. It's traditional to use everything and not waste. As Larry Campbell, a Swinomish elder who works in the Tribal Historical Preservation Office has said, "We must use things in moderation and use everything so there is no waste."

Taking Action

Musician, poet, and human rights activist Michael Franti writes music with a social or political message, including songs like "Hey World (Don't Give Up)" and "Nobody Right, No Body Wrong." In an interview, Franti spoke about one small change he made in an effort to reduce waste. This is an example of the kind of small effort our CCEAG members encourage tribal folks to put into practice. Below is an excerpt from his interview with Jeff Kant for *Planet Green*:

PG: What can…[we] do to unify people?

MF: The first thing is education. In terms of climate change, people have to know the facts, not just what they hear from politicians that are putting out a viewpoint to forward their argument, which is so tied to who's donating their money to them.… We all have to be as *well-informed* as we can possibly be, and be prepared to support businesses that are doing good practices and voting every day with our dollars and … be willing to make the small sacrifices that at the end of the day help a lot—and we feel better when we do them.

PG: So how do you try to educate people?

MF: The first thing is by example. We try to make our Power to the Peaceful festival and our tour as lean and green as we can. But one of the things that we did in our household is, about five years ago we decided that we weren't going to use any more plastic bags. It's something I think everybody can do.… We started carrying our own bags to the store.… I have a collection of reusable bags I keep in the back of my car. You really start to notice how often you go to a store and you buy some toothpaste and a roll of lip balm and you use this plastic bag to carry things to your car and into your house, and it's trash. (Michael Franti, interviewed by Jeff Kart, *Planet Green*, Discovery.com, March 22, 2010)

Following this model, CCEAG member Laura Kasayuli's daughter Ashley makes "upcycled" items to sell during community-wide events. Laura explained that "upcycled" crafts are artworks created from used clothes, linens, and so on. For example, an old T-shirt may be cut up and sewn into a new tote bag.

The challenge for CCEAG in moving forward beyond the Climate Change Initiative is to sustain the arduous and mindful task of translating the scientific information and emphasizing cultural sensitivity and meaningful approaches to education and awareness at the community level. In this manner we honor the teachings of our ancestors and uphold the natural law of protecting Mother Earth. As Diane Vendiola, a Swinomish elder and CCEAG member, put it, "The intention of CCEAG is to bring the people together and to follow the traditional teachings."

This process builds relationships and trust. Creating and sustaining good relationships, and continuing to partner with tribal departments, community programs, and sectors linked to outside organizations and institutions, such as the Coast Salish Sea Initiative, will open up more opportunities for sharing knowledge, building leadership skills, and developing technical resources.

Examples of these efforts include the following:

▶ Alliance-building beyond the tribe through grassroots mobilization: Shelter Bay/Swinomish Eco-Fair and Annual Shred Day, Swinomish Earth Enhancement Day, Skagit Beat the Heat, Tribal Canoe Journeys, and the La Conner Middle School Science Fair.

▶ Partnerships with institutions and organizations, including nDigiDreams, the Burke Museum, the Climate Impacts Group, the Evergreen State College's Northwest Indian Applied Research Institute, La Conner Middle School, Puget Sound Energy's It's Cool Campaign, Northern Cascade Institute's Climate Challenge Program, the Potlatch Fund, and so forth.

▶ Maintaining communication (to inform, educate, raise awareness, recognize, and promote) with the Swinomish *Kiyuuq^ws* (*Kee Yoks*) newspaper, the *Northwest Indian Fisheries Commission News*, the *Skagit Valley Herald*, KSVR Radio's Skagit Talks, Facebook, KANU *Native News*, and other outlets.

It is our hope is that the lessons learned through the efforts of CCEAG offer a model for youth leadership and development. An "organized generations" model of collective leadership, wherein youth complete a leadership development institute and these young leaders are then prepared to facilitate future community meetings and offer peer youth training events throughout the community and region, could be developed and implemented. However, the development of such a model largely relies upon adequate funding, qualified staffing, and appropriate support from the tribal leadership.

Another outcome, given support from the tribal leaders, would be to develop a strong cadre of young leaders who will work in partnership with the Swinomish community development planners, select tribal departments, CAPP, and the community at large to continually update and revise the climate change plans as needed. Continual planning for accelerated ecological degradation is a crucial process for maintaining our way of life in the face of this climate crisis.

Lastly, tribes must be mindful of other issues and solutions concerning the mitigation of climate crisis impacts, including the following:

► Controlling intellectual property rights and place-based stories and creating a curriculum for sharing important and appropriate cultural knowledge with tribal youth and others. One way that tribes can maintain control is through forming their own Tribal Information Review Board.

► Implementing sustainable tribal "transitions" in an effort to reduce waste and energy consumption. The transition should factor in sustainable economic development opportunities and models for clean energy.

► Strengthening institutional ties between tribes and organizations that are charting a new course—a clean, renewable, and reusable course. One Native organization making a significant shift from fossil fuels to clean energy production is Honor the Earth, which promotes renewable wind energy and sustainable projects in Native nations.

Essentially, tribes could create a vision of peace and wellness for our tribal communities and neighboring townships, thus opening a pathway toward preparedness and hope for our next generation of youth. We are the warriors of today; it is our responsibility as Indigenous people to remember where we come from and to do good things—for the well-being of ourselves, our families, our communities, and our nations. We must stay the course and focus on solutions! As the Indigenous Environmental Network declared in May 2010:

> Climate justice calls upon governments, corporations, and the peoples of the world to restore, revaluate, and strengthen the knowledge, wisdom, and ancestral practices of Indigenous Peoples, affirmed in our experiences and the proposal for "Living in a Good Way," recognizing Mother Earth as a living being with which we have an indivisible, interdependent, complementary, and spiritual relationship. (Indigenous Environmental Network 2010)

Resources

Community Alliance and Peacemaking Project: http://capp.web.officelive.com

Franti, Michael, interviewed by Jeff Kart, Planet Green, Discovery.com, March 22, 2010. http://video.planetgreen.discovery.com/feature/instrumental/michael-franti-spearhead-interview.html

Health Observatory: http://www.healthobservatory.org/library.cfm?RefID=72798

Honor the Earth: http://www.honorearth.org/

Indigenous Environmental Network. 2010. Four Principles for Climate Justice, May 4. http://www.ienearth.org

Institute for Agriculture and Trade Policy: http://www.iatp.org/

Inter-governmental Panel on Climate Change (IPCC). 2007. Climate Change 2007 Synthesis Report. Geneva: United Nations.

National Oceanic and Atmospheric Administration (NOAA). 2008. State of the Science Fact Sheet for Ocean Acidification. http://oceanservice.noaa.gov/education/yos/resource/01state_of_science.pdf

Pathways for Native Student Education. 2010. http://www.evergreen.edu/news/archive/2010/03/pathways.htm

Point Elliott Treaty. 1855.

http://washingtonhistoryonline.org/treatytrail/treaties/pdf/point-elliott-treaty.pdf

subiyay (Gerald Bruce Miller). 2000. Salmon Nation, Eco-Trust, May. http://www.salmonnation.com/voices/bruce_miller.html

Swinomish Climate Change Initiative: http://www.swinomish-nsn.gov/climate_change/about/about.html

Swinomish Planning and Community Development: http://www.swinomish.org/resources/planning.aspx

The 13 Moons of the Wsanec (Saanich People). JASON Project Curriculum Lesson. http://www.racerocks.com/racerock/firstnations/13moons/13moons.htm

Tsosie, Rebecca. Speech to the Salish Sea Gathering, 2009. A transcript is available from the Swinomish Planning Office.

U.S. Fish and Wildlife Service 2009. Endangered Species Bulletin (Summer). Washington Department of Ecology. 2009. News Release, January 7. http://www.fws.gov/endangered/bulletin/2009/bulletin_summer2009-all.pdf

Yelm History Project. 2010. Usual and Accustomed Places VI—The Boldt Decision, May 7. http://www.yelmhistoryproject.com/?p=928

Contacts

Ed Knight, Swinomish Planning and Community Development, 360-466-7280

Shelly Vendiola, CCEAG Communications Facilitator (Consultant), www.capp.web.officelive.com, msvendiola@gmail.com

GROUPS PRESS FOR TRIBE-FRIENDLY RENEWABLE ENERGY POLICIES

Rob Capriccioso

Indian Country Today, Jan. 23, 2009

Washington, DC – As more tribes explore and get involved in the renewable energy field, a network of tribal groups is asking President Barack Obama to support tribally owned and operated renewable energy projects, along with economic development initiatives that could reduce dependence on fossil fuels.

"The Obama economic stimulus plan that incorporates a green economy and green jobs portfolio must include provisions for access of these resources by our Native nations, our tribal education and training institutions and Native organizations and communities," according to a policy statement released jointly Dec. 17 by the Intertribal Council on Utility Policy, the Indigenous Environmental Network, the International Indian Treaty Council, and the Honor the Earth environmental group.

"When considering energy production, resource extraction, housing, and energy efficiency, it is essential that the incoming administration takes into account the disproportionate impacts of climate change and energy development on American Indian reservations and Alaska Native villages, and the potential for catalyzing green reservation economies."

The groups represent approximately 250 grassroots tribal organizations and tribes that want to ensure American Indian participation and prosperity in the green economy of the future.

The statement says that federal government subsidies for the nuclear, coal, gas, and oil industry should be rapidly phased out with a proportional ramp up of subsidies for renewable technologies and locally administered conservation and efficiency improvements.

Under current federal law, tribes are not directly entitled to credits provided to non-Native developers for renewable energy production. This has created a system where outside companies sometimes think twice about teaming with tribes on renewable energy projects, since, if they do so, the federal government does not allow for a full tax credit.

"Projects involving technologies like wind power could stand on their own if none of the energy sectors got [federal] subsidies or incentives, but there are already billions of dollars built into coal, gas, and coal subsidies," said Bob Gough, a leader with Intertribal COUP.

"To compete against them, renewable energy technologies require subsidies as well. You can't artificially keep the price of energy down, and then expect new kinds of technology to bear all the costs."

The groups are pressing for changes to subsidy laws to make them more tribe-friendly, and also say that any climate change legislation should not allocate funds for nuclear or clean coal technologies, as they believe those practices are often harmful to the Earth and to tribal interests.

The policy paper specifically asks that policymakers provide a renewable production refund for tribal projects that can't utilize current tax credits, as well as offer financial matching grants to capitalize renewable energy potential in tribal communities.

The organizations believe that a new crop of tribal renewable projects, which would be assisted by the legislative changes they seek, would provide dual benefits of low carbon power and green economic development where it is needed most.

Support for legislative action involving tribes and energy is based on the following research gathered by the groups:

► Tribal lands have an estimated 535 billion kWh/year of wind power generation potential.

► Tribal lands have an estimated 17,000 billion kWh/year of solar electricity generation potential, about 4.5 times the total U.S. annual generation.

► Investing in renewable energy creates more jobs per dollar invested than fossil fuel energy.

► Efficiency creates 21.5 jobs for every $1 million invested.

► The costs of fuel for wind and solar power can be projected into the future, providing a unique opportunity for stabilizing an energy intensive economy.

In sum, members of the tribal network believe that forward thinking energy and climate policy will have the ability to transform tribal and other rural economies, while also providing a basis for an overall economic recovery in the U.S.

Gough estimated that close to 100 tribes across the country have already assessed or are currently assessing the wind and/or solar energy resources that are available on their lands. Some of the tribes, including those in the Plains and Southwest regions, have found that their renewable energy resources rank among the most abundant in the U.S.

Tony Skrelunas, an America Indian program director with the Grand Canyon Trust environmental group, said that it will be important in 2009 for tribes to continue banding together to make their energy interests well-known to federal policymakers.

Skrelunas, a member of the Navajo Nation who used to manage the tribe's economic development operations, said many tribes are now savvy on energy issues, and have evolved to the point where they want strong federal policies put in place to help them harness their power. His group plans to help convene several tribal renewable energy players early in 2009 to focus on national strategies.

"There are a lot of issues that have to be worked out and laws that need to be clarified," Skrelunas said. "And the tribes have to be the ones championing this. The tribes have to be the ones going to Congress saying they want these laws changed."

Skrelunas said he is looking forward to what the Obama administration brings forward regarding tribal energy issues.

In terms of tribal economics, many energy experts say that renewable projects could create a more stable business model than, say, the development of casinos alone.

"One of the issues facing economic development with casinos is that you need to have a number of customers—you need to have a large population market to draw on, but that's not always the case for remote reservations," Gough noted.

"Whereas, it doesn't matter how many people want to use the electricity you're able to produce from a wind turbine in a rural area, you can serve thousands and thousands and thousands of customers from across a whole region."

Gough said he hopes to see dozens more tribal renewable energy projects up and running by the end of 2009.

A METHANE TO THEIR MADNESS:
Tribes and Farmers Come Together— Over Cow Manure

Lewis Kamb

Seattle Post-Intelligencer, April 22, 2003

Monroe, Wash.—Who knew the answers to so many problems lie in cow dung?

That the odorous piles could preserve farmland, save salmon, even build trust between traditional foes?

Through manure, an unlikely partnership has bloomed between Native Americans and farmers in Snohomish County—a partnership that is shattering a long-standing impasse through a common cause: harnessing the "green" power of methane.

In what they're calling a "historic agreement," the Tulalip Tribes have joined with local farmers, conservationists, and others to build a plant here that will convert cattle waste into energy.

The partners hope the new way of dealing with dung will help preserve the rural characteristics of the Skykomish Valley—and keep farmers farming and salmon spawning.

"I never knew how excited I could get over, you know, cow manure," said Kyle Taylor Lucas, government affairs liaison for the Tulalip Tribes. "But we really see this as a way to restore salmon runs and keep farms viable.

"And in terms of the long-term conflicts there have been between tribes and farmers, this partnership is absolutely unique."

Opposite camps

For years, tribes and farmers have battled over basic land-use issues.

Native Americans have complained that farming practices—including the tainting of groundwater, streams, and rivers along pastures by untreated manure—harm habitats for salmon upon which Indian fisheries have depended for decades.

Farmers have argued that continued restrictions on land—from Indian treaty rights and other limitations to farming along streams that traverse their fields—have strangled the economic options that keep them in business.

For the Tulalips and Snohomish County farmers, the dispute has meant years of standoff.

"Tribes and farmers have stayed in their opposing corners," Lucas said, "and no one talked."

Dairy cattle waste that previously would have contaminated salmon habitat now is used at Qualco Energy to produce methane energy.

But a funny thing happened about two years ago, when Tulalip Tribes Council Chairman Herman Williams Jr. realized that Indians and farmers have "so much more in common than not," Lucas said.

It happened during a celebration of farmer Dale Reiner's work to restore salmon habitat along the Haskell Slough that cuts through his 300-acre ranch near Monroe.

"He said some things that hadn't been said before," Reiner recalled. "He actually gave recognition to farmers, and that's a remarkable thing for a tribal member to do."

A series of informal conversations brought farmers and tribe members together in various living rooms. The conversations grew, ideas solidified, and common ground was reached.

"The Tulalips came to us with a pretty simple philosophy: 'We believe cows would be better in these valleys than condos,'" said Andy Werkhoven, a fourth-generation Skykomish Valley dairy farmer. "We, as farmers, couldn't agree more."

Soon, entities with a stake in rural Snohomish County—the Lower Skykomish River Habitat Conservation, Northwest Chinook Recovery, and the Washington State Dairy Federation, among them—joined the discussions.

The groups quietly hatched a strategy, then took their plans to Washington state's congressional delegation and the U.S. Department of Agriculture in Washington, D.C.

Earlier this month, the partners signed a formal agreement—a deal forged through cow manure.

Methane from manure

The plan involves constructing an anaerobic "digester" plant that pumps manure from local dairies. During a conversion process, methane is removed from the dung and used to fuel generators that produce electricity. The manure is broken down into a cleaner, eco-friendly grade of fertilizer and an effluent purged of the stench that permeates cow pies and cow towns alike.

The electricity can be sold on the market. Farmers can sell or use the fertilizer without fear that it will harm groundwater, streams, or salmon. And with a better way to deal with manure, farmers can expand their herds to better compete.

All of it, the partners hope, will mean that farming in the Skykomish Valley will remain economically viable, keeping farmers in business and keeping their land out of builders' hands.

They believe, as well, that by staving off creeping urbanization, Skykomish Valley salmon streams can be preserved—a matter of strong interest to tribes that have seen their salmon take dwindle over the years.

"If you lose the farms, you lose the fish," said John Sayre, director of Northwest Chinook Recovery, a member of the partnership. Armed with a recent $250,000 federal grant from the Department of Energy, Tulalips and local farmers are considering a site for the project at a state-owned "honor farm" for inmates near the Monroe State Reformatory.

No longer in use, the honor farm is expected to be placed on the state's surplus properties roll. It lies near four dairies that might fuel the digester. And as a former working dairy, much of the infrastructure needed for the project already is in place.

The Tulalips have asked the Legislature to hold the property from sale while the tribe uses its grant to study the project's feasibility and environmental issues, Lucas said. Tulalips, farmers, and conservationists traveled to Chino, Calif., last week for a two-day fact-finding mission at a biogas plant that began operating a year ago.

The Chino plant, funded entirely by state and federal grants, processes 225 tons of manure each day from ten nearby dairies, generating 500 kilowatts of electricity daily that is used to power turbines that remove salt from area groundwater.

"There are still lots of questions about this," said Werkhoven, whose farm is among the four that would fuel the prospective plant. "But we want to see if this can work to keep farmers in Western Washington."

If the project pans out, Tulalip leaders hope it will serve as a model for tribes and farmers throughout Washington. But even if it doesn't, the stalemate between traditional opponents has ended.

Already, Tulalips and farmers are working on other projects together, including one to identify salmon habitat they can preserve through land purchases and restoration.

"We've been at loggerheads for too long," Lucas said. "Who knew manure could bring us together?"

FINDING COMMON GROUND:
Qualco Biogas Project Brings Together Farmers, Natives while Helping the Salmon

James Careless

Manure Manager, September/October 2009

Historically, dairy farmers and members of the Tulalip Tribes in Washington's Tualco Valley have been at loggerheads. The farmers are pushing ahead with herd growth, thus increasing the risk of manure run-off into the Snohomish River.

This run-off could seriously hurt the river's ability to support salmon, an important resource for the Tulalip Tribes. Add local environmentalists, who are trying to preserve the Snohomish and the stage has been set for confrontation.

So why are these groups now working together, when they used to be fighting? One word: Biogas.

In a creative solution to the potential problem, local farmers belonging to the Sno/Sky Agricultural Alliance, Native Americans from the Tulalip Tribe, and a representative from the Northwest Chinook Recovery environmental group have banded together to form Qualco Energy. Qualco is a nonprofit group that runs a biogas electricity generator in the area. Currently, Qualco's biodigester is handling manure from 1,100 local cows, with capacity for double that, and 450 kilowatt/hours of electricity is being generated by Qualco regularly since it began generating in December 2008. That's enough to power 300 homes.

"Qualco Energy shows what can be done when people come together to solve problems to everyone's benefit," says Dale Reiner, Qualco's president. "It is truly a win-win for everyone affected—including the fish."

When he's not running Qualco, Reiner is managing his dairy herd on his 300-acre farm. It's been in the family for three generations, with frontage on the Snohomish River.

The Qualco Energy biogas plant is a nonprofit partnership of the Tulalip Tribes, Northwest Chinook Recovery, and the Sno/Sky Agricultural Alliance.

"Eight years ago, my brother and I had finished doing some waterfront restoration on our property, to help the local salmon population," he tells *Manure Manager*. "The fact that the Tulalip Tribes were doing the same kind of work got us to thinking: What if we could find a way to work together to restore the salmon, and deal with the potential run-off problem at the same time?"

Reiner kicked the idea around with local dairy farmers. They liked it, and he decided to approach the Tulalip Tribes. "We had to keep our meetings pretty hush-hush," Reiner says. "In those days, neither side trusted each other."

As people talked with each other, the distrust eased. Eventually the Sno/Sky Agricultural Alliance, Tulalip Tribes, and Northwest Chinook Recovery agreed to establish a biodigester in the area.

The suggested site was the former prison dairy known as the Monroe Honor Farm. Operated by the State of Washington from 1929 until 2001, the 277-acre farm is located close to local dairy farms, yet far enough away from subdivisions and other forms of housing that running a biogas facility wouldn't disturb anyone.

Working together, the three member groups of Qualco Energy received $250,000 from the 2003 Agriculture Appropriations bill to fund a feasibility study into the biogas facility. Once it was completed and their concept had been approved, Qualco had the basis to seek funding from the federal Biomass Research and Development program. As a result, a USDA grant amounting to $500,000 was received and used along with more than $2.9 million in acquired loans to fund the biogas project. Lobbying at the state level persuaded the legislature to give the farm to the Qualco Energy group. From that point, work on the biodigester began until it started operation late last year.

"It took a lot of hard work on everybody's part to get to this point, but we did it," says Reiner.

The System

The heart of Qualco Energy's biodigester is a large concrete digester tank. Located at the Monroe Honor Farm, the concrete covered tank is 198 feet long, 80 feet wide, and 17 feet deep. It has enough capacity to process waste from 2,200 cows at a time.

….The anaerobic bacteria within the manure grow within this environment, digesting the most toxic elements of the manure, releasing methane gas as a byproduct. This methane gas is captured from the tank, and then used to run a generator that produces electricity that is fed into the main grid. Meanwhile, the remaining processed manure is nearly odor-free. It has become nutrient-rich compost that is perfect for gardens and farms (as is the grey water left from the process).

"This is truly advanced recycling," Reiner says. "What used to be left in the fields to decompose—with a potential to overload the local water table with excess nutrients—is now creating clean electricity and natural fertilizer that lacks the pungent manure smell that people complain about. Meanwhile, the river is protected from the possibility of receiving excess dairy nutrients. This helps protect not just salmon, but the entire ecosystem."

Challenges

Currently, Qualco Energy's biodigester is operating at half capacity, with manure only coming from the nearby Werkhoven Dairy Farm. "Two farms have built pipelines to our site," says Reiner. "The first, owned by Werkhoven Dairy, is in operation. The second hasn't started production yet."

Why haven't other farmers started hauling manure to Qualco, at the very least? "It is a problem of economics," says Dale Reiner. "They want to use our facility, but they are having trouble affording to. Milk prices are down, while fuel prices are up. Combined, they make it expensive to haul manure to our site."

Even when the biodigester gets up to full capacity, other units will have to be built to make a real dent on local manure production. This will likely require more government assistance. However, given how completely biodigestion can reduce manure run-off, such funding should be politically popular with farmers, natives and environmentalists….

Speaking in broader terms, Reiner cites biodigestion as a technological solution that can reconcile the needs of farmers, natives, and environmentalists. "Qualco Energy is about more than dealing with manure constructively, although it is a big step in that direction," he says. "It is about bringing together conflicting interests in a positive way that really makes a difference. This is why the members of our company—the Sno/Sky Agricultural Alliance, the Tulalip Tribes, and Northwest Chinook Recovery—are so committed to this solution, and willing to keep at it until we have realized our collective dream. We all want farming to succeed here, because otherwise the land might end up being overtaken by condos. We all want the salmon to thrive, and we want to tackle the energy shortage in a clean, green manner."

"Biogas and biodigestion does all of these things," he concludes. "Everyone truly does win with this approach."

Qualco Energy can be found on the web at http://www.qualcoenergy.com

PART IV. POSSIBLE PATHS

POSSIBLE PATHS

The climate crisis poses a threat to Indigenous peoples and also an opportunity. For decades, forward-looking tribal members and leaders have tried to protect Native culture and language, bring back endangered species and traditional lifeways, and involve youth to practice their cultures. With the climate crisis, these changes have taken on a new urgency. Not only is accelerating these changes necessary for a decent life in Indigenous communities, but they are also increasingly necessary for the survival of the people.

Tribes, First Nations, and other Indigenous communities can lead the way with innovative solutions and provide models for non-Native neighbors. Rural non-Native communities are seeing their resources depleted, their kids leaving for jobs in the city, and their mom-and-pop stores replaced by corporate chains. With unresponsive governments, these communities can begin to learn from their Native neighbors and join forces to plan for the changing climatic conditions and more frequent emergencies. Indigenous sovereignty is not only healing for Native peoples but can also provide commonsense examples to non-Native peoples of a more healthy direction in relating to the Earth and fellow human beings.

Indigenous self-determination is not a matter of going backward to reclaim museum-preserved Native cultures, but a matter of applying traditional knowledge to solve twenty-first-century problems. Systems of Indigenous knowledge are not quaint folklore but ways of looking at the world that have proven more resilient and sustainable than the Western system, which brought about climate change in the first place. The Yuchi Muscogee educator Daniel R. Wildcat wrote in his 2009 book *Red Alert!: Saving the Planet with Indigenous Knowledge*:

> Saving the planet with indigenous knowledges will require the development of indigenuity—the ability to draw on insights found deep in tribal cultures—not only in North America, but around the world, and apply them in innovative ways to practical problems we face in our everyday lives. This will be crucial in the next several decades.
>
> Once we rid ourselves of the incredibly powerful but abstract notions of a linear temporal universal history, and its attendant conceptions of progress and civilization, the proposal to save

the earth with indigenous knowledges will be seen as a very sane suggestion.

Hopefully, modern humankind has reached a place and time where we can recognize that we must look beyond our notions of civilization and its presumed cultural superiority, and look outside...for solutions to problems that...our ideas of civilization have created.

Tribal water sampling at the Port Gamble S'Klallam community in Washington's Kitsap Peninsula.

KAUA E MANGERE—DO NOT BE IDLE

Maori Responses in a Time of Climate Change

Ata Brett Stephenson

Ata Brett Stephenson is of Te Kapotai, Ngapuhi, from Aotearoa/New Zealand.

Editors' note: This paper was delivered at the Leadership Conference of the United League of Indigenous Nations, Sep. 29–Oct. 1, 2008, at the National Museum of the American Indian, Smithsonian Institution, Washington, DC.

The word is out—people and communities have begun to embrace a deeper sense of acceptance and acknowledgement for the role of human impacts in climate change. Our debates can at last shift from the analysis of dramatic increases in atmospheric greenhouse gases (carbon dioxide, methane, and nitrous oxide) during the last 150 years of industrialization and its close coupling with surface temperatures, consistently above normal averages since the 1980s. The patterns of warming on a global basis have outpaced sensible, predictive trends that would merely reflect a moment of naturally occurring interglacial warming.

In New Zealand the National Institute of Water and Atmospheric Research (NIWA) has observed a warming of mean surface temperatures of 0.4C° since 1950 and a decrease in cold night or frost conditions by ten to twenty days a year (TVNZ News, Nov. 2007). The emerging climate change impacts for New Zealand are predicted to include an increase in the severity and frequency of storms and coastal flooding by 2050, an annual sea level rise of 5 mm per year, and average surface temperature increase up to 3.5°C by the year 2080. A similar set of climate-related projections appear in a 2008 Climate Change Statement from the Royal Society of New Zealand (RSNZ 2008).

While we recognize aspirations in human progress and, for some, the associated wealth and growth in national economies, it has come by way of accompanying resource losses in natural forests through burning and the consumption of carbon stores in coal and fuel oil. In a global context the heavily industrialized nations have historically resorted to the politics of blame-the-victim in the greenhouse gas emission debates by directing public attention to the constant land clearing by forest burning in tropical Amazonia and Southeast Asia. Equally, and perhaps again because of politics,

little has been made globally of the contemporaneous burning losses in northern boreal forests, which have more than doubled since the 1970s. Our inseparable human appetites for industrialization and consumerism are easily recognized sources of the problem. It should not go unacknowledged, however, that the mere combination of our global population growth and the limits on living space has required significant increases in land clearing, land use, and agricultural production, all of which contribute to greenhouse gas emissions and give it new qualities.

For some time now it has been part of the New Zealand national psyche to represent ourselves with a clean, green image—to represent our home as a place where the natural environment has been least affected by human habitation (Campbell 1999). This was the context for the New Zealand government's early adoption of the Kyoto Protocol on greenhouse gas emissions (see, for example, Ministry of Agriculture and Forestry 2004) in 2002, a move that was to cause increasing difficulties. It was only subsequent research that showed that high methane gas emissions from ruminant stock (such as sheep), favored by intensive pastoral production, could not be offset by total photosynthesis availability from forests and grasslands. As a matter of domestic climate change policy, the government chose to investigate and offer corrective measures by way of incentives, emission trading, and voluntary measures as a means "to achieve a permanent downward path for total gross [greenhouse] emissions by 2012" (Bosselmann, Fuller, and Salinger 2002).

Permanent Forest Sink Initiative

The photosynthetic capabilities of forest landscapes are significant in their ability to remove carbon dioxide from the atmosphere, and the Kyoto Protocol recognizes this as a mitigating factor against human-induced climate change. Carbon sink capability was recognized in the Kyoto Protocol in relation to the increase of carbon dioxide stored in a given area of forest between 2008 and 2012. As a result, a Permanent Forest Sink Initiative has been established in New Zealand (Ministry of Agriculture and Forestry 2004) and will enable qualifying landowners of young, post-1990s forest to benefit financially from emissions trading, a cornerstone to our climate change legislation, which was enacted in September 2008. While the key to the trading of an emissions unit is Crown (government) verification of the amount of increased carbon dioxide stored in the forest, and even though it is a dimension that will change as trees mature, it will in subsequent time periods (even beyond 2012, if allowed) be possible to harvest these forests on a continuous canopy basis. This term means that a selection of trees, about 20 percent, may be cut, providing sufficient canopy remains to cover ground space. The government has not yet specified how it will verify the changing carbon sink capacities of growing trees.

Conceptually, the carbon sink initiative would be of some advantage to Indigenous Maori who own undeveloped land but cannot afford to undertake its conversion to productive pasture. However, the history of British settlement since the signing of the 1840 Treaty of Waitangi (this is the foundation document of agreement with Maori chiefs to establish a British Crown Colony in Aotearoa/New Zealand) has generally not supported Maori interests in and values concerning land and forest. After 1840, what had been significant stands of large native trees under Maori control were clear-cut for settler farms or logged for economic returns—little of which was of any benefit to a Maori community. These native tree species include, in northern forests, the dominant Kauri (*Agathis australis*) and mixed groups of endemic podocarps such as Rimu (*Dacrydium cupressinum*), Miro (*Prumnopitys ferruginea*), Totara (*Podocarpus totara*), Tanekaha (*Phyllocladus trichomanoides*), and Kahikatea (*Dacrycarpus dacrydioides*).

Stripped lands were reduced to "marginal land" once any space suitable for agricultural use had been "claimed" and occupied by colonial settlements. From that time, economically unviable marginal spaces (those with mixed vegetation stands of regenerating native species and introduced weeds) were left to recover and often by way of confiscation to become Crown land (owned by the government).

Some of this "marginal" land has returned to Maori control thanks to claims against the Crown brought before the Waitangi Tribunal, to which Maori *iwi* [tribes] apply for control over areas they have lost since the 1840 Treaty. Waitangi Tribunal claims take the style of tribal research papers such as "Te Runanga o Ngati Awa" (1994) or as reports on evidence presented to the Crown in tribal applications for a redress of historical grievances such as that given in the *Muriwhenua Land Report* (Waitangi Tribunal Report 1997). In these instances, the presentation of historical information is highly significant in that it stems from culturally held sources (M. Stephenson 2008) and employs analytical

pathways akin to those presented by Smith (1997) in describing *Kaupapa Maori* theory.

When the Crown has been fully persuaded and it has the legal capacity to return forest and land to Maori claimants under redress claims, it has often done so. Thus in recent times Maori have become "new" owners and managers of existing forests, but paradoxically the vast majority had been mostly planted prior to the 1990s or were relics of Indigenous forest, and therefore these forests do not qualify under the permanent sink initiative for emissions trading. As Mead (2003) explains, the ancestral and living components of Maori social structure are fundamentally based on relationships to land, including land recovered through Crown treaty settlements.

At a late 2000s *marae* [Maori community center] debate, the issue of forest vegetation within the domain of our Te Kapotai *hapu* [subtribe] was re-examined. Tribal leader Hautautari Hereora (personal communication, June 10, 2008) explained that our people had resisted a government plan in the 1980s to replant a previously milled forest relic and its margins with the exotic Monterey pine (*Pinus radiata*), a species that has become the hallmark of commercial forestry in New Zealand. Of considerable significance to Maori leaders was the character of forest biodiversity, in our case the natural re-emergence of forest podocarps (Kauri, Rimu, Totara, and Kaihikatea), and its ecological importance for native habitat, wetland, and river protection. Our history and experience gave us sufficient reasons to reject government proposals, even though the government had suggested that Maori would benefit in long-term financial gains after the redevelopment of that land.

Of equal interest in the current debate over planting pine as a carbon sink was the situation of adjacent subtribes. Those which had undertaken exotic tree planting were finding twenty-five to thirty years later that the monocultural *Pinus radiata* planting had brought unacceptable outcomes in the shape of losses to cultural resources and water catchment, as well as downstream alterations to stream flow and aquatic species. Additionally, the pine management program required repeated spraying against fungal infections, leading to the accumulation of chemical residues that contaminated ground areas and waterways to the point of bringing further losses in a once biodiverse landscape. Members of those *hapu* now have to visit "our" forest to continue gathering resources, particularly plant materials for *rongoa* [pharmaceutical uses] and *raranga* [weaving].

Water

Water has come to the forefront in our thinking, and in recent times of unprecedented climatic extremes there are few who have not experienced situations of drought or flood in New Zealand. In 2008, after double low pressure storms, a week of rainfall in northern/central regions of the North Island caused flooding of the Waipa and Waikato river watersheds, inundating farm pastures that had only just been restored after a prolonged summer drought (TVNZ News, Aug. 4, 2008). To add to the problem, the principal water storage reservoir, Lake Taupo, feeding the Waikato River had nearly reached its maximum operating level, and its overflow into a downstream system of hydro storage dams could not be curtailed. It should not go unnoticed that the river, its floodplains, and its wetlands constitute a water systems network which should not have become dismembered by embankments, drainage, roadways, and human settlement under the guise that it can be sensibly "controlled."

Like other Indigenous people, Maori traditionally recognize the significance of landscape and because of that have a philosophy that they "belong to" the physical and biological world rather than "own" it. We know the mood of the river and behave accordingly; our expecta-

tion is to make use of existing circumstances. As previously indicated (B. A. Stephenson 2006), the centrality of the marae to Maori communities is absolute in terms of cultural practice, learning, and development. In the local context the issue of climate change preparedness forms part of the cultural dimension. *Ko Waikare te awa*—Indigenous Maori commonly speak of the rivers as part of their landscape, marking boundaries and access routes to tribal territory and providing water to living organisms. The river denotes a sense of spiritual strength that underpins our identity and well-being. Floodplains and wetlands associated with the river have always been held in high regard, as they were the source of significant food and plant resources and the location of faunal nurseries and nesting sites. Significantly, they also contain a region of transitional surface water that can mitigate the effects of drought. Wetlands also help to clean water because the river's waterborne sediment load is deposited there and altered by fungal and bacterial action before it is released.

Because of their tradition, Maori are considerably guarded in response to an emerging debate on Crown agency in the management and control of our waterways. To some extent government policy is increasingly focused on halting the decline of water quality, particularly in mitigating the effects of intensive land use and urban runoff. But there are concerns, equally, for flood protection and soil conservation in response to

Waikare River in flood.

the weather extremes caused by climate change. In this context Maori would contend that financial reward and consumerism have played too large a role in societal acceptance of "risk factors" in environmental management. The Ministry for the Environment reported on its website in 2008 on a number of strategies that are progressing with government approval under a *Sustainable Water Programme of Action*. What is largely absent from these debates is any clear message about impacts on natural ecosystems and the allocation of water for organismic life.

A "sustainability principle" (Williams and Nolan, 1997) governing water allocation lies within the 1991 Resource Management Act (RMA), but people disagree about its meaning, interpretation, and implementation, leading to conflicting outcomes for contemporary Maori (Harmsworth 2002). The standards and boundaries for community care can be challenged, mitigated, or ignored—the consistency of a particular moral or ethical response is questionably superficial. Our notion of "sustainability" becomes even more confused when combined with "words like growth, management, use, economy, society," because it brings "a wide range of interpretations, many of which are contradictory" (Williams and Nolan 1997). Regional councils, which by and large have a responsibility for the management of RMA issues, are also charged with taking into account Maori advocacy in preparing and formulating action on district plans. As part of future planning, a Cultural Health Index, utilizing the knowledge of tribal leaders and Maori communities, has been established for water assessment (Tipa and Teirney 2003). In this, traditional knowledge and experience become excellent guidelines to modern-day management practices, and Maori informants and the community share in the responsibility. In a *Te Tepu* (a popular TV program) roundtable debate between tribal informants on Maori Television (Sep. 8, 2008), two interesting points were offered on issues of heavy rainfall and water:

▶ "The *mana* [cultural influence] of the water remains under Maori community control, not with the environmentalist or a government."

▶ "When it rains, wheels are no use but we have our horses. They know how—and we can always walk."

Both remarks indicate that, particularly for rural Maori, traditional views and values are to some degree a normal part of their lifestyle. Like many others, the community at Waikare has addressed the issue of modern facilities for water supply and storage. The Waikare River and numerous associated aquifers that emerge from an otherwise well saturated groundwater system

provide access to natural water sources. The name *Waikare* [trickling waters] is significant in our ancient repertoire of place names (Hancy 2007). In addition, we hold some reassurances about the catchment in the face of the increasing drought circumstances predicted for Northland and East Coastal lands. One key part of the catchment is the Russell State Forest, which is generally inaccessible to the public; any road or walkway is restricted to its fringes (Howard Hereora, personal communication, August 25, 2008). It has been a little more than sixty years since logging ceased here, and plant recovery within this relic forest creates warm, moist microclimate circumstances with high levels of evapotranspiration. There is also sufficient elevation in topography created by the five principal mountain peaks that are the cornerstones of Waikare. These, together with their forest cover, create the circumstances for condensation fog and cloud formation. In the foreseeable future this catchment is likely to remain intact. Any proposals for change would be vigorously resisted, because the catchment is a source of our survival. In the alternative climatic extreme, flooding, Te Kapotai is also well accustomed to short-phase, high flood levels in the Waikare River and its marginal floodplain. The sensible management of people and property protection is part of the lifestyle here.

While global and national surges in demands for natural resources are generally anticipated, the deeper concerns over who will manage and control these resources are less prevalent for water than for hydrocarbons, productive farming, and homes. Historically

Native kauri endemic forest in Aotearoa / New Zealand, threatened by development and diseases.

we have all adapted, or constructed, a degree of water self-sufficiency reliant on "free" and available rainfall. Climate change will rapidly and severely affect the understanding of "sufficiency" and "adequate storage." Our predictive bases for water allocation are proving unsustainable, and various forms of restriction are already operating for some communities. In New Zealand a high degree of competition has been emerging among the applicants for groundwater uses and river diversions associated with farm irrigation, hydro storage, public/tourist recreational space, and the conservation of natural resources. For urban populations alone, the current rates of demand in water use almost outstrip the capacity to plan and construct new points of collection and storage.

A related debate about the human ethical responsibility for water should also underpin the planning and management model. Discussions of UNESCO's Water Ethics Working Group set some useful parameters in considering the trade-off among interests in water allocation (Acreman 2001). While in a generic sense a principle of conserving biodiversity has become embedded within national RMA legislation, our attempts to implement the principle are persistently dwarfed by competing human wants and needs. More is required on the science of "minimum flow"—the estimate that would determine the water allocation to an ecosystem. This problem is additionally dependent on the human recognition of other organisms, species' needs for water to complete reproductive stages in a life cycle, and the extent of water sufficiency for community biodiversity in an ecosystem (Acreman 2001). From the position of an ecosystem, water allocation has almost become entirely dependent on the ill-equipped human species to make such decisions.

Ecosystems are unfailing in their response to human intervention. What is often misunderstood or ignored by modern societies generally, but commonly acknowledged by Indigenous peoples, is the degree of reciprocity that underpins the relationship. Those closely associated with the science of environmental management recognize that applied biodiversity involves other beneficial species. The quality of water can be improved through treatment of waste and sewage water. An additional use of these wastes includes the recovery of sediments, which is fundamental to good soil formation and instrumental in the reduction of pollution. Managed water systems are also essential to irrigation, hydro storage, and aquaculture. Management of storm water and floods takes similar pathways, but the excess water may be detrimental to existing ecologies or, if redirected into other channels, affect water quality

and quantity in other communities. Because of a commitment to reciprocity, Maori generally favor a natural approach with less of a "need to control" as a means to resolve these issues.

Seed Banking

One of the current directions in environmental guardianship that arises out of climate change preparedness is seed banking. We cannot underestimate the potential value of seed banking as a device to protect existing species or enhance new crop varieties in a world of climate change. The New Zealand Plant Conservation Network (managed by the Crown) recognizes seed conservation as an insurance policy for endangered Indigenous plant species (Ministry for Agriculture and Forest Research 2007). The Department of Conservation announced in 2008 that a native seed bank in Rangitane/Ngati Kuri Maori territory has received the first seeds collected from the herb *Atriplex hollowayii*, which grows on the coastal dunes of Northland and is endangered nationwide. Some commercial or economic interests, however, will argue that seeds should be selected with a consistent phenotypic outcome in mind. This perspective holds little regard for the diverse genotypic opportunities of a gene pool. Clearly, this preference for genetic modification has characterized some of the commercially sponsored research programs, which are at odds with genotypic diversity, as evidenced in an announcement of the successful production of "superior carbon-capture" pine trees from genetically modified seeds extracted from immature pine cones (NZ Yahoo News, Aug. 9, 2008).

The agonizing alternative to extinction is to entomb or isolate seed materials from their respective gene pools, frozen from evolution. Only through human intervention (such as farming) can these materials be used in their active context. Instead, the genetic information of seeds is subject, like water, to the imponderables of human decision making, and more than likely governed by social, political, or economic demands that have little relevance to species and the ecology of the natural environment. Without questioning the disparities between rhetoric and reality in social decision making, it nonetheless cannot be suggested that in a future time of crop failure our human behaviors will be any less confrontational. Moreover, access to seed banks may be restrictive, and the extraction and transfer of genetic materials for commercial food crops will more than likely follow what has become the general practice in genetic modification—the releases of modified seeds or plants that will not produce viable offspring.

Practices not unlike seed banking are widespread amongst Pacific peoples in cultivating vegetative root crops like *kumara* [sweet potato] (Yen 1963) and taro (Ferdon 1988). What is insurmountably different between "traditional" and modern seed banking, however, is the notion of seed isolation from all other means to life by *kaitiaki* [guardians], from other environmental sources, from soil landscapes and the physical world. In "traditional" and even contemporary Indigenous practice, no seed or plant can be left on a bare soil with the blind expectation that it would simply perform as a food source. There is an underlying interconnectiveness, shaped by knowledge and experience, inclusive of spiritual and ecological factors, that determines the outcomes of crop production. Isolation would bring ecological separation from both the soil character essential to growth and the evolved interrelationship and interdependence with other organisms.

In the transfer and outcropping of plants essential to conservation and local resource supply, wetlands were used by Maori as a plant nursery, and without care and succession of community structure a plant would simply behave as a "mongrel," without human care or management (Howard Hereora, personal communication, no date). Like many other tribal areas, Te Kapotai maintained extensive gardens for immediate food crops and as a bank for future generations (Hancy 1970). The Indigenous question for modern isolated seed banks is whether those seeds can survive in a later environment with which the species has lost its connection.

Climate Change Preparedness

Without doubt, rapid and unpredictable climate change, along with historically determined and survivable risk limits, will continue to be challenging for centuries after high levels of greenhouse gases are stabilized or reduced. The time-lag relationship among atmospheric composition, shifts in solar radiation effects, and any corrective reversal of deep oceanic circulation is a major factor in this crisis. Quite understandably, much of the present focus has been concentrated on the effects of climate change on the human quality of life, but since we have an absolute reliance on other organismic life and non-living resources, we cannot sensibly examine a future in anthropomorphic isolation.

Climate change is predicted to operate differently in particular regions from the pattern national and global trends would suggest. Thus regional natural resource wealth and community interrelationships will to some extent determine how well populations will manage and survive. But in a generic scenario on the global scale, the infrastructures for relief and redevelopment will be quite inadequate. Current evidence suggests that in large-scale, repetitive disasters the availability

of national and global aid and the nature of voluntary assistance can be determined by the culture, religion, and social class of the displaced people.

For Indigenous people, generally, it is the community that holds and shapes people. The community skill base and social organization can be promptly initiated and trusted because community members are known or genealogically linked to each other. For Maori a high cultural value is placed on *manakitanga* [hosting and care giving] and an associated responsibility to provide without limitation for those who seek to use a community place like *marae* (Mead 2003). Management committees of most marae have well-established plans for, among other things, coordinated relief and care in emergency situations. Community strengths, rather than an individual reliance on what *might* emerge from political governance and voluntary arrangements, may be pivotal to a way forward in responding to climate change.

References

Acreman, M. 2001. Ethical aspects of water and ecosystems. *Water Policy* 3:257–265.

Bosselmann, K., J. Fuller, and J. Salinger. 2002. *Climate Change in New Zealand: Scientific and Legal Assessments.* New Zealand Centre for Environmental Law Monograph Series vol. 2. Auckland: University of Auckland.

Campbell, J. 1999. Managing Environments. In *Explorations in Human Geography: Encountering Place*, edited by R. Le Heron, L. Murphy, P. Forer, and M. Goldstone, pp. 237–260. Auckland: Oxford University Press.

Ferdon, E. N. 1988. A Case for Taro Preceding Kumara as the Dominant Domesticate in Ancient New Zealand. *Journal of Ethnobiology* 8(1):1–5.

Hancy, I. M. K. 2007. *Iwi and Hapu Studies and Long-Term Survival and Prosperity of Te Kapotai.* Unpublished Report, personal papers from Wananga, examined May 5, 2008. Copies of these papers are held in the office of the Waikare Marae, Kawakawa, Northland.

Harmsworth, G. 2002. Indigenous Concepts, Values and Knowledge for Sustainable Development: New Zealand Case Studies. Conference paper presented at International Centre for Cultural Studies, Nagpur, India, Nov. 22–24.

Mead, H. M. 2003. *Tikanga Maori: Living by Maori Values.* Wellington, New Zealand: Huia Publishers.

Ministry of Agriculture and Forestry. 2004. The Permanent Forest Sink Initiative: An Initiative to Support Reforestation Carbon Sinks. Media release, Indigenous Forestry Unit, Christchurch, May 19.

———. 2007. New Zealand Plant Conservation, August 16. http://www.govt.nz/templates/news

Royal Society of New Zealand. 2008. Climate Change Statement. Press release, July 10. http://www. royalsociety.org.nz/news/press release/2008

Smith, G. H. 1997. The Development of Kaupapa Maori: Theory and Praxis. PhD dissertation, University of Auckland.

Stephenson, B. A. 2006. Global Climate Change—The Implications for Indigenous Practices. In *Climate Change and Pacific Rim Indigenous Nations*, edited by Alan Parker and Zoltán Grossman, pp. 29–38. Olympia, WA: Northwest Indian Applied Research Institute.

Stephenson, M. S. 2008. Timeless Projects: Remembering and Voice in the History of Education. *History of Education Review* 37(2):3–13.

Te Runanga o Ngati Awa. 1994. *Ngati Awa Me Ona Karangarangatanga*. Research Report No. 3. A report prepared in support of claim WAI-46 to the Waitangi Tribunal, Wellington. Te Runanga o Ngati Awa, Whakatane.

Tipa, G., and L. Teirney. 2003. *A Cultural Health Index for Streams and Waterways; Indicators for Recognising and Expressing Maori Values*. Wellington, New Zealand: Ministry for the Environment.

Waitangi Tribunal. 1997. *Muriwhenua Land Report, WAI 45*. Wellington, New Zealand: GP Publications.

Williams, D. A. R. and D. Nolan, eds. 1997. *Environmental and Resource Management Law in New Zealand*, 2nd ed. Wellington, New Zealand: Butterworths.

Yen, D. E. 1963. The New Zealand Kumara or Sweet Potato. *Economic Botany* 17:31–45.

POTENTIAL PATHS FOR NATIVE NATIONS

Laural Ballew and Renée Klosterman

Masters in Public Administration—Tribal Concentration, The Evergreen State College (Olympia, Washington). Laural Ballew is a member of the Swinomish Tribe.

We, the Indigenous Peoples, walk to the future in the footsteps of our ancestors.
—*Kari-oca Declaration, Brazil (1992)*

Indigenous peoples share significant experiences as a result of colonialism, such as the loss of land, natural resources, and subsistence; the abrogation of treaties; and the imposition of psychologically and socially destructive assimilation policies. Non-Indigenous societies typically do not relate to the land, sea, and sky in the same manner as Indigenous peoples, instead extracting natural resources and demanding absolute control of nature itself.

Over the decades, Indigenous peoples of the Pacific Rim have observed changes in the environment, including both local phenomena such as changing land management practices and global alterations such as shifts in bird migration or water levels. Tribal histories note that the times, indeed, "are a' changin.'" One example in Aotearoa/New Zealand illustrates the larger trends: "Images of eel have appeared on Maori artifacts and throughout oral histories. According to the oral histories, never has the eel migration been so erratic. Eel migrations correspond with changes in local water temperature and rainfall. But New Zealand's weather has been increasingly volatile, throwing the eel off" (Dawson 2006).

Historically when environmental shifts occurred, the Indigenous people shifted habitation to another location, returning later if the climate permitted. With the colonization of Indigenous lands (driven by consumerism exploding at a global scale), there simply is no longer a place to go—and most likely no longer a place to return. Aotearoa, Australia, and other countries on the Pacific Rim are running out of land that can sustain nature. In Indigenous terms, "Nature" is viewed as a garden, distinct from cultivated fields and yards but not a "wilderness," as in Western thinking. The Native garden requires tending with prescribed burns, tree and brush thinning, weeding, planting of preferred cultural plants, and management of culturally important animals. In short, Indigenous peoples have long practiced what Western society today calls "sustainability." In his 2006 song "After the Garden is Gone," Neil Young sings,

> After the garden is gone
> What will people do?
> After the garden is gone
> What will people say?
> After the garden is gone …

In *Native Science*, Native educator Gregory Cajete writes, "Native science's view on sustainability has begun to take effect. The survival of Planet Earth may be contingent on Western science, governments and private industry to relinquish their monopoly on accepted 'knowledge,' and listen to the voices of the indigenous cultures" (2000:269).

Indigenous cultures are locally based but maintain a universal view of nature that can be applied anywhere in the world. The phrase "think globally, act locally," now more imperative than ever, has special resonance in the Indigenous world. Perhaps this is the true vision of "homeland security," developing local strategies in Indigenous communities to change the course for planet Earth. Climate change awareness is relatively low throughout developed countries, and the general understanding of the phenomenon is still quite poor. Issues of "global warming" are so complex that many people feel helpless or frustrated. For some Indigenous people and tribes, the terms "global warming" and "climate change" might not be in their vernacular.

In improving people's understanding of climate change and empowering them to do something about it, the local context is most important to consider: what is affecting them locally? Chemical changes caused by climate change may result in warmer waters, which may lead tuna or anchovy (instead of sockeye salmon) to traditional fishing areas. In the words of Terry Williams, the Fisheries and Natural Resources Commissioner of the Tulalip Tribes in Washington state, the "People of the Salmon" may become the "Mahi-Mahi People."

If most tribal members in the Pacific Northwest of North America were asked about the effects of climate change on the environment of the Pacific Rim, they would likely not know how to respond. However, if the question concerned the natural resource management process, chances are they would be quick to recognize climate change's environmental impact on their locale. This connection is evident with regard to the sockeye salmon diversion, and the devastating effects of the increasingly intense El Niño and La Niña cycles, including storms, floods, and droughts—to name just a few.

Since time immemorial, Pacific Northwest Indian tribes have depended on natural resources for cultural, spiritual, and economic necessities. It is not coincidental that the tribes have been situated in each major watershed in the region and have been able to adapt quickly to the changes of the ecosystems. Tribes hold thousands of years of knowledge and experience to share with non-Native people and organizations that are willing to work with them.

The Treaty tribes in Washington state have the power to successfully preserve, protect, and restore natural resources for the benefit of all. For example, climate change and urban development have severely affected water resources and aquatic ecosystems in western Washington state. The Northwest Indian Fisheries Commission (NWIFC), with a membership of twenty tribes in Washington state, has recognized the need for a comprehensive assessment of the quantity and quality of water resources in western Washington as the starting point for informed management of them.

This assessment plan involves gathering information about the quantity and quality of fresh water available in western Washington. It can produce the scientific information useful for a wide range of tribal resource management, administrative, and legal activities. NWIFC recognizes the importance of managing water resources while protecting tribal rights, and working in collaboration with non-Native citizens and agencies to protect and repair our river basins, such as those in the Nisqually watershed.

Lummi Nation

The NWIFC is not alone in developing a response to the effects of urbanization and climate change on water supplies. The Lummi nation, located near the city of Bellingham, Washington, has established a Water Resources Division to promote the protection of treaty rights to water and develop and implement a comprehensive water resource management program. The program has several ongoing projects—from water quality monitoring to flood damage reduction planning to water rights negotiation and litigation on the reservation. These programs support the safeguarding of reservation water and land resources against degradation.

The water quality monitoring program, which has been ongoing since 1993, is intended to allow the Water Resources Division to document a baseline for the conditions of Lummi nation waters. This information is important to protect groundwater, tribal tidelands, and sensitive shellfish-growing areas. Protecting groundwater is necessary because it is the source of almost all of the water consumed on the reservation, and a sufficient supply of quality water is important to fulfill the purpose of the reservation to sustain life for future generations.

The Pacific Northwest tribal response to the tsunami threat can serve as a model for their response to climate change. Tsunami emergency planning has been an afterthought to most non-Native communities until recently. The Lummi nation plans to install two tsunami warning towers, technically called All Hazard Alert Bulletin (AHAB) radios, on the reservation. A third AHAB on the reservation was provided by the local county government. The Lummi nation has signed a Memorandum of Agreement with the state of Washington emergency response by local activation or remote activation in the event of an earthquake. This proactive response to the threat of tsunamis (set off by earthquakes) can help guide the tribes' response to climate change related disasters by focusing attention on emergency preparation, community resilience, and collaboration with neighboring communities.

Global and Local Responses

To limit the effects of human-induced climate change, the global Indigenous movement must work closely with the national and regional intertribal groups and local tribes, through communication and education initiatives. International and national Indigenous groups need the support of local tribes or local Indigenous groups (*iwi* in Aotearoa), which can empower people by educating them about how they can make a difference in their community. Currently, numerous organizations are bringing together elders, youth, tribal leaders, environmentalists, and concerned citizens to address environmental concerns. Other Indigenous groups are working in the international arena for human rights, land, and treaty issues and incorporating climate change as a new challenge facing Indigenous cultures and self-determination.

An example of an organization working toward an Indigenous role in climate change responses is the Inuit Circumpolar Council (ICC). The council has consultative status in the United Nations. Every four years the ICC holds a general assembly that is the driving force for the organization. The assembly offers the opportunity for sharing information, discussing common concerns, debating issues, and reinforcing the bonds linking all Inuit. Delegates from the Inuit Circumpolar Youth Council and the International Elders Council are present, encouraging communication and partnerships.

The ICC was established to promote the protection of human rights of all 150,000 Inuit in Alaska, Siberia,

Canoe arrives at the Lummi Nation in Washington in 2007, after a journey of more than 400 miles from the Bella Bella First Nation in British Columbia.

Canada, and Greenland. In order to thrive in their circumpolar motherland, the Inuit had the vision to speak with a united voice on issues of common concern and to unite their forces and abilities to protect and support their way of life.

In 2006, the ICC took firm action on climate change by submitting a petition to the Washington, D.C.–based Inter-American Commission on Human Rights, seeking relief from infringements on Inuit human rights because greenhouse gas emissions were causing global warming. ICC Chair Sheila Watt-Cloutier stated in her annual report: "We have alerted the world that we will not become a footnote to the onslaught of globalization by finalizing and filing a complaint at the Inter-American Commission on Human Rights to defend our human rights against the impacts of climate change" (*Indian Country Today*, January 5, 2006).

The ICC petition asked the commission to recommend that the United States implement mandatory limits to its emissions of greenhouse gases, and cooperate with the global community of nations to "prevent dangerous anthropogenic interference with the climate system," which is the intention of the U.N. Framework Convention on Climate Change (UNFCCC). The petition also appeals for the commission to affirm that the United States has a responsibility to collaborate with the Inuit to implement a plan to help them adjust to the inevitable impacts of climate change and to consider the impact of its emissions on the Arctic before endorsing any key government actions.

"This petition is not about money," Ms. Watt-Cloutier said. She continued:

It is about encouraging the United States of America to join the world community to agree to deep cuts in greenhouse gas emissions needed to protect the Arctic environment and Inuit culture and, ultimately, the world. We submit this petition not in a spirit of confrontation—that is not the Inuit way—but as a means of inviting and promoting dialogue with the United States of America within the context of the climate change convention. Our purpose is to educate not criticize, and to inform not condemn. I invite the United States of America to respond positively to our petition. As well, I invite governments and non-governmental organizations worldwide to support our petition and to never forget that, ultimately, climate change is a matter of human rights. (Quoted in Norrell 2006b)

Gregory Cajete observes, "We need an impassioned 'mistica'—mission, passionate desire and empowered need—to strive for ecological personhood. Our very survival as a species depends on our ability to make such a transformation" (2000:266). Creating an action agenda, a "mistica," is what international Indigenous organizations can do. The following model, described by Terry Williams, the Fisheries and Natural Resources Commissioner of the Tulalip Tribes, is one approach in creating this agenda:

Local/ Regional	Data	Setting goals
National	Organization	Cultural sustainability
Global	Declaration	Action plan

Questions in developing this agenda must include:

➤ How do we sustain our people at home?

➤ Are climate extremes beyond the traditional limits?

➤ What is the role of tribal government?

➤ What is the role of media—getting out the story, identifying what can be changed and what is possible, not always focusing on just the negative?

Communication

Media projects are critical to increasing public understanding of climate change. "Community begins with communications," said Govinda Dalton of the Earth Cycles Radio Project in Calpella, California (quoted in Norrell 2006a). Films such as *An Inconvenient Truth* and the Discovery TV series, *Global Warming: What*

You Need to Know, portray climate change not as a "theory" but as a reality.

International Indigenous groups have access to information and communication technologies now more than ever. In his article "An Indigenous World: How Native Peoples Can Turn Globalization to Their Advantage," Moises Naim (2003) argues that without discounting the atrocities and damage that globalization-from-above has inflicted on Indigenous people worldwide, globalization-from-below has also had a powerful benefit—the ability to communicate and share information globally. The increased reach of the global environmental movement, which has experience in organizing politically and the ability to mobilize international media and governments, has made Indigenous issues visible to a larger global audience than ever before. As a result, Naim says:

Indigenous people have a louder voice that can be heard internationally, and increased political influence at home. More fundamentally, globalization's positive impact on indigenous peoples is also a surprising and welcome rejoinder to its role as a homogenizer of cultures and habits. When members of the Igorot indigenous tribe in northern Philippines and the Brunca tribe from Costa Rica gather in Geneva, their collaboration helps to extend the survival of their respective ways of life—even if they choose to compare notes over a Quarter Pounder in one of that city's many McDonald's. In short, globalization's complexity is such that its results are less preordained and obvious than what is usually assumed. As the Maori, the Mayagnas, and the Tlicho know, it can also be a force that empowers the poor, the different, and the local. (Naim 2003)

The media have enormous power to determine what issues are important and to set the public agenda. They have enormous power to shape the meaning of these issues and as a result strongly influence people's ideas and values, including their ideas about Indigenous peoples. Native Americans, Maori, First Nations peoples, and the Aboriginal peoples of Australia are in a position to provide information about environmental concerns to Indigenous and non-Indigenous people through the use of radio, television, print, internet, and film.

As Cajete notes, every generation "must develop and add a social ecological imperative to perennial truths; these are the foundations of life we pass down to subsequent generations" (2000:267). He adds, "The accumulated knowledge of the remaining Indigenous groups around the world represents an ancient body of thought, experience and action that if honored and preserved as a vital storehouse of environmental wisdom, can form the basis for evolving the kind of cosmological reorientation so desperately needed" (281).

Education

The power and energy of youth, combined with the knowledge and experience of elders, is a powerful tool in pursuing the mission to combat global climate change. Increasing the creativity of organizations at all scales will educate young and old alike. It's Getting Hot in Here is a website devoted to "dispatches from the Global Youth Climate Movement," using creative and humorous means to grab the attention of and get the word out to young people. More traditional means can also be used to educate Indigenous youth. Maori *Tangata Whenua* (people of the land) transmit their environmental knowledge, which encompasses historical knowledge passed down through the generations. This accumulated wisdom can assist in the reconstruction of long-term climate trends (Stephenson 2006). Aleut elders are similarly identifying and educating youth in their culture to prepare for climate change in southwestern Alaska (see Merculieff, Part I).

Numerous Canadian First Nations youth groups are involved in the environmental movement. At the World Youth Forum in Vancouver, B.C., on June 23, 2006, the attendees issued the Ayateway Declaration. The preamble states: "The Ayateway Declaration is a living document. These words are not set in stone, they have been planted and fed by many minds and will continue to grow as we share them with our communities. We

Evergreen's Climate Change and Pacific Rim Indigenous Nations program meets with Tulalip natural resources staff in 2006: (left to right) Daryl Williams, Terry Williams, Alan Parker, Brett Stephenson, Renée Klosterman, Laural Ballew, Jill Bushnell, and Preston Hardison.

hope that this declaration will allow for the gathering of thoughts, beliefs and concerns of the Indigenous youth globally" (Indigenous Peoples Youth Caucus 2006).

If the message of Native resilience is to be carried into the future, it is important to include the younger generations. They may be more willing to hear the message of climate change and come up with ideas about how to combat the climate change crisis. Youth can also contribute fresh ideas that could be beneficial to this cause. Youth participation is now a prominent goal of international organizations as they work toward solving problems that will affect children's experiences in the future, such as environmental degradation (Blanchet-Cohen, et al. 2003). Researchers and policymakers have recently begun to pay more attention to the implications of carrying out research with children and youth and to the variety of ways that they can be meaningfully involved in matters that affect their lives. It is now commonly recognized that working with young people requires adapting conventional research methods and creating new ones. This call for creativity has encouraged researchers to explore different approaches and media, such as action research and other participatory methods (Christensen and James 2000; Greig and Taylor 1999; Punch 2002).

The Pacific Northwest

In order for Northwest tribes to take climate change and how it affects the region seriously, the message needs to be aired continually. Doing so includes frequent visits of tribal members to conferences and gatherings, as well as to tribal schools and colleges.

Housed on the Lummi reservation is Northwest Indian College (NWIC), which is the only accredited tribal college in the Washington, Oregon, and Idaho region. Its primary goal is to serve the educational and training needs of Pacific Northwest tribes and their members. The main branch is located on the Lummi Reservation, with several campuses and learning centers on Swinomish, Tulalip, Muckleshoot, Port Gamble S'Klallam, and Nez Perce lands.

NWIC collaborated with faculty from the Huxley College of Environmental Studies at Western Washington University to offer the Tribal Environmental and Natural Resources Management (TENRM) program, which was devised to produce tribal managers who know the land and its resources firsthand. It was funded by the National Science Foundation and several other collaborating institutions. The program has been structured in a thoroughly Native way and designed to meet the needs expressed by leaders of twenty-six Pacific

Northwest tribes (Tribal Environmental and Natural Resources Management [TENRM] 2001).

The TENRM program created the framework used to design the current Native Environmental Science (NESD) bachelor's degree program offered by Northwest Indian College. The college offers a two-year associate's degree, which is the first step toward a four-year degree program. The NESD program is intended to produce effective Native American leaders and environmental scientists who will be knowledgeable in both culture and science. This program was implemented with extensive contributions from Pacific Northwest tribal elders, leaders, environmental managers, educators, and students. The program highlights and surveys the connection between Native ways of knowing and traditional ecological knowledge and contemporary science. The NESD program prepares graduates who aspire to work with tribal communities to support environmental conservation and revitalization.

This type of program presents the best opportunity for organizations to make Native students aware of the local effects of climate change. Communication is vital to enlighten the public, and the tribal college provides an ideal forum. Not only can the tribal colleges reach the younger generation, but also the older, non-traditional student population as well. Starting such a program in one school can fuel the fire to inspire other colleges as their leaders gather at annual conferences for all tribal colleges throughout the nation. Tribal college students also meet regularly at trainings sponsored by the American Indian Alaskan Native Climate Change Working Group.

At the same time, is important to include the elders of this region. The elders are highly respected among their communities, and they are able to exert more leverage in encouraging official tribal participation in climate change forums. It is important to include the elders together with the youth groups to carry the message of climate change crisis around Indian Country.

Legal Strategies

From an environmental and spiritual perspective, Maori see the world as a unified whole, in which all elements, including *tangata whenua*, are connected. Emphasis is placed on maintaining the balance of cultural and spiritual values in the environment when using resources for social and commercial purposes. Such a perspective can be defended legally and extended to cover climate change. Thus, Maori demand that the New Zealand government maintain and develop policies to combat climate change to live up to its promises in the 1840 Treaty of Waitangi.

In the United States, there are also useful legal precedents for defending U.S. tribes' environmental rights, including the 2004 U.S. Supreme Court decision in the case of *South Florida Water Management District v. Miccosukee Tribe of Indians*. The court reaffirmed federal trust responsibility to U.S. tribes under the Clean Water Act. The ruling stated, "The interests being threatened here, including the threat to the Miccosukee Tribe's homeland, sovereignty, economic integrity, resources, and its right to conduct its religious and cultural practices, are precisely the interests the United States is duty bound to protect" (*Oneida Nation v. County of Oneida*). Further research on how federal laws and policies interact with tribal jurisdiction on climate change issues needs to be conducted, so tribes can pressure the federal government to fulfill its trust obligation to tribes threatened by climate change.

In the context of a shifting climate, international legal strategies will be a key part of any political effort. Indeed, "International Indigenism" is a growing movement, according to researcher Ronald Niezen. In his 2003 book *The Origins of Indigenism—Human Rights and the Politics of Identity*, Niezen examines how the relatively recent emergence of an internationally recognized identity of "Indigenous peoples" merges with the development of universal human rights laws and principles. Together, the focus of these Indigenous groups is on human rights laws and policies and the ability of international organizations to resist and change the political, cultural, and economic sanctions of individual states, through actions such as the Inuit petition to the Inter-American Commission on Human Rights. Niezen argues that from a new position of legitimacy and influence, these Indigenous groups are striving for greater recognition of collective rights, in particular their rights to self-determination, in international law.

These efforts are, in turn, influencing local politics and encouraging more ambitious campaigns for autonomy in Indigenous communities worldwide, encapsulated in the 2007 United Nations Declaration on the Rights of Indigenous Peoples (see Grossman, Part III). The 2006 Bemidji Statement on Seventh Generation Guardianship states, "From the smallest unit of society to the largest unit of government, we can protect, enhance, and restore the inheritance of the Seventh Generation to come" (Indigenous Environmental Network 2006:1).

National Indigenous organizations are leading the call to develop policies addressing global climate change. In the Assembly of First Nations in Canada, the Environmental Stewardship Unit developed a flyer to further educate First Nation leaders to recognize the enormous impacts of climate change on the land and people. Many First Nations citizens reside in the northern boreal forest, the place where climate change is already having massive negative effects (see Williams, Part I). The flyer encourages finding ways of reducing greenhouse gas emissions.

Besides the Assembly of First Nations, other national and regional intertribal organizations that have passed climate change resolutions or implemented climate change programs include the National Congress of American Indians (NCAI), National Tribal Environmental Council (NTEC), Affiliated Tribes of Northwest Indians (ATNI), Northwest Indian Fisheries Commission (NWIFC), Aboriginal Fisheries Commission in Canada, and Maori iwi (tribal) meetings in Aotearoa.

Conclusions

Traditionally, Indigenous cultures have had the capacity to mitigate or adapt to climate change. Many Indigenous communities can redeploy and reestablish these forms of resource management, derived from their traditional knowledge base, in response to contemporary climate change. Local indicators, or what is locally appropriate, will vary by region. Cajete writes, "Only by truly touching the Earth can we honor and enable the vision and action necessary to recapture the feeling and understanding that we have always been a part of the living and 'conscious' Earth" (Cajete 2000:267). Indigenous governments can determine the goals that are entailed in this vision and take action in the following ways:

► Develop an international partnership to promote communication about climate change issues, using print, television, film, radio, internet, and other media.

► Conduct research into how national laws protecting tribal rights can be used to make changes at local and national levels. At the same time, they can collaborate with state, provincial, and federal government agencies to ensure proper funding is available for Indigenous communities to carry out research and climate change responses.

► Become involved in establishing carbon sinks as part of the carbon-trading market. However, it should be noted that these sinks have been deemed unacceptable by many Indigenous groups. Under the proposal "land" or "forests" could become carbon "credits" that can be traded between participating countries, encouraging the spread of fast-growing monocrop tree plantations. Further research and consultation on the socioeconomic and cultural context of carbon trading is necessary prior to the acceptance of this idea by Indigenous governments.

► Form youth/elder coalitions to educate Indigenous people at local, national, and international levels and encourage youth and elders to collaborate in preparing their communities for climate change and reducing its threat.

► Disseminate valuable environmental information contained in the oral history and traditions of Indigenous communities to assist in understanding of the long-term weather changes by both Native and non-Native communities. Indigenous groups collaborating with Western scientists may lead to mitigation of climate change.

► Develop one voice in the numerous international organizations through the United Nations or other large-scale organizations. The United League of Indigenous Nations is one step toward this cooperation across international boundaries.

► Think globally, act locally.

The intricate web of international Indigenous organizations has the ability to work toward weaving a future for the generations to come, sustaining the current generation and respecting the ancestors. It is clear that action is necessary to change policies, laws, and education. As Cajete explains, "Human beings consist of seventy percent salt water and thirty percent minerals of the earth. We are the earth and her waters made more animate. Our understanding of this essential natural connection is one of understanding our relationships to all things of the Earth" (Cajete 2000:279).

Resources

An Inconvenient Truth. 2006. Directed by Davis Guggenheim. Paramount Home Entertainment. http://www.climatecrisis.net

Blanchet-Cohen, Natasha, ed., 2003. Children Becoming Social Actors: Using Visual Maps to Understand Children's Views of Environmental Change. *Children, Youth, and Environments* 13(2).

Cajete, Gregory. 2000. *Native Science*. Santa Fe, NM: Clear Light Publishing.

Council of Energy Resources Tribes (CERT): http://www.certredearth.com

Dawson, Durrel. 2006. Early Signs. *Living on Earth*, June 23. Archived at www.LivingonEarth.org

Environmental Protection Agency (U.S.), American Indian Environmental Office: http://www.epa.gov/indian

Hardison, Preston. 2006. Guest speaker in the Impacts of Global Warming on Pacific Rim Indigenous Nations course. The Evergreen State College, Olympia, Wash., July.

Havemann, Paul, and Helena Whall. 2002. *The Miner's Canary: Indigenous Peoples and Sustainable Development in the Commonwealth.* http://www.commonwealthadvisorybureau.org/fileadmin/CPSU/documents/Projects/Indigenous_Rights/2002_The_Miner_s_Canary_Indigenous_peoples_and_sus_devt_2002.pdf

Indigenous Environmental Network. 2006. The Bemidji Statement on Seventh Generation, Protecting Mother Earth Conference, Bemidji, Minn., July 6.

Indigenous Peoples Youth Caucus. 2006. Ayateway Declaration. Issued at the World Urban Forum III, Union of B.C. Indian Chiefs, Vancouver, B.C., Canada, June 23. http://www.ubcic.bc.ca/News_Releases/UBCICNews06260601.htm

Inuit Circumpolar Council (ICC). http://www.inuitcircumpolar.com

Maori Environmental Science, National Climate Centre, National Institute of Water and Atmospheric Research (NIWA) Science: http://www.niwa.co.nz/our-science/climate

Naim, Moises. 2003. An Indigenous World, How Native Peoples Can Turn Globalization to Their Advantage. *Global Policy Forum.* http://www.globalpolicy.org/component/content/article/161/28019.html

National Tribal Environmental Council: http://www.ntec.org

New Zealand Climate Change/Rerekitanga Ahuarangi o Aotearoa. Climate Change, the Way Ahead. http://www.climatechange.govt.nz

Northwest Indian College: http://www.nwic.edu

Niezen, Ronald. 2003. *The Origins of Indigenism: Human Rights and the Politics of Identity.* Berkeley, CA: University of California Press.

Norrell, Brenda. 2006a. Indigenous Environmental Network Brings Regeneration to Warriors. *Indian Country Today,* June 24. http://groups.yahoo.com/group/NatNews/message/43541

———. 2006b. Indigenous Peoples Voice Urgency On Global Warming. *Indian Country Today,* January 5. http://groups.yahoo.com/group/worlds-indigenous-people/message/6351

Northwest Indian Fisheries Commission. 2006. Comprehensive Tribal Natural Resource Management. A Report from the Treaty Indian Tribes in Western Washington. Olympia, Wash.: Northwest Indian Fisheries Commission.

National Wildlife Foundation (NWF): http://www.nwf.org

Stephenson, Brett. 2006. Guest speaker in the Impacts of Global Warming on Pacific Rim Indigenous Nations course. The Evergreen State College, Olympia, Wash., July.

Tribal Environmental and Natural Resources Management (TENRM) Handbook. 2001. Bellingham, Wash.: Northwest Indian College. http://faculty.wwu.edu/gberardi/links/TENRMHandbook.pdf

Williams, Terry. 2006. Guest speaker in the Impacts of Global Warming on Pacific Rim Indigenous Nations course. The Evergreen State College, Olympia, Wash., July.

World Urban Youth and Forum: http://www.eya.ca/splash.php

Teach Us, and Show Us the Way
(Chinook Blessing Litany)

We call upon the earth, our planet home, with its beautiful depths and soaring heights, its vitality and abundance of life, and together we ask that it Teach us, and show us the Way.

We call upon the mountains, the Cascades and the Olympics, the high green valleys and meadows filled with wild flowers, the snows that never melt, the summits of intense silence, and we ask that they Teach us, and show us the Way.

We call upon the waters that rim the earth, horizon to horizon, that flow in our rivers and streams, that fall upon our gardens and fields and we ask that they Teach us, and show us the Way.

We call upon the land which grows our food, the nurturing soil, the fertile fields the abundant gardens and orchards, and we ask that they Teach us, and show us the Way.

We call upon the forests, the great trees reaching strongly to the sky with the earth in their roots and the heavens in their branches, the fir and the pine and the cedar, and we ask them to Teach us, and show us the Way.

We call upon the creatures of the fields and forests and the seas, our brothers and sisters the wolves and deer, the eagle and dove, the great whales and dolphin, the beautiful Orca and salmon who share our Northwest home, and we ask them to Teach us, and show us the Way.

We call upon all those who have lived on this earth, our ancestors and our friends, who dreamed the best for future generations, and upon whose lives our lives are built, and with thanksgiving, we call upon them to Teach us, and show us the Way.

And lastly, we call upon all that we hold most sacred, the presence and power of the Great Spirit of love and truth which flows through all the Universe, to be with us to Teach us, and show us the Way.

NO LONGER THE "MINER'S CANARY"

Indigenous Nations' Responses to Climate Change

Zoltán Grossman

Member of the Faculty in Geography/Native American and World Indigenous Peoples Studies, The Evergreen State College, Olympia, Washington.

Editors' note: A previous version of this chapter was published in the *American Indian Culture and Research Journal* 32:3 (2008):5–27.

It's getting hotter, harder to breathe,
Why should I calm down, I know I've been deceived.
Like oceans of regret, all these questions rise.
Will they drown with our mistakes,
or will they learn to fly?

She said it's over, overwhelming.
We're past the breaking point, the breaking point again.

It's getting hotter, and harder to see.
Balancing the contradictions, how
much do we really need?
Standing on the broken edges of apathy,
Occupied by your destruction, your
waves crashing over me.

So restless, she's shaking,
Can you feel her temperature rising?
We're so complacent and apathetic,
while she's given us everything …

—Blackfire (Diné Nation), "Overwhelming," on [Silence] is a Weapon. Reproduced by permission of Blackfire (Klee, Jeneda, and Clayson Benally.)

Climate change is usually portrayed as a process of "global warming" that is so large that it can be addressed only by national governments or international agencies. We are told that we can only respond to climate change in a personal way—by changing our light bulbs or automobiles—and that we cannot change the industrial policies that generate most greenhouse gases (also called carbon emissions). We are made to feel powerless and fatalistic in the face of the biggest threat to the Earth's well-being in its history.

Yet increasingly, the most effective solutions to global warming are not being seen at the national or international scale, but at the local scale. European cities have taken steps to curb their greenhouse gases, and U.S. cities are beginning to follow their example, rather than follow the inaction of their federal government. Municipal, state, and provincial governments are also beginning to respond, particularly on the west coast (Flesher 2007).

On one hand, Indigenous peoples are on the front line of climate change—the first to feel its effects, with subsistence economies and cultures that are the most vulnerable to climate catastrophes. Paraphrasing Felix Cohen, Paul Havemann and Helena Whall observe, "Indigenous Peoples are like the miner's canary. When their cultures and languages disappear this reflects the profound sickness in the ecology" (Havemann and Whall 2002). In this view, the fate of Native peoples provides an early warning to the fate of all humanity.

On the other hand, Indigenous peoples have certain advantages in responding to the challenge of climate change, compared to non-Native neighbors or local governments. They want their continuing lifeways (not their death) to provide some direction to the rest of humanity. Perhaps the "miner's canary" formulation could be revised, so the canary escapes the cage, flaps its wings, and shows the hapless miner a safe way out of the toxic mine. Indigenous knowledge and experience has the potential to help the rest of humanity get more out of harm's way.

Native peoples have faced massive ecological and economic changes in the past—from colonialism, genocide, epidemics, industrialization, and urbanization—yet many Native cultures have survived against overwhelming odds. The climate crisis is the latest, and perhaps the ultimate challenge, but this history may make Native peoples better equipped than the non-Native society that is reliant on chain grocery stores and shopping malls. It is critical that tribal climate change discourse not only warn Native communities of the dangers in climate instability, but empower them around inherent tribal strengths:

▶ Traditional Ecological Knowledge: Indigenous cultures have centuries of experience with local natural resources, so they can recognize environmental changes before Western scientists detect them and can develop ways to respond to these changes.

▶ Political Sovereignty: Because tribes have a unique status as nations, they can develop their own models of dealing with climate change and managing nature in a sustainable way.

▶ A Sense of Community: In contrast to much of the non-Native population, Indigenous peoples still have community. Native peoples still have extended families that care for each other, assume responsibility for each other, and extend hospitality in times of need.

The Treaty of Indigenous Nations builds that sense of community, by including other tribal nations in the community, even those who live on the other side of imposed colonial borders or on the other side of the ocean. Indigenous peoples have survived the effects of colonialism and environmental destruction only by cooperating with each other. It is no longer just a good idea to build these relationships; *climate change makes them much more urgent.* This article explores some of the relationships that are already being built or have the potential to be built among Indigenous nations, then with local governments, national governments, and international agencies.

Cooperation among Indigenous Governments

In the coming years of climate change, intertribal cooperation will become more important, in order for Indigenous cultures and communities to survive. Many recent reports and articles have examined in depth the threat of climate change to Indigenous peoples and cultures, but precious little has been written about possible responses.

The Northwest Indian Applied Research Institute (NIARI) has described the effects of climate change on the Pacific Rim region in its report *Climate Change and Pacific Rim Indigenous Nations* (NIARI 2006). NIARI later published a companion *Community Organizing Booklet*, with the technical language translated into understandable English for tribal members (McNutt 2010). NIARI's Climate Change and Pacific Rim Indigenous Nations Project has developed and presented recommendations to tribal governments (NIARI 2009).

On August 1, 2007, indigenous nations from within the United States, Canada, Australia and Aotearoa (New Zealand) signed a treaty to found the United League of Indigenous Nations. The Treaty of Indigenous Nations offers a historic opportunity for sovereign Indigenous

governments to build intertribal cooperation, outside the framework of the colonial settler states. Just as the Pacific Rim states have cooperated to limit Native sovereign rights and build polluting industries, Indigenous nations can cooperate to decolonize ancestral territories and protect their common natural resources for future generations.

The treaty process has involved indigenous political alliances such as the National Congress of American Indians (NCAI), the Assembly of First Nations (AFN) in Canada, and Ngati Awa Maori Confederation. The treaty identifies four main areas of cooperation: increasing trade among Indigenous nations, protecting cultural properties, easing border crossings, and responding to the urgent threat of climate change (United League of Indigenous Nations 2007).

Indigenous peoples of the Pacific Rim already share much in a common natural region, have similar fishing cultures, and have been in contact with each other for many centuries. In the treaty, the second Mutual Covenant commits the signatory nations to "collaborating on research on environmental issues that impact indigenous homelands, including baseline studies and socio-economic assessments that consider the cultural, social, and sustainable uses of indigenous peoples' territories and resources" (ULIN 2007). The United League of Indigenous Nations (ULIN) can help facilitate this collaboration by helping to build an Indigenous nations' climate change network. This network could include different working committees, which would include representatives from:

▶ Tribal or First Nation sovereign governments

▶ Indigenous community members, traditional harvesters, and spiritual leaders

▶ Researchers, educators, and students

Exchanges within this network (working with existing Indigenous organizations) could help implement practical projects to adapt to (or mitigate) survivable climate changes and develop joint responses to more destructive climate changes. These exchanges could include sharing information, connecting tribal youth, training harvesters of shifting plant and animal species, and ensuring access to food, water, and power.

Sharing Information

In order to survive climate change, Indigenous communities will have to share information with each other about the effects of global warming, as well as share different responses. But the first priority is to share information *within* each community. With Indigenous governments' success in establishing their own environmental departments, many tribal members assume that staff members will take care of all natural resource issues. Some tribal natural resources staff are already working on climate change–related issues. But the challenge of climate change will not be met by tribal government officials alone; it is simply too huge of a problem and needs to involve the entire community.

The tribal government can get tribal members gathered together to share information, and tribal members can request their elected officials to respond to their concerns about the effects of climate change. The first step is to bring together tribal members to discuss how climate change might be affecting tribal life and culture and what can be done about it within the tribe, or in cooperation with other governments. In 1998, NASA funded a Native Peoples/Native Homelands national conference, which included workshops of tribal members documenting the effects of climate change on their regional cultures (Maynard et al. 2001).

In November 2009, the second conference drew tribal college and university students and White House Council on Environmental Quality staff to Shakopee (Minnesota) to discuss "Indigenous Responses to the Challenge" of climate change (Native Peoples/Native Homelands Climate Change Workshop II 2009).

A series of regional and national conferences have been held on tribal climate change issues. The Tribal Lands Climate Conference held at the Cocopah Nation in Arizona in 2006 was an example of intertribal cooperation to document global warming (National Wildlife Federation 2006). A water rights group held a climate change conference hosted by Washington's Squaxin Island Tribe in 2007 (Center for Water Advocacy 2007). Around the same time, the University of Colorado Natural Resources Law Center published a report on the effects of the climate crisis on tribes in different U.S. regions: *Native Communities and Climate Change: Protecting Tribal Resources as Part of National Climate Policy* (2007).

In 2008, the emphasis began to switch from documenting the effects of climate change to discussing tribal responses. A conference was held in Boulder on "Planning for Seven Generations: Indigenous and Scientific Approaches to Climate Change," by the American Indian and Alaska Native Climate Change Working Group (UCAR Community Building Program 2008). Tribal Climate Change Forums in Oregon drew tribal agency representatives from around the Northwest starting in 2009 (Sustainable Northwest 2009). Northern Arizona University started a Tribes and Climate Change website to gather information of effects and responses (Institute for Tribal Environmental Professionals 2009).

Because their environment is being so drastically altered by climate change, Indigenous peoples in the Arctic and Subarctic are leading the way in sharing information about its effects. For example, one Inuit community in Nunavut held large community discussions and produced a video (International Institute for Sustainable Development and Hunters and Trappers Committee of Sachs Harbour 2000). An educational book was cooperatively produced from interviews with hunters and fishers from twenty-six Inuit and Cree communities around Hudson Bay (McDonald et al. 1997). In Alaska, according to Aleut leader Larry Merculieff, Aleut villages have held community meetings of harvesters to discuss changes in the resources. Merculieff stresses the importance of including both elders and youth in these discussions and of collecting field samples and observations together. He says, "As species go down, the levels of connection between older and younger go down along with that" (Rosen 2004).

Connecting Youth

It will be especially important to share information with youth in Indigenous communities, to make them more aware of climate change and get them energized and involved in the issue. Through practicing their culture with their elders, they can learn traditional ecological knowledge and be more able to understand changes in weather patterns or in plant and animal species. Through working with each other, young people can also learn about climate change and educate their entire communities about the issues. The urgency of responding to climate change is being incorporated into tribal youth conferences and can become a key part of exchanges among Indigenous nations.

Tribal leadership can encourage middle school, high school, and college-age youth to form their own groups to get active. First Nations youth were among the activists outside the UN conference on climate change held in Montreal in 2005, as part of the Energy Action's youth climate movement It's Getting Hot in Here. Native youth organized by the Indigenous Environmental Network have represented the Campus Climate Challenge and attended Powershift student climate conferences and rallies, such as the 2009 Green Jobs rally in front of the U.S. Capitol (Tribal Campus Climate Challenge 2009). The Alaska Youth Environment Action sent delegates to Japan and Iceland to attend the International Youth Eco-Forum on Climate Change and Renewable Energy and collected thousands

Makah children present a paddle dance at Makah Days in Neah Bay in 2005.

of signatures on a climate change petition, which they presented to their congressional delegation at the U.S. Capitol (AYEA 2009).

But youth also can become involved in their local communities. A model already exists among B.C. First Nations, who have trained Aboriginal young people to map their territories, in order to protect natural resources and strengthen land claims. The youth in the Strategic Watershed Analysis Teams (SWAT) interview elders and other harvesters, gather field data with GPS units, and produce maps (Collier and Rose 2000). Starting in the Tribal Canoe Journey of 2008, First Nations and tribal youth have been involved in water sampling to investigate effects of climate change in the Salish Sea, as part of an initiative sponsored by the Coast Salish Gathering (Kapralos 2008). Similar youth teams could also participate in tribal Hazard Identification and Vulnerability Assessments to examine how to "harden" their communities against destructive climate change, or help tribal planning departments develop long-term plans for survival. Since they will be around to see the full effects of global warming, tribal youth deserve a role in planning for the future.

Training for New Species

As the weather becomes warmer farther north, we will be seeing more species shift out of their usual habitats and into other regions. In the Pacific Northwest, this will both mean that some plant and animal species will move from south to north and also that they will shift up mountain slopes. The most endangered species are those that cannot shift quickly—such as trees—and shifting land-based species that are blocked by a body of water (such as the Salish Sea), high elevation (the tree line), or high latitude (the tundra) and cannot migrate

any further. Droughts could also severely hurt species just when they are vulnerable, unless urgent measures are taken to protect their habitats.

Plant, animal, and marine species will shift into new areas, where tribal harvesters may not be familiar with them, and they may not fit into local Indigenous cultural and spiritual systems. Indigenous communities are already thinking about the implications of traditional resources moving out of their historic territories. Some fish runs, for example, may disappear, and other fisheries may be replaced partially or entirely by new species coming from the south. Whether or not Indigenous harvesters can adapt to these new species may determine whether tribal economies survive. New pests and diseases (such as the spruce bark beetle infestation) already threaten tribal health and economies. In either case, Indigenous nations that choose to adapt to the new species can draw on the expertise of neighbors further south.

Species have migrated before in the past (even if not as suddenly), so tribal ancestors must have helped each other adapt. Indigenous governments can help facilitate a series of exchanges between tribal communities, so they can teach each other about unfamiliar species and train each other how to harvest (or avoid) them. At the Tribal Lands Climate Conference, a Haudenosaunee (Iroquois) woman reported that she had visited relations to her south to learn what was coming into her territories, then visited communities to her north to let them know what may be coming their way.

In the Pacific Northwest, we can learn from people in California about their species, and people in British Columbia and Alaska may have to learn about our species. Basketweavers from coastal tribes have "stated that they often purchase bark from gatherings or trade with other tribes" further north (Papiez 2009). The existing relationships and family bonds among different nations will be immensely helpful in preparing for the arrival of new species and in learning techniques to harvest them.

Cooperation in Food Security

Another important area in which tribes can cooperate is in securing access to food in times of shortage. We have all grown used to going down to the supermarket to get our basic essentials. But all it takes is one storm power outage, or one flood or landslide, to remind a community of "the old days" when food did not come only out of a grocery bag. Indigenous peoples know (better than most non-Native neighbors) how fishing, hunting, gardening, and so on can supply needed food in hard times.

A growing movement for Native agriculture and food systems is emphasizing locally grown, traditional foods that revitalize tribal cultures and a sense of community and local control (First Nations Development Institute 2004). Traditional foods are also healthier for tribal members than the colonial white flour and sugar diet that has created an epidemic of diabetes (Severson 2005). Because traditional crops and animals are historically more locally adapted, they also can be more resilient to climate changes. Some tribes are researching and adopting more deep-rooted or drought-resistant "ancient seeds for modern needs" into their food systems (Native SEEDS/Search 2009). Other tribes are reintroducing bison and other locally adapted livestock (Inter-Tribal Buffalo Cooperative 2009).

But what if climate change also affects traditional foods, creating a shortage of fish and wild game, or drying up farm crops or gardens? Tribes need to think ahead to these situations when basic needs cannot be met with local foods. Some tribes have food storage facilities, but storage for both perishable and non-perishable crops are needed for food security. Intertribal cooperation will become essential, since some tribes lack suitable conditions or enough land for sustainable agriculture, while other tribes have adequate land, food crops, and livestock herds. Intertribal agreements could set up a trade network that takes food security into account, particularly within a single region (since fuel shortages may disrupt long-distance transport). A precedent can be seen in the growing network of remote agricultural tribes supplying Native foods (such as bison and salmon) to tribal casinos in more urbanized areas.

Whether they decide to cooperate about food security, harvesting new species, or sharing information, it is to the advantage of Indigenous governments to make agreements with each other *now*, when they have funds and resources available, rather than in response to a local climate change crisis, in which resources may be scarce and funding prioritized for other communities. The same situation holds for cooperation between tribes and their neighboring non-Native towns, especially since the relationships between tribal and local governments are often tenuous or even tense. Building a positive relationship *before* an emergency hits will enhance mutual understanding and cooperative bonds that can help each other survive.

Tribal and Local Governments

In some areas, Indigenous and local governments have begun to overcome their differences over jurisdiction and work together for the common good (Northwest Renewable Resource Center 1997). This cooperation

between Native and non-Native neighbors is going to become more crucial as climate changes intensify. The most important ways to survive climate change—adequate food, water, shelter, power, and so forth—are most efficiently and cheaply found in our own local areas. Cut off from help by floodwaters or mudslides or lacking aid from unreliable national agencies (remembering the Federal Emergency Management Agency [FEMA] during Hurricane Katrina), we will all have to rely on our neighbors. When push comes to shove, all that we will have is each other.

Climate change adaptation is usually presented as a sad or scary topic, but it can also be viewed as an unprecedented opportunity. Climate change adaptation can be effectively used a reason to quickly make fundamental changes in our environmental, economic and cultural practices that otherwise may take years to implement. Adaptation is a good excuse to make necessary changes that we should be making anyways for a healthier future, and making the changes more quickly than we otherwise would have. For example, regulating stormwater runoff has become more urgent to offset increasing acidification (linked to climate change) of the Salish Sea. A sense of community and respect for the land are no longer just good ideas, but they are absolutely necessary to survive the troubles ahead.

Climate disasters are often used to centralize political control and privatize economies, as described in Naomi Klein's 2007 classic *The Shock Doctrine*. As a flip side of the shock doctrine, communities can use climate change adaptation to increase awareness of ecological ways to prevent future disasters, the need to share resources among neighbors, and deepen cooperation between communities—beyond the immediate sandbagging of a river. Some of the most important "green jobs" for youth may be in rural and urban planning, disaster prevention, and emergency response, not only to heal our communities but to make them more humane and sustainable than they were before the disasters.

Renewable Energy Partnerships

Tribal and local governments can cooperate to build renewable energy projects that can reduce their dependence on dirty fossil fuels. Tribes can have access to federal funds and "Green Tag" Renewable Energy Credits to start up their own energy projects (Native Energy 2009). Perhaps the most promising direction for Indigenous nations in combating climate change is in adopting renewable energy technologies that reduce Native dependency on the colonial economy and the centralized electrical grid, at the same time as providing a model for non-Indigenous communities of reducing fossil fuel use. As a Michigan tribal resolution states:

> Tribal lands represent a vast amount of renewable energy potential, including wind and solar power that can meet the energy needs of both local tribes and surrounding communities; wind power blowing through Indian reservations in just four northern Great Plains states could support almost 200,000 MW of power, enough to reduce output from coal plants by 30% and reduce our electricity base global warming pollution by 25%, and Great Lakes Indian nations could similarly produce alternative non-polluting renewable energy for our tribal communities and for export (Little Traverse Bay Bands 2005).

Renewable energy projects on Native American reservations enable tribes to tap into federal funds and use their sovereignty to shift their energy economies away from the centralized, fossil fuel dependent model. On the Hopi Reservation, for example, NativeSUN has been installing photovoltaic panels on tribal homes to harvest the Southwest's abundant solar power (LaDuke 1999). Other U.S. tribes are involved in energy projects that tap into geothermal, tidal, or wind energy sources. They have asked Congress for renewable energy production tax breaks, because the export of renewable power can build a sustainable reservation economy that brings revenue and jobs to rural communities.

The Intertribal Council On Utility Policy (COUP) provides a tribal forum for policy issues dealing with energy operations and services. It asserts that U.S. tribes have "tremendous untapped energy potential in reservation wind resources" such that the Northern Great Plains could become the "Saudi Arabia" of renewable wind energy (Intertribal COUP 2009). In partnership with Native Energy, Intertribal COUP is developing an 80 MW distributed wind project, hosted in 10 MW clusters at eight different reservations (Native Energy 2009). With access to a predictable revenue stream from renewable energy, the tribes can sell power at a profit through the federal energy grid and at the same time reduce dependency on incoming power through the same grid (Native Wind 2009). A successful tribal effort to convert to renewable energy can become the prototype for non-Native communities that also wish to develop decentralized energy economies and reduce fossil fuel use. Northwest Tribes have a *Northwest Energy Planning Guide* to identify the technologies most appropriate to their locations (NWSeed 2009).

The concept of Native American renewable energies is slowly being combined with the commitments of U.S. cities to shrink their carbon footprint. Mayors from 180 U.S. cities symbolically signed the Kyoto Protocol, committing their governments to reducing carbon emissions even though the federal government had not ratified the protocol. In November 2005, a Native Renewable Energy Summit was held in Denver to discuss ways that U.S. cities and tribes can partner to achieve their economic and environmental goals. The summit was one step toward tribes and cities working together to reduce fossil fuel consumption while generating sustainable energy employment. At least twenty-nine Local Governments for Sustainability have expressed interest in such partnerships, including Minneapolis, Salt Lake City, New Haven, and Brooklyn (Gough 2005). As *Indian Country Today* noted, "The many cities that have pledged to reduce their dependence on carbon-producing power share a common ground with the tribes. Tribes could lead the way by showing their commitment to clean air and water and creating the potential to expand the distribution of power" (Melmer 2005).

In Washington state, for example, the Tulalip Tribes are working with local dairy farmers on the Qualco Energy methane project, which creates biogas power from cattle waste. The waste is thus also kept out of the Snohomish River, aiding in tribal salmon recovery efforts (Kamb 2003). The Makah Tribe was part of a consortium operating the Makah Bay Offshore Wave Energy Pilot Project, studying how to use special buoys to turn the motion of ocean waves into electricity for the county utility, at least until the project's major investor withdrew in 2008 (Renewable Energy Access 2006). In British Columbia, the Gitga'at and other First Nations are exploring options for small-scale hydroelectric dams that would not endanger salmon runs (Sustainable Communities 2003).

Joint Land-use Planning

Joint land-use planning by tribal and local governments can prevent some of the most disastrous effects of climate change and build more self-sufficiency. Together, governments also have to anticipate the effects of climate changes, such as preventing hillside erosion, maintaining alternate road access, growing local food crops, and preventing new pests and diseases from getting a local foothold.

One of the most important areas for cooperation is to ensure a supply of fresh water, which may be in demand as glaciers melt and streams and rivers dry up in the summer months. Tribal and local governments can work together to protect and treat their water supplies, conserve water use, and store the glacial runoff in reservoirs or underground aquifers (Ghoghaie 2011). Tribes can use their federally recognized senior water rights (under the Winters Doctrine) to secure access to fresh water, as it becomes a commodity as valuable as oil or gold.

One of the main threats of global warming is rising sea levels, from both melting polar ice and the warmer, expanding oceans. Coastal communities that already face dangers from storms, floods, and tsunamis now also have to contend with the new threat of rising ocean levels, which make the more familiar threats much more dangerous. The Pacific Northwest coast is particularly vulnerable to these risks (Erzen 2006). Tribal and local governments need to build and retain wave barriers, prevent shoreline erosion, and build new homes above the floodplains. As one innovative example of planning ahead, the Nisqually Tribe signed an agreement with the City of Olympia in 2008 to switch its drinking water source from McAllister Spring to a wellfield on higher ground. The proactive move avoids possible saltwater intrusion from sea-level rise and restores water flow to Medicine Creek, ironically the site of the signing of the 1854 Treaty (Iyall 2008).

Joint Emergency Planning

If catastrophe occurs, collaboration by tribal and local governments can also prevent loss of life and community wealth. Many tribes now work with local governments on emergency services, such as acquiring fire trucks or sharing EMT services. Deeper relationships will be needed in case of climate change disasters (such as windstorms, floods, droughts, and landslides) to keep them from wiping out communities and their livelihood.

Tribes can lead the way by being models to neighboring local governments. For example, during a June 2005 tsunami warning, Washington coastal tribes quickly evacuated their reservations, while non-Native citizens were angered that their own local governments did not respond as quickly. In the December 2006 windstorm blackouts, some Washington tribes opened their emergency shelters and health clinics to adjacent towns. After the December 2007 megastorm devastated Lewis County with mudslides and flooding (making Interstate 5 inaccessible for one week), the Chehalis Tribe offered preference for flood victims in job openings for its new resort. The December 2008 blizzard (and resultant January 2009 floods) and January 2012 ice storm also brought together tribal and local governments in their responses. Whether these particular storms were caused by climate change is a matter of scientific debate,

but because many residents have experienced isolation during the more intense winter megastorms, they are thinking more about cooperation with their neighbors.

The Swinomish Tribe has taken the lead among Native American nations by starting the Swinomish Climate Change Initiative. The tribe is working closely with local non-Native governments in the Skagit Delta, and has a Community Engagement Group for tribal members to get involved in the adaptation planning process. Chairman Brian Cladoosby hopes the initiative serves as a model for other Northwest tribes to account for climate change in their joint planning with their neighbors (Swinomish Indian Tribal Community 2009).

Tribal and local governments can develop Hazard Identification and Vulnerability Assessments not only to deal with short-term emergencies or to develop evacuation procedures. They will also need to look toward pooling their resources for longer-term periods without electricity, gas, or access to supermarkets. People tend to come together in disasters, and sharing will become more essential in the future to meet daily needs of food, water, heat, and power.

However, tribal-local cooperation only works if local governments respect the inherent sovereignty of Indigenous nations and understand how tribal sovereignty can actually benefit them—by pressuring state and federal governments into action (Grossman 2005). By slowly turning local governments from adversaries into allies, tribal governments can strengthen their own sovereignty (Zaferatos 2004).

Native Nations and Federal Governments

Indigenous nations in different countries have many varied relationships to their national governments—from treaty relationships to autonomous territories and (in the United States) federal trust responsibility. In the United States, the Bush administration was reluctant to meet with tribes or to meet international standards. President Obama has been more open to a stronger trust relationship with sovereign tribes, giving an opportunity to tribes to shift gears and go from the defensive to the offensive when it comes to protecting natural resources and economies from climate change. A major obstacle to climate action at the federal level is the climate-skeptic sentiment that has grown during the Obama Administration, shaping its reluctance to take a stand for change at UN climate summits.

One cartoon of the 2009 Copenhagen summit (seen in USA Today) sums up the problem. The conference screen displays the desperately needed measures to lessen greenhouse gas emissions: "Preserve rainforests,

Sustainability, Green jobs, Livable cities, Renewables, Clean water, air, Healthy children." A perturbed white man turns to a woman of color and asks, "What if it's a big hoax and we create a better world for nothing?" (She merely looks back at him, annoyed.)

In September 2009, Secretary of the Interior Ken Salazar ordered his staff to pay attention to tribal concerns about climate change:

> Climate change may disproportionately affect tribes and their lands because they are heavily dependent on their natural resources for economic and cultural identity. As the Department has the primary trust responsibility for the Federal government for American Indians, Alaskan Natives, and tribal lands and resources, the Department will ensure consistent and in-depth government-to-government consultation with tribes and Alaskan Natives on the Department's climate change initiatives. Tribal values are critical to determining what is to be protected, why, and how to protect the interests of their communities. The Department will support the use of the best available science, including traditional ecological knowledge, in formulating policy pertaining to climate change (Salazar 2009).

Reforming Federal Laws

An important tribal tool in the United States has been the Treatment-As-State (TAS) status recognized by the Environmental Protection Agency (EPA). The U.S. Congress amended the Clean Air Act in 1977, adding Prevention of Significant Deterioration provisions allowing a governmental entity to "redesignate" its air quality to a higher standard. The Northern Cheyenne tribe in Montana was the first tribe to use the amendment to secure "Class I" air quality over its reservation (Small 1994). In 1990, Congress again amended the Clean Air Act to authorize the EPA to treat tribes as "states" whenever tribes are capable of carrying out state-like regulatory and enforcement authority.

TAS status and sovereign environmental standards have been an even stronger tribal environmental tool when they were used to protect a more localized and trackable natural resource: water. In 1987, congressional amendments to the Clean Water Act allowed the EPA to treat "qualified" tribes as states for regulatory and enforcement purposes. The act allowed tribes designated by the EPA to have the same powers as states in setting EPA-approved water quality standards that would govern upstream polluters inside and outside reservation boundaries. Isleta Pueblo in New Mexico,

for example, successfully secured TAS status in order to force Albuquerque to stop dumping municipal wastes into the Rio Grande upstream from the reservation (Gordan 2000).

The power to enhance tribes' own air and water quality standards represented a new and potentially powerful tool to protect traditional resources and reservation environments, but it has stimulated strong resistance from state and local governments (Galloway 1995). As of now, TAS standards can counter threats to air and water that (in the words of one tribal environmental coordinator) are "very close, very big, very nasty" (Forest County Potawatomi Tribe 1995). But in the emerging political landscape, tribal governments and their allies could begin to lobby for a change in the law to cover impacts on tribal air and water from more distant sources, such as coal plants, to address even more severe threats, including acid rain and climate change.

Using Treaties to Protect Habitat

Climate change is an environmental violation of treaty rights. Emitting greenhouse gases into the atmosphere alters the climate and so alters or eliminates habitat for species that tribes were guaranteed access to in the treaties. Since it may destroy habitat for tribal resources, climate change can be seen as a violation of treaty rights. Pacific Northwest tribes have used treaty rights to get a seat at the table to decide resource policy covering treaty resources such as salmon, shellfish, wild game, and medicinal plants. The Northwest Indian Fisheries Commission (NWIFC), Columbia River Inter-Tribal Fish Commission (CRITFC), and tribal resource departments use the treaty powers to protect habitat and in doing so are already dealing with issues affected by climate change. When melting glaciers and a reduced mountain snowpack reduce stream flow, salmon and other aquatic life may die. But the treaties only recognize tribal rights within ceded lands and waters, so what happens when the species shift outside the treaty boundaries?

In the Pacific Northwest, the Boldt II process opened up the possibility of tribes using treaty rights in federal court to force states and private interests to protect or restore fish habitat, and to force effective management of natural resources. The prospect would seem to hand tribes an unprecedented legal trump card to protect the environment. After the 1980 Orrick Decision, Northwest tribes used treaty rights as a political and legal wedge to defeat proposals that threatened fish habitats (Cohen 1986; U.S. District Court 1980). Yet tribes have been reluctant to pull out their "treaty card" in federal environmental cases. Using the treaties can open tribal sovereignty to unfavorable rulings by federal courts, which have at times interpreted a tribal share in the resources to include a share in the diminishment of the resources.

Despite this tribal reluctance, resource companies were terrified by the implications of the Boldt II process and anticipated that the tribes would continue their string of federal court victories from harvest allocation issues to habitat issues. The industries' fears in fact provided one more reason for the tribes to not pursue Boldt II in the courts; in short, the tribes did not have to. Industries and state agencies were willing to come to the negotiating table with the tribes, simply out of fear of the drawn-out, expensive, and economically paralyzing lawsuits that would result if they did not. The outcome in Washington state was the present system of tribal–state co-management. The main point is that tribes did not necessarily need a court victory to bring industry and governments to the table.

Yet the tribes are now beginning to win federal court victories to protect salmon habitat. In 2007, the U.S. District Court of Western Washington found in favor of treaty tribes who had sought restrictions on culverts that blocked salmon, and directed the state to repair or replace existing culverts and impose "fish-friendly" conditions on new culverts. The federal court declared "that the right of taking fish, secured to the Tribes in the Stevens Treaties, imposes a duty upon the State to refrain from building or operating culverts under State-maintained roads that hinder fish passage and thereby diminish the number of fish that would otherwise be available for Tribal harvest" (U.S. District Court 2007). The Culvert Case may provide a precedent for other cases that affect treaty harvesting of natural resources, perhaps one day even resources diminished by climate change.

Protecting Coastal Communities

As mentioned in the previous section, rising sea levels are emerging as one of the main threats from climate change. An ocean level rise of a one to four feet may seem gradual, but it can make a huge difference in coastal erosion and storm damage, and Intergovernmental Panel on Climate Change (IPCC) scientific projections go higher—up to twenty feet by 2100—as Greenland's ice cap melting increases (Borenstein 2007; Rosenthal and Revkin 2007). Tribal and local governments can shore up beaches against higher waves and protect their fresh water supplies from saltwater intru-

sion, but they can only do so much. Federal involvement is needed when entire villages are endangered. On the Washington coast, federal–tribal cooperation has been enhanced by the establishment of the Olympic Coast National Marine Sanctuary, which is managed jointly by coastal tribes, the state, and the National Oceanic and Atmospheric Administration (Preston 2007).

Several Washington tribes have gained federal support to relocate their coastal housing out of floodplains to higher ground. The Quileute Reservation is moving tribal structures in La Push to higher ground, out of the path of tsunamis (like the ones that struck the West Coast in 1964 and 2011). The tribe closed a trail into a National Park Service beach to put pressure on the federal agencies (Kowal 2006). Congress finally passed a land transfer bill in February 2012 (Hotakainen 2012). The Hoh tribe has also acquired higher land from neighboring governments to move housing and government offices, through a 2010 congressional bill (LaCorte 2006, Mapes 2010). The Skokomish tribe plans to move housing out of a low, marshy area (created by the Cushman hydroelectric project), partly to help clean up the Hood Canal (Allen 2007). The Makah, Quinault, and Lower Elwha Klallam tribes are similarly planning to shift new housing plans to higher ground. Though all these moves have not been taken specifically because of climate change, rising sea levels make the tribal goals far more urgent.

Affirming Trust Responsibility

As an ultimate goal, tribes could begin to pressure the federal government to curb carbon emissions as part of fulfilling the federal trust responsibility to protect reservation air and water. In 2006, the Ninth Circuit Court of Appeals reversed a federal permit for a geothermal plant on land sacred to California's Pit River Tribe, saying that federal "agencies violated their duties ... and their fiduciary duties to the Pit River Tribe by failing to complete an environmental impact statement" (Reuters 2006).

Many other similar cases in the United States have addressed federal trust responsibility to ensure the health and well-being of tribes, but generally on a local scale. A compelling exception is the Ninth Circuit ruling in 2006 that held the Teck Cominco mine in British Columbia responsible for violating U.S. laws by discharging mine wastes downstream in the Columbia River that eventually contaminated the Colville Reservation (U.S. District Court of Appeals, Ninth Circuit 2006; see also *Pakootas v. Teck Cominco* in Center for

The Nisqually Tribe has led collaborative salmon habitat restoration projects in Washington's Nisqually River Basin. Here, the Tribe has worked with the Town of Eatonville to clear banks of weeds, plant trees to prevent erosion, and insert log jams to create pools and shelters for salmon.

Global Law and Policy 2007). The Colville Confederated Tribes had the backing of the EPA and the state of Washington against the Canadian mining company, and a 2004 district court ruling held that U.S. environmental laws apply to pollution regardless of where it originates (Environmental News Service 2004). The case was put on the U.S. Supreme Court's docket, and a ruling against cross-border, point-source pollution may provide a loose precedent for climate change litigation.

In February 2008, the Alaskan Iñupiat community of Kivalina sued Exxon Mobil Corporation, eight other oil companies, fourteen power utilities, and one coal company in a federal lawsuit, claiming that the greenhouse gases they emit threaten the community's existence. The lawsuit estimates the cost of the community's relocation at $400 million. It sought to recover "monetary damages for defendants' past and ongoing contributions to global warming" and "damages caused by certain defendants' acts in furthering a conspiracy to suppress the awareness of the link between these emissions and global warming." The lawsuit was dismissed by the Northern California district court in September 2009, and an appeal was filed (*Native Village of Kivalina v. ExxonMobil Corp. et al.* 2009).

The Winters Doctrine (from the *U.S. v. Winters* case) recognizes tribal rights to sufficient water for a reservation; but it is not clear if these laws can be used

to seek relief when climate change dries up the rivers and streams. The jury is still out on the question of whether Native treaty rights can be used to protect natural resource habitat from a threat as global and pervasive as climate change. Sovereign tribal governments could make appeals to the federal government to cooperate with international agencies in curbing greenhouse gases, but the effectiveness of this appeal is also unclear.

Coordinated International Strategies

At the very least, U.S. tribal governments, First Nations in Canada, Maori tribal nations, and other Indigenous nations could consider a joint, coordinated strategy, to have a voice and presence at the international level. A united Indigenous nations delegation to the United Nations Framework Convention on Climate Change (UNFCCC) is one appropriate vehicle for such advocacy, but certainly not the only one.

Indigenous governments could also consider putting pressure on Asia-Pacific Economic Cooperation (APEC) Forum states to reduce carbon emissions. APEC has brought together states around the Pacific Rim to improve economic and political ties. APEC members include some of the national governments that have been most resistant both to Indigenous sovereignty and to carbon emission reductions. We can often see how federal Indian policies in the United States and Canada are first "tested" in Australia or New Zealand, and vice versa (Galloway 2007). These former British colonies are coordinating their efforts against both Native rights and greenhouse gas reductions; the responses of Indigenous nations should also become more coordinated.

Because climate change is perhaps the most urgent challenge facing Native peoples today, it is critical that Indigenous nations' leaders do not wait for a certain critical mass of nations to sign the United League of Indigenous Nations treaty. The most effective climate change cooperation among the nations will not come bureaucratically from above but organically from below, in the direct cross-border relationships among tribal nations themselves. This kind of bilateral and multilateral cooperation has begun to develop across the colonial boundaries in the Salish Sea and the Great Lakes, and it can develop climate change responses to serve as models for other nations.

Conclusions

The most promising avenues for Indigenous climate change advocacy appear to bypass the established global system of sovereign states, by asserting Native sovereignty in other areas. By not including the settler states, the Treaty of Indigenous Nations recognizes that the sovereignty of First Nations does not stem from their relationship with a federal government but is rather inherent, and stems from their existence before the arrival of the colonial powers.

The treaty also recognizes that the powers of Indigenous nations are not simply legally confined within the Western system of laws, but are also social, economic, cultural, and spiritual. Even if the United States, Canada, and other countries are not responsive to Indigenous concerns, tribal leadership has a responsibility to safeguard the health and well-being of the tribal community by working with other Indigenous peoples, allies, and neighbors.

Indigenous nations can begin to exercise the sovereign right to survive climate change by getting engaged with all levels of government—sharing information within their own communities (especially youth and elders), training and assisting each other to meet the challenges of shifting species, working with neighboring governments to coordinate local responses and planning, challenging industries and governments that contribute to global warming, getting involved directly in the international regulatory process, and much more. U.S. tribes, in particular, have an important role in the middle of the world's largest emitter of greenhouse gases. Only if U.S. policy shifts dramatically will the possibility exist of coordinated international action.

The development of renewable energy systems in Indigenous communities can not only protect the environment from fossil fuel burning but also develop tribal economies and build a new web of economic relationships with non-Native local governments and communities. These innovative and creative approaches may be initially reliant on national funding but can help build a de facto sovereign reality on the ground for Indigenous nations. At the same time, they can provide a model to non-Native communities that they do not have to be reliant on centralized, corporate control of the energy economy—the status quo that generated the climate change crisis.

The most important Indigenous responses to climate change will not be in tribal government offices or negotiations over political rights with other governments, but in the ability of tribal members to pass on cultures that respect the land. Tribes have survived conquest, wars, epidemics, poverty, and resource shortages before but have persevered through keeping the cultures strong. The late Ojibwe environmental leader Walter Bresette proposed a Seventh Generation Principle as language for state, tribal, and national constitutions: "The right of the people to use and enjoy air, water, sunlight, and other natural resources determined to be common property, shall not be impaired, nor shall such use impair their availability for future generations" (LaDuke 1999).

The United League of Indigenous Nations is one vehicle for tribal nations to help each other survive the changes ahead and exercise their sovereignty to meet the challenge of global warming, instead of simply asking the colonial system to take action. As Haudenosaunee leader Oren Lyons told the historic 2007 treaty gathering at the Lummi Nation: "*Sovereignty is the Act Thereof.*"

Resources

Alaska Youth for Environmental Action. 2009. http://www.ayea.org

Allen, Chester. 2007. Tribe's New Housing Plan Aims to Help Struggling Canal. *The Olympian*, March 4.

Borenstein, Seth. 2007. Global Warming Unstoppable, Report Says. *Washington Post*, February 2. http://www.washingtonpost.com/wp-dyn/content/article/2007/02/02/AR2007020201093_pf.html

Campus Climate Challenge. 2009. The Story of Energy Action Coalition and The Challenge. http://www.campusclimatechallenge.com

Center for Water Advocacy. 2007. 4th Annual Northwest Tribal Water Rights Conference.

Climate Change: Impacts to Water, Fish, Cultures, Economies, and Rights. Hosted by Squaxin Island Tribe, Kamilche, Washington, October 24–25. http://turtletalk.wordpress.com/2007/10/20/squaxin-island-4th-annual-tribal-water-rights-conference-climate-change-impacts-to-water-fish-cultures-economies-and-rights/

Center for Global Law and Policy. 2007. "*Pakootas v. Teck Cominco Metals, Ltd.*: Liability for Transboundary Environmental Damage?" Santa Clara University School of Law. http://law.scu.edu/international/pakootas.cfm

Cohen, Fay G. 1986. *Treaties on Trial: The Continuing Controversy over Northwest Indian Fishing Rights*. A Report Prepared for the American Friends Service Committee. Seattle: University of Washington Press.

Collier, Russell, and Martine Rose. 2000. The Gitxsan Model: A Vision for the Land and the People. ESRI Conservation Program website, December 7. http://www.conservationgis.org/native/native2.html

Environmental News Service. 2004. Teck Cominco Will Appeal Columbia River Pollution Ruling. *Mines and Communities*, November 10. http://www.minesandcommunities.org/Action/press480.htm

Erzen, Alex. 2006. Climate Change and Coastal Vulnerability in Washington (draft). Ocean Policy Work Group, March 21. http://courses.washington.edu/oceangov/OPWG_Docs/3_15Drafts/ClimateChange3_21_06.pdf

First Nations Development Institute. 2004. Native Agriculture and Food Systems Initiative (NAFSI), guide to the Native Food Summit, Milwaukee, Wisconsin, September 9–11. http://www.docstoc.com/docs/20682827/Summit-Report

Flesher, John. 2007. Governors Address Climate Change. *Washington Post*, July 21. http://www.washingtonpost.com/wp-dyn/content/article/2007/07/21/AR2007072101160.html

Forest County Potawatomi Tribe. 1995. Class I air quality request to the U.S. Environmental Protection Agency. Wisconsin, June.

Galloway, Gloria. 2007. Did Australia Demand Reversal on Natives? *Toronto Globe and Mail*, June 9.

Galloway, William C. 1995. Tribal Water Quality Standards Under the Clean Water Act: Protecting Traditional Cultural Uses. *Washington Law Review* 70:177–202.

Ghoghaie, Nahal. 2011. Native/non-Native Watershed Management in an Era of Climate Change: Freshwater Storage in the Snohomish Basin. Master's Thesis, Master of Environmental Studies program, The Evergreen State College, Olympia, Wash.

Gordan, Linda I. 2000. Water Quality Standards for Arsenic in the Rio Grande: Isleta Pueblo Water Quality Protection and the Clean Water Act. In *Policy Conflicts and Sustainable Water Resources Development in New Mexico's Rio Grande Basin*, Michael E. Campana, ed. University of New Mexico Water Resources Program Publication No. WRP-2 (February). Albuquerque: University of New Mexico Water Resources Program. http://www.unm.edu/~wrp/wrp-2.pdf

Gough, Robert. 2005. Western Regional Air Partnership Meeting on Climate Change: Tribal Climate Change Initiatives. PowerPoint presentation, Palm Springs, Calif., Dec. 14–15. http://www.wrapair.org/forums/cg/meetings/050826climate/gough.pdf

Grossman, Zoltan. 2005. Unlikely Alliances: Treaty Conflicts and Environmental Cooperation between Native American and Rural White Communities. *American Indian Culture and Research Journal* 29:4:21–43.

Havemann, Paul, and Helena Whall. 2002. *The Miner's Canary: Indigenous Peoples and Sustainable Development in the Commonwealth*. London: Indigenous Rights in the Commonwealth Project, University of London Institute of Commonwealth Studies. http://www.commonwealthadvisorybureau.org/fileadmin/CPSU/documents/Projects/Indigenous_Rights/2002_The_Miner_s_Canary_Indigenous_peoples_and_sus_devt_2002.pdf

Hotakainen, Rob. 2012. Quileutes' land transfer gets Congress' OK. The Olympian (February 15). http://www.theolympian.com/2012/02/15/1990789/quileutes-land-transfer-gets-congress.html

Institute for Tribal Environmental Professionals. 2009. Tribes and Climate Change. Flagstaff: Northern Arizona University. http://www4.nau.edu/tribalclimatechange

International Institute for Sustainable Development and Hunters and Trappers Committee of Sachs Harbour. 2000. *Sila Alangotok: Inuit Observations on Climate Change* (video). http://www.iisd.org/media/2000/nov_16_2000.asp

Inter-Tribal Buffalo Cooperative (ITBC). 2009. http://itbcbuffalo.com/

Intertribal Council On Utility Policy (COUP). 2009. http://www.intertribalCOUP.org

Iyall, Cynthia. 2008. City, tribe team up on clean water project. *Seattle Daily Journal* (June 26). http://www.djc.com/news/en/11202008.html

It's Getting Hot in Here. 2009. *Dispatches from the Youth Climate Movement.* www.itsgettinghotinhere.org

Kamb, Lewis. 2003. A Methane to Their Madness: Tribes and Farmers Come Together—over Cow Manure." *Seattle Post-Intelligencer*, April 22. http://www.seattlepi.com/news/article/A-methane-to-their-madness-1112846.php

Kapralos, Krista J. 2008. Science is Hitching a Ride on Tribal Canoe Journey. *Indian Country News*, March 3. http://www.heraldnet.com/article/20080303/NEWS01/855603740

Kowal, Jessica. 2006. In a Bid for Higher Ground, a Low-Lying Indian Tribe Raises the Stakes. *New York Times*, July 30. http://travel2.nytimes.com/2006/07/30/us/30beach.html?partner=rssnyt&emc=rss&pagewanted=print

LaCorte, Rachel. 2006. Quileute Tribe holds beach access as leverage for land exchange. Seattle Times November 24. http://community.seattletimes.nwsource.com/archive/?date=20061124&slug=dispute24m

LaDuke, Winona. 1999. *All Our Relations: Native Struggles for Land and Life.* Boston: South End Press.

Little Traverse Bay Bands of Odawa Indians. 2005. Resolution #051505-01: Adoption of Kyoto Protocol and Renewable Energy Standards.

Mapes, Lynda V. 2010. Congress OKs land transfer to Hoh Tribe. Seattle Times (December 14). http://seattletimes.nwsource.com/html/localnews/2013679641_hoh15m.html

Maynard, Nancy, ed. 2001. *Circles of Wisdom: Native Peoples–Native Homelands Climate Change Workshop Final Report.* Washington, D.C.: U.S. Global Climate Research Program. http://www.usgcrp.gov/usgcrp/Library/nationalassessment/native.pdf

McAllister Wellfield Partnership. 2008. Nisqually Tribe–City of Olympia Agreement, May 13. http://olympiawa.gov/city-utilities/drinking-water/McAllister.aspx

McDonald, Miriam, ed. 1997. *Voices from the Bay: Traditional Ecological Knowledge of Inuit and Cree in the Hudson Bay Bioregion.* Ottawa, ON: Canadian Arctic Resources Committee and the Municipality of Sanikiluaq. http://www.carc.org/pubs/v25no1/voices.htm

McNutt, Debra, ed. 2010. *Northwest Tribes Meeting the Challenge of Climate Change.* Olympia, WA: Northwest Indian Applied Research Institute. http://academic.evergreen.edu/g/grossmaz/climatechangebooklet.pdf

Melmer, David. 2005. Renewable energy may bring economic boom. *Indian Country Today*, Nov. 18. http://www.bluefish.org/reneboom.htm

Mystic Lake Declaration. 2009. Native Peoples/Native Homelands Climate Change Workshop II: Indigenous Perspectives and Solutions, Shakopee Mdewakanton Dakota Community, Minn., November 21.

National Wildlife Federation. 2006. Tribal Lands Climate Conference. Hosted by Cocopah Indian Nation, Yuma, Arizona, December 5–6. http://www.tribalclimate.org

Native Energy. 2009. How We Are Different. http://www.nativeenergy.com/who-we-are

Native Peoples/Native Homelands Climate Change Workshop II. 2009. Shakopee Mdewakanton Sioux Community, Minnesota, November 18–21. http://www.nativepeoplesnativehomelands.org

Native Seeds/SEARCH. 2009. About Us. http://www.nativeseeds.org

Native Village of Kivalina v. ExxonMobil Corp. et al. 2009. U.S. District Court of Northern California No. C 08-1138 SBA, September 30.

Native Wind. 2009. www.nativewind.org

Northwest Indian Applied Research Institute (NIARI). 2006. *Climate Change and Pacific Rim Indigenous Nations.* Olympia, WA: The Evergreen State College. http://academic.evergreen.edu/g/grossmaz/IndigClimate2.pdf

———. 2009. Climate Change and Pacific Rim Indigenous Nations, PowerPoint presentation, http://academic.evergreen.edu/g/grossmaz/IndigClimate2009.ppt

Northwest Renewable Resources Center. 1997. *Building Bridges: A Resource Guide for Tribal-County Intergovernmental Cooperation.* Seattle, WA: Northwest Renewable Resources Center.

Northwest Sustainable Energy for Economic Development (NWSeed). 2009. *Northwest Energy Planning: A Guide for Northwest Indian Tribes.* Seattle, WA: Northwest Sustainable Energy for Economic Development. http://www.nwseed.org/documents/NWSEED_Tribal GB_Final.pdf

Obama, Barack. 2009. Remarks by the President During the Opening of the Tribal Nations Conference and Interactive Discussion with Tribal Leaders, November 5. http://www.whitehouse.gov/ the-press-office/remarks-president-during-opening-tribal-nations-conference-interactive-discussion-w

Papiez, Chelsie. 2009. Climate Change Implications for Quileute and Hoh Tribes of Washington: A Multidisciplinary Approach to Assessing Climatic Disruptions to Coastal Indigenous Communities. Master's thesis, The Evergreen State College. http://archives.evergreen.edu/mastderstheses/Accession86-10MES/Accession86-10E-Theses.htm

Preston, D. 2007. Marine Sanctuary Council Formed. *NWIFC News* (Spring 2007). http://access.nwifc.org/ newsinfo/documents/newsletters/2007_1_Spring. pdf

Renewable Energy Access. 2006. Wave Energy Project Gets Environmental OK. October 30. http://www.renewableenergyworld.com/rea/ news/article/2006/10/wave-energy-project-gets-environmental-ok-46408

Reuters News Service. 2006. U.S. Court Backs Tribe in Fight Over Calpine Plant. *Klamath Forest Alliance*, November 6. http://www.klamathforestalliance.org/ Newsarticles/newsarticle20061104.html

Rosen, Yereth. 2004. Warming Climate Disrupts Alaska Natives' Lives. Reuters, April 20. http:// omega.twoday.net/20040420

Rosenthal, Elisabeth, and Andrew Revkin. 2007. Panel Issues Bleak Report on Climate Change. *New York Times*, February 2. http://www.nytimes. com/2007/02/02/science/earth/02cnd-climate.html

Salazar, Ken. 2009. Addressing the Impacts of Climate Change on America's Water, Land, and Other Natural and Cultural Resources. Interior Department Secretarial Order No. 3289, September 14. http://www.nps.gov/sustainability/documents/ Quick-Links/SecOrder3289[1].pdf

Severson, Kim. 2005. Native Foods Nourish Again. *New York Times*, November 23. http:// www.nytimes. com/2005/11/23/dining/23nati.html

Small, Gail. 1994. The Search For Environmental Justice in Indian Country. *News From Indian Country*, March. http://nativenet.uthscsa.edu/ archive/nl/9404/0029.html

Sustainable Communities. 2003. Hartley Bay, British Columbia: Gitga'at First Nation Community Energy Planning. http://www.pembina.org/bc/communities

Sustainable Northwest. 2009. Tribal Climate Change Forum. http://www.sustainablenorthwest.org/ events/tribalclimateforum

Swinomish Indian Tribal Community. 2009. Swinomish Climate Change Initiative. Adaptation Strategy Toolbox and Draft Impact Assessment Technical Report. http://www.swinomish-nsn.gov/ climate_change/project/reports.html

Tribal Campus Climate Challenge. 2009. Indigenous Environmental Network. http://www.ienearth.org/ tccc.html

UCAR Community Building Program. 2008. Planning for Seven Generations: Indigenous and Scientific Approaches to Climate Change, Boulder, CO, March 19–21. http://www.cbp.ucar.edu/conferences/ seven_generations

United League of Indigenous Nations. 2007. United League of Indigenous Nations Treaty. http://www. indigenousnationstreaty.org

U.S. District Court of Western District Washington. 1980. *United States v. State of Washington.* 506 Federal Supplement 187. Washington, DC: Government Printing Office.

U.S. District Court of Western District Washington. 2007. Culvert Case Summary Judgment. *United States of America et al v. State of Washington.* Case No. 9213RSM. Subproceeding No 01-01. http:// www.scribd.com/doc/259364/Culvert-Case-Summary-Judgment

U.S. District Court of Appeals, Ninth Circuit. 2006. *Pakootas v. Teck Cominco Metals, Ltd.*, 452 F.3d 1066. http://www.ca9.uscourts.gov/datastore/ opinions/2011/06/01/08-35951.pdf

University of Colorado Natural Resources Law Center. 2007. *Native Communities and Climate Change: Protecting Tribal Resources as Part of National Climate Policy.* http://www.colorado.edu/law/ centers/nrlc/publications/ClimateChangeReport-FINAL _9.16.07_.pdf

White Earth Land Recovery Project. Mino-Miijim (Good Food) Program. www.nativeharvest.com/ node/2

Zaferatos, Nicholas Christos. 2004. Tribal Nations, Local Governments, and Regional Pluralism in Washington State: The Swinomish Approach in the Skagit Valley. *Journal of the American Planning Association* 70(1) (Winter). http://faculty.wwu.edu/zaferan/ Zaferatos%20-%20JAPA%20Winter%2004.htm

RECOMMENDATIONS TO NATIVE GOVERNMENT LEADERSHIP

Alan Parker

Director of the Northwest Indian Applied Research Institute; faculty in the Master of Public Administration program, The Evergreen State College, Olympia, Washington

Recommendations for Indigenous leadership regarding Climate Change Impacts and Tribal Communities

This section of the anthology is addressed primarily to the leadership of U.S. tribal nations and to First Nations, Maori, and other Pacific Rim Indigenous nation leaders. Through our research and consultations with tribal officials, we have determined that climate change impacts Indigenous peoples in distinctive ways that demand distinctive responses.

Climate change brought about by human activity has already resulted in irreversible global warming. It is impacting Native peoples in the Arctic and Subarctic by permanently disrupting their lifestyles centered on subsistence hunting and fishing. Over many thousands of years the Inuit, the Iñupiat, and the Yup'ik have lived in Arctic climates by adapting to frigid weather conditions and seasonal changes, based on long winter nights and long summer days. In a matter of a few short years these conditions have been disastrously altered.

Global warming changes are much more intense in the Arctic than they are elsewhere. Whereas globally we have seen average mean temperature increases of 1 percent, Arctic peoples have experienced an increase of up to 8 percent. In many areas the permafrost has melted and ice packs have retreated at a rate that is beyond the adaptation abilities of wildlife and ocean life. *For Indigenous people who live on the land and the water, climate change is already a disaster.*

In the temperate zones where the great majority of Pacific Rim tribal peoples live, evidence of disastrous climate change has already been documented, although it has escaped the attention of the U.S. public (which depends on commercial media for information).

In the Pacific Northwest, glacier-fed rivers and streams have permanently warmed because of the decline in winter snowpack and the retreat of high mountain glaciers. Global warming means that these glaciers will not reappear, and the fish and wildlife that depend on clear, cold water are disappearing. These fish and wildlife are necessary supplements to the diet of Northwest tribes and essential to the practice of their tribal cultures.

The Critical Importance of Community

When we contemplate the urgent need to plan for climate change impacts, an important advantage that Indigenous peoples have in contrast to the non-Indigenous population is that *we still have community!* Our tribal communities are comprised of extended families who care for each other, who keep track of each other, and who insist that the collective family, the tribe, do everything possible to take care of the tribal community.

Our colleague from the Maori nation points out that the foundation of Maori community is the *marae*. As we understand it, a marae may be thought of as a building and is comparable to Native American communities that are organized around a longhouse. But the marae is not simply a building or a structure that serves as a meeting place for those families that belong to a particular marae. It is also a sacred space that serves as the center of community ceremonial life. We were informed that the members of a marae, historically a tribe or subtribe, assume a responsibility for community members who experience family losses or tragedy, as well as to extending hospitality to ensure that guests of the marae are fed and housed.

In this respect, Maori custom is typical of that of Indigenous people worldwide, who place great value on maintaining community as the focal point for cultural and social values and practices. As we contemplate the potentially disastrous consequences of climate change upon our communities, it seems clear that we must look to the structures and institutions of community as the means to prepare for and deal with these consequences.

As one contemplates acceleration of climate change in the next five to ten years, our tribal communities must adapt to these changing conditions at a pace that will stress their social, economic, and cultural fabrics. But, we cannot afford to join our fellow Americans in massive denial. *The time to plan and adapt is now.*

Recommendations

1. *Gather information* on the impacts of climate change in your region and *make it available to your tribal community.* As tribal people who have survived against all odds in the past, we will survive the changes associated with global warming—if we prepare. We cannot even begin to prepare if we are uninformed or unaware of the facts as they pertain to our own regions and localities. Moreover,

this information must be shared within the entire community if leadership is to have the support and understanding to allocate tribal resources to the different steps discussed below. Our project is intended to assist tribal leadership begin the process of informing their respective communities, but it must be supplemented by information specific to the different regions and locales that comprise Indian Country in the United States and Indigenous communities in the Pacific Rim region.

2. **Secure sources of water** (fresh water and drinking water) for drought-impacted regions. In North America we have many reports that global warming will result in severe weather change taking the form of extended droughts. In the near future we could face situations in which drinking water supplies for metropolitan areas (derived from reservoirs and other storage facilities and underground aquifers) literally dry up. For those tribal communities living adjacent to such localities, prior planning could be extremely crucial. Even in rural areas there may be demand for agricultural or industrial water use that far exceeds supply, and tribal communities may be forced to pay exorbitant rates unless they have already secured water sources. Under existing federal law the Winters right (based on the *U.S. v. Winters* case) recognizes tribal rights to sufficient water to fulfill the purposes of the federal reservation. Not all tribes are in a position to assert such a federal right, and even those that have already established such rights may find them overridden in a time of extreme shortage.

3. **Secure sources of food** that will continue to be available in disaster or emergency conditions:

    ▶ Plan storage facilities for both perishable and non-perishable foods

    ▶ Negotiate agricultural production capabilities for food crops that will be able to adapt to changing climate conditions through agreements among tribes

    This recommendation contemplates that tribes will need to develop a level of self-sufficiency because of a climate change-induced breakdown in the commercial food production and supply infrastructure. Even if such an extreme situation is avoided in the future, prices will no doubt increase in relation to constraints on supply. Having a supplementary food production and storage capability could be crucial to maintaining a healthy diet for the community. For example, different root crops may be more adaptable to drought conditions or might produce much more volume from the same amount of land that is in production.

    Many small land-based tribes do not have food production resources, while some large land-based tribes have excessive production capabilities. If the smaller tribes are planning ahead, they may consider proposing joint agreements that involve the investment of capital resources into food production systems with tribes possessing a larger land base.

4. **Prepare for impacts on plant and animal species**; determine whether species adaptation will provide alternative sources of culturally important plant and animal species. As climate change impacts our landscapes, inevitably we will see the disappearance of fish runs and entire fisheries. Perhaps in some cases such fisheries will be replaced by other runs as species adapt by shifting north or south.

    Drought conditions will also influence plant and animal species that have sustained Indigenous peoples as subsistence food sources or as essential elements of their ceremonial life. To the extent that they can, tribes should be studying these phenomena in order to adapt as the plants and animals adapt.

5. **Develop relationships with neighboring governments and communities** regarding emergency plans for the more disastrous impacts of climate change. It is predictable that local governments will be more responsive to climate change than state or national governments. They are by definition closer to the people, who will no doubt demand that government take whatever action is possible, and their smaller size makes them more flexible. Although the residents of jurisdictions under local governments are not in communities as close-knit as tribal communities, they still possess the ability to work together as neighbors, having common interests in local impacts on their homes and neighborhoods. It is recommended that tribal leadership look to these municipal and county governments to search for common ground in addressing climate change impacts.

6. **Consider political alliances** with municipal, county, and other local governments **to build a renewable energy policy** on local, regional, and national levels with the aim of

    ▶ developing carbon emission reduction and control measures;

    ▶ developing renewable energy policies and practices;

► advocating for responsive and proactive positions by national governments.

Many local government leaders have taken initiative, and west coast mayors asserted leadership among U.S. cities in adopting carbon emission control measures that would apply to city government. California has adopted stringent carbon emission standards that would apply to the world's third-largest economy.

Tribal governments should consider joining municipal governments in these efforts, as discussed at the 2005 Native Renewable Energy Summit. The value of such local initiatives with respect to carbon emission controls is not that they could impact global warming directly but that they will have a cumulative effect, and they keep the issue on the public agenda until such time as our federal government is forced to respond.

In contrast to their relative inability to affect carbon emission levels, U.S. tribes are in a posi-tion to participate meaningfully in programs of renewable energy. Indigenous nations can begin convert from dependence on fossil fuels and the centralized electrical grid to cleaner and more locally controlled renewable energies, and select the energy systems most appropriate for their natural region, whether it is wind, solar, biomass, wave, tidal, and so on. These energy sources can also provide a source of income and trade and serve as an example to non-Indigenous communities.

The Tulalip Tribes in Washington state have embarked on a biomass energy development project involving a number of large dairy farmers. The project will benefit both parties. Through making use of diary farm waste to produce methane (which is then burned to generate electricity), the Tribes not only are acquiring a source of renewable energy but also are providing for disposal of diary farm waste that would otherwise add to the pollution of the Snohomish River, a major salmon fishery for the tribe. The dairy farmers acquire a

Suquamish canoe family from the Salish Sea attending the Treaty of Waitangi commemoration in Aotearoa/New Zealand, February 6, 2011.

means of waste disposal to relieve them of potential liability and can thereby expand to meet increased demand for their dairy products. Other tribes have embarked on wind, wave energy, small hydroelectric, thermal, and biomass projects.

Native American communities through their tribal governments are uniquely positioned to become influential and credible advocates for public education and much-needed government initiative on global warming. This is an opportunity to address a real issue that has already began to impact their vital interests in distinctive manner, and by taking the lead tribes will gain public support and even admiration.

7. ***Consider strategies to unite Tribes and First Nations around the habitat protection*** needed to defend treaty rights to fish and game, and to seek enforcement of U.S. trust responsibilities to protect tribal homelands through U.S. adoption of carbon emission control and mitigation. Treaty signatory tribes, particularly those in the Pacific Northwest, have been engaged in many ways and on many levels in asserting their rights, guaranteed under treaty, that fish and game habitat be protected. As we have already seen, climate change impacts habitat in devastating ways.

By creating a national program to unite tribes around treaty-defined habitat protection issues, and by combining the treaty rights protection claims with claims that the United States is breaching its trust duty to protect tribal homelands, tribes may strengthen their voices in a unique way. Through the strategy of placing such tribal advocacy in the context of climate change and the lack of U.S. response to evidence of climate change, U.S. tribes raise a new argument that only tribes are entitled to make, thereby generating yet another level of public pressure and public education.

8. ***Consider active involvement as sovereign Indigenous governments in global climate change negotiations***, including formal Indigenous government representation at the annual Conference of the Parties of the United Nations Framework Convention on Climate Change (UNFCCC). The special and unique concerns of Indigenous nations have so far only been represented by non-governmental organizations (NGOs) at this UN forum.

It is recommended that U.S. tribal governments, First Nations based in Canada, Maori tribal

nations, and other Indigenous nations in the Pacific Rim consider a joint, coordinated strategy, perhaps through the structure of the United League of Indigenous Nations, to have a voice and presence at the international level. The UNFCCC is the most appropriate venue for such advocacy, but certainly not the only one. Pacific Rim Indigenous nations should also consider putting pressure on Asia-Pacific Economic Cooperation (APEC) Forum states to reduce carbon emissions.

9. ***Get youth involved in cultural education and defending the future of their nation*** from harmful climate change. By making the youth of Indigenous communities more aware of climate change issues, tribal leaders can energize and inspire a level of activism and involvement that will benefit future generations. Young people are often more idealistic than their parents' generation and willing to commit to a cause. However, they need the elders' traditional ecological knowledge, including awareness of climatic cycles and species. Support youth who want to set up groups of secondary school and college-age Indigenous youth around these issues.

10. ***Work with other Indigenous nations*** across imposed colonial boundaries on the basis of being part of the same natural region (such as the Pacific Rim). Share skills and knowledge about animal and plant species, as their habitats shift due to climate change. Exchange ideas and training about community adaptation, climate change mitigation, and renewable energies.

Pacific Rim Indigenous nations can use the Treaty of Indigenous Nations process as a way to build political alliances and practical programs that are based on cooperation and joint responses to climate change impacts. As the Pacific Rim states have cooperated historically in order to colonize Indigenous lands and increase industrial growth, Indigenous nations need to cooperate to decolonize ancestral territories and protect our common property (the air and water) for future generations.

PART V. NATIVE CLIMATE CHANGE RESOURCES
AUTHOR BIOGRAPHIES

NATIVE CLIMATE CHANGE RESOURCES
Examples of Model Projects and Groups, Organized around NIARI Recommendations

PROJECT

Northwest Indian Applied Research Institute (NIARI)

Founded in 1999 to expand the services that The Evergreen State College can offer Northwest tribes, assisting them to meet their economic, governance, and resource goals. At the same time, the Institute has provided additional, real-life learning opportunities for Evergreen students.

http://nwindian.evergreen.edu

Alan Parker, Director, parkeral@evergreen.edu

Climate Change and Pacific Rim Indigenous Nations Project

NIARI project initiated in 2006 to document effects of climate change on Native homelands, describe examples of Indigenous nation responses at local and international levels, and recommend future paths for Indigenous nation governments to consider.

Project page with updated links, PowerPoint, and community organizing booklet available for download.

http://academic.evergreen.edu/g/grossmaz/climate.html

Zoltán Grossman, grossmaz@evergreen.edu

Faculty website: http://academic.evergreen.edu/g/grossmaz

Evergreen Native Programs

Native American and World Indigenous Peoples Studies (NAWIPS) undergraduate Olympia program, Reservation-Based / Community-Determined Program, Master of Public Administration / Tribal Concentration, and Longhouse Education and Cultural Center.

http://www.evergreen.edu/nativeprograms

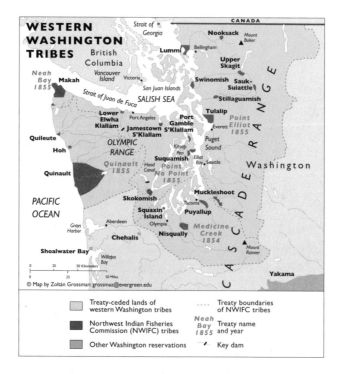

I. SHARING INFORMATION

Educate tribal membership on the present and future effects of climate change on tribal homelands.

Native Peoples/Native Homelands Climate Change Workshops
NASA-funded gatherings of Native peoples to exchange information and responses on climate change in their regions. The 1998 conference released the Circles of Wisdom report, and the 2009 conference released the Mystic Lake Declaration.

http://www.usgcrp.gov/usgcrp/Library/nationalassessment/native.pdf (1998)

http://www.nativepeoplesnativehomelands.org (2009)

(605) 747-4097 or (605) 441-8316

gough.bob@gmail.com or nancy.g.maynard@nasa.gov

Native Communities and Climate Change
A 2007 report by the Natural Resources Law Center at the University of Colorado School of Law, including comprehensive overviews of effects in each U.S. region.

http://www.colorado.edu/Law/centers/nrlc/publications/ClimateChangeReport-FINAL%20_9.16.07_.pdf

(303) 492-1287

hannajm@colorado.edu or mark.squillace@colorado.edu

Indigenous Environmental Network (Native Energy and Climate Program)
Strengthens and builds the capacity and political power of Indigenous peoples to address the impacts of fossil fuel development in Indigenous communities and advocates the creation of sustainable and clean energy and climate policies at all levels of governance.

http://www.ienearth.org/climatejustice.html and http://www.ienearth.org/nativeenergy.html

PO Box 485, Bemidji, MN 56619

(218) 751-4967 or (928) 214-8301

ienenergy@igc.org or ien@igc.org

Tribes and Climate Change
Website of the Institute for Tribal Environmental Professionals at Northern Arizona University, providing information, resources, and contacts tailored to helping Native people gain a better understanding of climate change in their communities.

http://www4.nau.edu/tribalclimatechange

ITEP, PO Box 15004, Flagstaff, AZ 86011

(928) 523-1488 or (928) 714-1906

Susan.Wotkyns@nau.edu or Dennis.Wall@nau.edu

American Indian and Alaska Native Climate Change Working Group
Because of this organization's recognition that the lifeways of Indigenous people hold tremendous lessons for all of humankind to consider, it takes steps to ensure tribal peoples will have the expertise within their own communities to make good decisions.

http://www.haskell.edu/climate/index.html

Red Alert!: Saving the Planet with Indigenous Knowledge (Dr. Daniel Wildcat)

http://www.fulcrum-books.com/productdetails.cfm?PC=6120

AIS, Haskell Indian Nations University, 155 Indian Ave., Lawrence, KS 660446

(785) 832-6677 office • (785) 832-6643 fax

nabis1@cresis.ku.edu and dwildcat@sunflower.com

Pacific Northwest Tribal Climate Change Network
Formed in 2009 to provide guidance and assist with outreach on tribal climate change efforts.

http://tribalclimate.uoregon.edu/network

(503) 808-2018 or (541) 346-5777

edonoghue@fs.fed.us or kathy@uoregon.edu

Planning for Seven Generations
March 2008 conference in Boulder, CO, that approached climate change from perspectives rooted in Indigenous experiences and present-day science, to point to shared strategies for understanding, adapting to, and mitigating climate change.

http://www.cbp.ucar.edu/conferences/seven_generations

(785) 832-6677

dwildcat@sunflower.com

Global Climate Change Impacts in the United States
A 2009 report from the U.S. Global Change Research Program, documenting climate change effects in each U.S. region, using accessible language and graphics.

http://www.globalchange.gov/publications

National Wildlife Federation (Tribal Lands Program)
A forum for tribes to share first-hand, on-the-ground accounts of climate change and its impacts on their economic, natural, and cultural resources. Has organized a series of conferences on tribal energy and climate issues since 2006.

http://www.tribalclimate.org

(303) 786-8001 or (303) 441-5153

voggesser@nwf.org or wilensky@nwf.org

Climate Impacts Group
University of Washington interdisciplinary research group studying the regional impacts of climate change and working to increase the resilience of the Pacific Northwest in the face of fluctuations in climate.

http://cses.washington.edu/cig

UW Center for Science in the Earth System, Joint Institute for the Study of the Atmosphere and Ocean, Box 355672, Seattle, WA 98195-5672

(206) 221-0222 or (206) 616-5350

aksnover@u.washington.edu or karpov@u.washington.edu

International Institute for Indigenous Resource Management

Internationally based legal scholars and researchers working on cutting-edge projects designed to empower Native peoples by examining the role the law can play in establishing and enhancing Indigenous peoples' control over and management of their lands and resources. Institute teams also study ways Indigenous peoples can control the impacts of science and technology on their societies and help build and strengthen Native legal, technical, management, and other systems and institutions.

http://www.iiirm.org/

444 South Emerson Street, Denver, CO 80209-2216

(303) 733-0481

mervtano@iiirm.org

California Nevada Applications Program (CNAP)—California Climate Change Center

Climate information for California and Nevada decision makers.

http://meteora.ucsd.edu/cap

Climate Research Division, Scripps Institution of Oceanography,

University of California, San Diego, 9500 Gilman Drive, La Jolla, CA 92093-0224

(858) 534-4507

mtyree@ucsd.edu

Pacific Climate Impacts Consortium

Regional climate service centre on the physical impacts of climate change in B.C. and Yukon.

http://www.pacificclimate.org

PO Box 3060, Stn CSC, U. Vic., Victoria BC V8W3R4 Canada

(250) 721-6236

http://www.pacificclimate.org/contact

Alaska Center for Climate Assessment and Policy (ACCAP)

Alaska climate information and climate change impacts.

http://ine.uaf.edu/accap

ACCAP, University of Alaska–Fairbanks, PO Box 755910, Fairbanks, AK 99775

(907) 474-7878

sarah.trainor@alaska.edu or accap@uaf.edu

Alaska Native Tribal Health Consortium (ANTHC)—Center for Climate and Health

Works to understand local changes and to develop strategies that encourage wellness, resilience, and sustainability. Performs local and regional surveys of health effects related to climate change; provides consultation on climate adaption and mitigation.

http://www.anthc.org/chs/ces/climate

(907) 729-2464

akaclimate@anthc.org

Cold Climate Housing Research Center

Facilitates the development, use, and testing of sustainable, durable, healthy, and cost-effective building technologies for people living in the circumpolar north.

PO Box 82489, Fairbanks, AK 99708

http://www.cchrc.org

(907) 457-3454

http://www.cchrc.org/contact-us

Unikkaaqatigiit: Inuit Perspectives on Climate Change (Canada)

The Inuit, living in the vast Arctic regions, are feeling the first and substantial effects of global warming. Forces mostly outside of the Arctic have caused climate change, manifested in the Arctic by changing sea ice, tundra, and wildlife patterns.

http://www.itk.ca/publication/unikkaaqatigiit-perspectives-inuit-canada

Teacher Handout: http://www.climatechangenorth.ca/section-LP/pdf/LP_07_print.pdf

Māori Environmental Knowledge

Māori environmental knowledge of weather and climate, include the naming and classification of climate phenomena; the oral recording of weather- and climate-based events and trends; and the use of environmental indicators to forecast and predict.

http://www.niwa.co.nz/publications/wa/vol14-no2-june-2006/understanding-local-weather-and-climate-using-maori-environmental-knowledge

Te Kūwaha o Taihoro Nukurangi/National Institute of Water and Atmospheric (NIWA) Research/Māori Research and Development Unit, P.B. 14901, Kilbirnie, Wellington, NZ

64 (4) 386 0511 or 64 (27) 246 6177

c.severne@niwa.co.nz

Tuanuku Climate Change Network

This group has been established as a tool to assist Maori and their allies across Aotearoa, to keep them informed about the latest developments in climate change, disseminate information in their local community, assess the immediate and midterm risks, and identify ways to "future proof" the children, taonga [treasures], community assets, and security by developing local adaptation plans. Food security, water security, and energy security are three key pillars in the notion of Tino rangatiratanga (self-determination).

http://www.facebook.com/group.php?gid=162984541246

rangiwhero@gmail.com

U.S. Environmental Protection Agency, Region 10 (Tribal Trust and Assistance Unit)

Committed to working with federally recognized Tribes in the Pacific Northwest and Alaska on a government-to-government basis, to protect, restore and preserve the environment for present and future generations.

http://yosemite.epa.gov/R10/tribal.nsf

Tribal Leaders' Summit (2012)
http://www.grandronde.org/tls
Promoting Generations of Self-Reliance: Stories and
Examples of Tribal Adaptation to Change (8/12/11)
http://ine.uaf.edu/accap/documents/epa_tribal_
adaptation.pdf
(907) 271-3434
davis.michellev@epa.gov

Alaska Native Knowledge Network
*Resources for compiling and exchanging information
related to Alaska Native knowledge systems and ways of
knowing.*
http://www.ankn.uaf.edu
c/o University of Alaska-Fairbanks, Bunnell Bldg., Rm.
117, Fairbanks, AK 99775
(907) 474-5897
fyankn@ankn.uaf.edu

Climate Change Adaptation Resource
*A wide collection of knowledge, lessons and experience
from five countries across the Northern Periphery,
including twelve communities working to develop their
capacity to adapt to the impacts of current and future
climate changes, under the themes of transport, energy,
risk management and tourism.*
http://www.climatechangeadaptation.info
clive.d.bowman@gmail.com or carlo.aall@vestforsk.no

**C3 - Climate Change Collaboration in the Pacific
Northwest (U.S. Fish and Wildlife Service)**
*Better organizes, integrates, and focuses the federal com-
munity's efforts to address the effects of climate change on
natural resources in the Pacific Northwest region; fosters
collaborative efforts between research, management,
and regulatory agencies and programs ("knowledge-
to-action"); and provides a portal to the federal climate
change community in the Pacific Northwest, for states,
academic organizations, tribal organizations, and others.*
http://www.c3.gov
*National Fish, Wildlife & Plants Climate Adaptation
Strategy*
http://www.wildlifeadaptationstrategy.gov
(206) 220-4616
lief_horwitz@usgs.gov and pat_gonzales-rogers@fws.gov

Climate Science Centers
*U.S. Geological Survey (USGS) regional Climate Science
Centers to provide scientific information, tools, and
techniques to anticipate, monitor, and adapt to climate
change.*
http://www.doi.gov/csc/index.cfm

2. WATER SECURITY

*Secure sources of fresh water now to meet
future needs of tribal communities located
in drought- and flood-impacted areas.*

Pacific Northwest Coast Sea-Level Rise Report
*In 2007, the National Wildlife Federation looked at a
number of sea level rise scenarios for the region and
found that even a relatively moderate scenario of sea level
rise of just over two feet by 2100 will have a significant
impact on coastal habitats.*
http://www.nwf.org/sealevelrise/Maps_of_the_Pacific_
Northwest_Coast.cfm

Coast Salish Water Quality Project
*Coast Salish leaders, canoe skippers and USGS scientists
partnering to collect & analyze Salish Sea water quality
data during Tribal Canoe Journeys.*
http://www.usgs.gov/features/coastsalish
Canoe Journey Science Coordinator, SITC, 11404
Moorage Way, La Conner, WA 98257
(360) 466-1236
sakin@swinomish.nsn.us and egrossman@usgs.gov

Indigenous Water Initiative
*Promoting better understanding of Indigenous perspec-
tives on water and development among water professionals and enhancing dialogue between indigenous political
and spiritual leaders and the agents of water resources
development.*
http://www.indigenouswater.org
1021 Camino Santander, Santa Fe, NM 87501
(505) 992-0309
dgroenfeldt@indigenouswater.org

Center for Water Advocacy
*Non-profit public interest law firm that intends to
promote the long-term sustainability of water resources
in the western United States for the benefit of fish and
wildlife populations, habitat, aesthetics, recreation, and
traditional and cultural activities.*
http://www.centerforwateradvocacy.org
Squaxin Island conference 2007: http://
centerforwateradvocacy.org/sitebuildercontent/
sitebuilderfiles/finalconferenceagenda2007.pdf
76 S. Main Street, Suite 20, Moab, UT 84532
(435) 259-2958 office • (435) 259-0708 fax
waterlaw@uci.net

Western Water Assessment
*Information about natural climate variability and
human-caused climate change, forecasts, and outlooks for
Colorado, Utah, and Wyoming.*
http://wwa.colorado.edu
NOAA Earth System Research Laboratory, R/PSD 325
Broadway, Boulder, CO 80305
(303) 497-6449
wwa@noaa.gov

Black Mesa Water Coalition (Arizona)

Formed in 2001 by a group of young intertribal, interethnic people dedicated to addressing issues of water depletion, natural resource exploitation, and health promotion within Navajo and Hopi communities.

http://www.blackmesawatercoalition.org

PO Box 613, Flagstaff, AZ 86002

(928) 213-5909 office • (928) 213-5905 fax

blackmesawatercoalition@yahoo.com or chelsea.rc@ gmail.com

North Australian Indigenous Land and Sea Management Alliance (Australia)

The Indigenous Water Policy Group (part of the North Australian alliance) aims to increase the awareness of Indigenous people about the government's current policies for water reform and to engage in research relating to Indigenous rights, responsibilities, and interests in water resources.

http://www.nailsma.org.au/

Bld Purple 12.3.27, Charles Darwin University, Darwin, NT 0909 Australia

08 8946 7691 office • 08 8946 6388 fax

Jessica.LewFatt@cdu.edu.au

3. AGRICULTURE and FOOD SOVEREIGNTY

Secure a future source of food stocks, long-term storage capacity, and production capabilities (including intertribal trade) for crops that can adapt to climate change.

Native Agriculture and Food Systems Initiative

Initiative of the First Nations Development Institute for tribes and Native communities as they seek to strengthen the food system in their communities, improve health and nutrition, and build food security.

http://www.firstnations.org/node/217

Native Food and Agriculture Resource Manual

http://firstnations.org/book/export/html/70

FNDI, 703 3rd Avenue, Suite B, Longmont, CO 80501

(303) 774-7836 office • (303) 774-7841 fax

info@firstnations.org

Intertribal Agriculture Council (American Indian Foods Program)

Founded in 1987 to pursue and promote the conservation, development, and use of agricultural resources for the betterment of Indigenous people. The harmonies of man, soil, water, air, vegetation, and wildlife that collectively make up the American Indian agriculture community influence their emotional and spiritual well-being.

http://www.indianaglink.com

100 North 27th Street, Suite 500, Billings, MT 59101

(406) 259-3525 office • (406) 256-9980 fax

info@indianaglink.com

InterTribal Buffalo Cooperative

Restoring the buffalo to Indian Country to restore our historical, cultural, traditional, and spiritual relationship for future generations. ITBC has a membership of fifty-seven tribes with a collective herd of over 15,000 bison.

InterTribal Buffalo Cooperative

http://www.itbcbuffalo.com

2497 West Chicago St., Rapid City, SD 57702

(605) 394-9730 office • (605) 394-7742 fax

jstone@itbcbison.com

White Earth Land Recovery Project (Ojibwe, Minnesota)

Facilitating recovery of the original land base of the White Earth Indian Reservation, while preserving and restoring traditional practices of sound land stewardship, language fluency, community development, and strengthening our spiritual and cultural heritage.

http://www.nativeharvest.com

607 Main Avenue, Callaway, MN 56521

(888) 274-8318

http://nativeharvest.com/contact

Traditional Native American Farmers Association (New Mexico)

Based in the Indigenous communities in New Mexico but with projects as far afield as Belize, the association is a leading voice for food sovereignty, with many successes in getting farmers back on the land, farming organically and traditionally.

http://nativeharvest.com/tnafa

PO Box 31267, Santa Fe, NM 87594

(505) 983-4047

cbrascoupe@yahoo.com

Native Seeds/SEARCH (Arizona)

Promotes the use of ancient Southwestern crops and their wild relatives by gathering, safeguarding, and distributing their seeds to farming and gardening communities. We also work to preserve knowledge about their uses.

http://www.nativeseeds.org

526 N. 4th Ave., Tucson, AZ 85705

(520) 622-5561 store • (520) 622-5591 fax

info@nativeseeds.org

Mvskoke Food Sovereignty Initiative (Oklahoma)

Works to enable the Mvskoke people and their neighbors to provide for their food and health needs now and in the future through sustainable agriculture, economic development, community involvement, and cultural and educational programs.

http://www.mvskokefood.org

208 W. 6th St., Okmulgee, OK 74447

(918) 756-5915 office • (918) 756-5918 fax
mvskokefood@gmail.com

Tohono O'odham Community Action (Arizona)

A community-based organization dedicated to creating a healthy, sustainable, and culturally vital community on the Tohono O'odham nation.

http://www.tocaonline.org
PO Box 1790, Sells, AZ 85634
(520) 383-4966 office • (520) 383-5286 fax
information@tocaonline.org

Gitiganing Garden Restoration Project (Bad River Ojibwe, Wisconsin)

A collaborative effort of at least a dozen active community members, the VISTA volunteers they arranged to work for them, an elected board, and more than sixty families that take advantage of the community field's individual plots and services for home gardeners.

http://feeds.uwex.edu/gardens/gardendetails.
cfm?gardenid=92
PO Box 275, Odanah, WI 54861
(715) 685-2784 or (715) 682-2601
badrivervistas@yahoo.com

Tsyunhehkwa Farm (Oneida, Wisconsin)

An agricultural community and culturally based program of the Oneida Nation of Wisconsin with three main components: agriculture, canning, and retail. Our primary focus is on self-sustainability and food security.

http://www.oneidanation.org/Tsyunhehkwa
PO Box 365, Oneida, WI 54155
(920) 869-2141 office • (920) 869-2147 fax
tsyunheh@oneidanation.org

Dream of Wild Health Farm (Minnesota)

A 10-acre organic farm growing Indigenous seeds—primarily corn, beans, and squash—donated by tribes from the region; some of these seeds are hundreds of years old. It grows a vegetable garden for a farmers' market close to a Twin Cities Native community.

http://dreamofwildhealth.org
16085 Jeffrey Avenue, Hugo, MN 55038
(651) 439-3840
diane@dreamofwildhealth.org

Hawai'i SEED (Hawai'i)

Working on five islands to educate the public about the risks posed by genetically engineered organisms. We are dedicated to promoting diverse, local, healthy, and ecological food and farming that supports real food security for the Hawaiian Islands.

http://www.hawaiiseed.org
(808) 331-1211
PO Box 2352. Kealakekua, HI 96750
hawaiiseed@hawaiiseed.org

Native American Natural Foods (South Dakota)

Marketing Tanka Bars, based on traditional bison meat and pemmican.

http://www.tankabar.com
287 Water Tower Road, Kyle, SD 57762
(605) 455-2019 or (605) 455-2187
customerservice@tankabar.com

4. PLANT AND ANIMAL SPECIES

Prepare for impacts on culturally significant wild plant and animal species and teach each other about both harvestable and invasive species that are shifting northward.

State of the Salmon

EcoTrust program to build knowledge across borders, linking a greater understanding of Pacific salmon to their improved management and conservation around the Pacific Rim.

http://www.stateofthesalmon.org
721 NW Ninth Avenue, Suite 200, Portland, OR 97209
(503) 227-6225 office • (503) 227-1517 fax
inquiries@stateofthesalmon.org

Center for World Indigenous Studies (Center for Traditional Medicine)

Conducts research, education, and clinical treatment that integrates traditional systems of Indigenous healing with complementary and integrative medicine. The center promotes social change through its activities by and for Native and non-Native peoples worldwide.

http://www.cwis.org and
http://centerfortraditionalmedicine.org/
PMB 214-1001 Cooper Pt. Rd. SW, Suite 140, Olympia,
WA 98501
(360) 586-0656
chair@cwis.org or lekorn@cwis.org

Climate Change Impacts on Tribal Resources

Tulalip Tribes flyer on climate change impacts on natural resources.

http://www.tulalip.nsn.us/pdf.docs/FINAL%20CC%20
FLYER.pdf

Shadow of the Salmon

The video tells the story of a young man from Lakota nation who comes to the Pacific Northwest to visit his Coast Salish relatives. The Northwest Indian Fisheries Commission film received three Regional Emmy Award nominations.

http://salmondefense.org/projects/shadow-of-the-
salmon-video

Teacher's Resource Guide: http://education.wsu.
edu/nativeclearinghouse/achievementgap/
ShadowoftheSalmonCurriculumGuide(4).pdf

NWIFC, PO Box 38, 7205-A Martin Way East,
Olympia, WA 98516-5535

(360) 528-4351

water4fish@comcast.net

Hongoeka Development Trust Ltd. (Aotearoa/New Zealand)

A community-based aquaculture project on Māori land at Plimmerton, NZ, using abalone polyculture technologies to develop economically sustainable systems that make aquaculture more accessible to Māori.

http://www.niwa.co.nz/our-science/te-kuwaha/
research-projects/all/accessible-aquaculture---
hongoeka-development-trust-ltd

Te Kūwaha o Taihoro Nukurangi/National Institute
of Water and Atmospheric (NIWA) Research/
Māori Research and Development Unit, P.B. 14901,
Kilbirnie, Wellington NZ

64 (4) 386 0511 or 64 (27) 246 6177

c.severne@niwa.co.nz

5. RELATIONSHIPS WITH NEIGHBORS

Develop relationships with neighboring governments and communities regarding land use planning to prevent worsening storm effects and to make emergency plans for the more disastrous impacts of climate change.

Swinomish Climate Change Initiative

A comprehensive and long-term effort to address climate change issues, starting with a two-year project to identify probable impacts and develop an action plan. The Swinomish Tribe has partnered with regional science experts and invited neighboring jurisdictions to participate in providing for a healthy and resilient community.

http://www.swinomish-nsn.gov/climate_change/
project/project.html

SITC Planning and Community Development, 11404
Moorage Way, LaConner, WA 98257

(360) 466-7280

eknight@swinomish.nsn.us

McAllister Wellfield Partnership

In 2008, the Nisqually Tribe and the City of Olympia formed an historic partnership and Stewardship Coalition to replace McAllister Springs (threatened by saltwater intrusion), with wells on higher ground, thereby restoring water flow to Medicine Creek.

http://olympiawa.gov/city-utilities/drinking-water/
McAllister.aspx

(360) 456-5221 ext. 1112 or (360) 753-8495

cushman.joe@nisqually-nsn.gov or rhoey@ci.olympia.
wa.us

Sustainable Northwest

Uses collaboration to bridge rural and urban interests, encourage entrepreneurship, and build trust in sustainable natural resource management and utilization. Coordinated a series of tribal climate change policy training seminars (course materials available on its website).

http://www.sustainablenorthwest.org

813 SW Alder, Suite 500, Portland, OR 97205

(503) 221-6911 ext.109

apomeroy@sustainablenorthwest.org

Pacific Northwest Research Station

Generates and communicates scientific knowledge that helps people understand and make informed choices about people, natural resources, and the environment, and builds relationships between the U.S. Forest Service and tribal governments.

http://www.fs.fed.us/pnw

PO Box 3890, Portland, OR 97208

beav@fs.fed.us

Community Alliance and Peacemaking Project (CAPP)

Grassroots group using facilitation and peacemaking strategies and techniques, including traditional models of peacemaking. Promotes values such as respect for self and honoring the circle of life and Mother Earth.

http://capp.web.officelive.com/default.aspx

PO Box 535, Anacortes, WA 98221

360-293-7197

msvendiola@gmail.com

Nisqually River Council

Encouraging and supporting sustainability in the Nisqually River Watershed in order to steward our resources in perpetuity and build a model for harmonious living.

http://nisquallyriver.org

12501 Yelm Highway SE, Olympia, WA 98513
360-438-8715

info@nisquallyriver.org

Skagit Watershed Council

A "big tent" community-based partnership of organizations that came together in 1997 to work to protect and restore salmon habitat in the Skagit and Samish watersheds.

http://skagitwatershed.org

PO Box 2856 Mount Vernon, WA 98273

360-419-9326

council@skagitwatershed.org

Skagit Climate Science Consortium

Working to understand climate-related changes occurring in Washington's Skagit Basin.

http://www.skagitclimatescience.org

(206) 293-4741

cmacilroy@gmail.com

ICLEI-Local Governments for Sustainability
A membership association of local governments committed to advancing climate protection and sustainable development. Website has regional contacts and resources for climate change mitigation and adaptation planning.
http://www.icleiusa.org
180 Canal Street, Suite 401, Boston, MA 02114
617-960-3420
iclei-usa@iclei.org

U.S. Environmental Protection Agency (State and Local Climate and Energy Program)
Provides technical assistance, analytical tools, and outreach support to state, local, and tribal governments. Listserv shares news of important developments in climate change and clean energy policies, programs, and opportunities.
http://www.epa.gov/statelocalclimate/

New Zealand Ministry for the Environment
Report on 2007 Regional Consultation Hui (meetings) with Māori on Climate Change in Atearoa/New Zealand.
http://www.mfe.govt.nz/publications/climate/consultation-maori-hui-report-nov07/index.html

Alaska Sea Grant Marine Advisory Program
Supports coastal communities through research, education, and extension. Supports marine and coastal research, provides education and extension services, and distributes information about Alaska's seas and coasts.
http://seagrant.uaf.edu/map/climate
Alaska Climate Change Adaptation Planning Tool:
http://seagrant.uaf.edu/map/climate/docs/adaptation-planning-tool.pdf
1007 West 3rd Ave, Suite 100, Anchorage, AK 99501
(907) 274-9695
terry.johnson@alaska.edu

Olympic Coast Intergovernmental Policy Council
Created in 2007 by the Hoh, Makah, Quileute tribes, and Quinault nations, the state of Washington and the National Oceanic and Atmospheric Administration (NOAA) Office of National Marine Sanctuaries.
http://olympiccoast.noaa.gov/management/intergovernmentalpolicy.html
First Stewards: Coastal Indigenous Communities Climate Change Symposium at the National Museum of the American Indian in Washington, DC; hosted by Hoh, Makah, Quileute and Quinault nations, July 17-20, 2012.
http://www.firststewards.org
Olympic Coast National Marine Sanctuary, 115 East Railroad Ave., Ste. 301, Port Angeles WA 98362
(360) 457-6622 or (360) 276-8211 x. 368
ejohnsto@quinault.org

Coquille Indian Tribe Climate Change Committee
Established in 2008 engage tribal government, tribal members, and natural and cultural resource managers in the development of a climate change action plan.
Coquille Indian Tribe, 3050 Tremont, North Bend OR 97459
http://www.coquilletribe.org
http://www4.nau.edu/tribalclimatechange/tribes/northwest_coquille.asp
(541) 290-3552
billsnyder@cedco.net or donivy@coquilletribe.org

Puget Sound Partnership
A community effort of citizens, governments, tribes, scientists & businesses working together to restore & protect Puget Sound.
http://www.psp.wa.gov
Salish Sea Ecosystem Conference (2011)
http://www.salishseaconference.org
326 East D St., Tacoma, WA 98421
(360) 464-1232
info@psp.wa.gov

Columbia Basin Trust
Supports efforts by the people of the Columbia Basin in BC to create a legacy of social, economic and environmental well-being, and to achieve greater self-sufficiency for present & future generations.
http://www.cbt.org
http://www.cbt.org/Contact_Us

6. RENEWABLE ENERGIES

Consider alliances with local governments to build and market renewable energy capacity, such as wind and biogas power.

Energy Planning: A Guide for Northwest Indian Tribes
Renewable energy booklet by Northwest Sustainable Energy for Economic Development (NWSEED), an organization whose goal is to establish a clean, diverse, and affordable energy system based on efficient use of renewable resources, with maximum local control and ownership of energy.
http://www.nwseed.org/documents/NWSEED_Tribal%20GB_Final.pdf
1402 3rd Ave., Suite 901, Seattle, WA 98101
(206) 328-2441 office • (206) 770-6570 fax
info@nwseed.org

Qualco Energy
Biogas plant built by the Tulalip Tribes in partnership with Sno/Sky Agricultural Alliance and Northwest Chinook Recovery to turn dairy farm cattle waste into

methane energy, thereby keeping the waste out of the Snohomish River salmon fishery.
http://www.qualcoenergy.com
18117 203rd St SE, Monroe, WA 98272
(425) 344-2940 or (360) 651-4399
darylwilliams@qualcoenergy.com

Makah Bay Offshore Wave Energy Project
A 30-MW Wind Power Pilot Project begun in 2006 on the Makah Indian nation, with the aim of selling the power to a local utility (project postponed in 2008 by the withdrawal of an investor).
http://teeic.anl.gov/documents/docs/library/
　MakahBayOffshoreWave.pdf
Makah Tribal Planning, PO Box 115, Neah Bay, WA
　98357
(360) 645-3281 office • (360) 645-2033 fax
mtcbud@centurytel.net

Native Energy
By helping finance construction of Native American, family farm, and community-based renewable energy projects, our customers help communities in need build sustainable economies, with significant Native American ownership.
http://www.nativeenergy.com
(800) 924-6826
support@nativeenergy.com or pat.spears@
　nativeenergy.com

Intertribal Council on Utility Policy (COUP)
A forum for discussing utility issues from regulatory and economic perspectives among ten Northern Plains member tribes. We provide policy analysis and recommendations, as well as workshops. The Native Wind project emphasizes wind energy development.
http://www.intertribalcoup.org
PO Box 25, Rosebud, SD 57570
(605) 945-1908 or (605) 280-7999
patspears25@gmail.com or gough.bob@gmail.com

Honor The Earth
This organization's Energy Justice Initiative aims to develop the community infrastructure essential to a sustainable future in Indian Country, addressing energy policy as a means to democratize power production and create systemic change that advances environmental and social justice.
http://www.honorearth.org/
2104 Stevens Avenue South, Minneapolis, MN 55404
(612) 879-7529 or (218) 375-3200
http://www.honorearth.org/contact

U.S. Department of Energy (Tribal Energy Program)
Promotes tribal energy sufficiency, economic growth, and employment on tribal lands through the development of renewable energy and energy efficiency technologies. One of its projects is Wind Powering America.

http://www.eere.energy.gov/tribalenergy
Wind Powering America
http://www.windpoweringamerica.gov/
　nativeamericans/index.asp
1000 Independence Avenue SW, Washington, DC
　20585
(303) 275-4727
lizana.pierce@go.doe.gov

National Renewable Energy Laboratory
NREL Tribal Energy Program, Office of Energy Efficiency and Renewable Energy
http://www.nrel.gov
1617 Cole Rd., Golden, CO 80401
 (303) 275-4335
kimberly.craven@nrel.gov

Energy Justice Network
This organization's energy agenda includes supporting communities threatened by polluting energy and waste technologies. Taking direction from its grassroots base and Principles of Environmental Justice, it advocates a clean energy, zero-emission, zero-waste future for all.
http://www.energyjustice.net
catalyst@actionpa.org

Green Native Council
Educates and empowers American Indian nations to create sustainable solutions to the severe housing crisis in reservation communities; this organization also builds on the strengths of each tribe's unique traditions and works with all generations to achieve positive family and community development.
http://www.greennativecouncil.com
Bend, OR 97701-9033
(541) 306-7425
tmonroe@greennativecouncil.com

Indigenous Community Enterprises (Navajo, Arizona)
Developing community and economic development opportunities that respect and incorporate traditional culture, foster responsible stewardship of the land, and maintain and enhance the well-being and self-reliance of communities and individuals.
http://www.cba.nau.edu/ice
2717 N. Steves Blvd Suite 8, Flagstaff, AZ 86004
(928) 522-6162 office • (928) 522-6163 fax
hjames@icehome.org

Kumeyaay Wind Energy Project (California)
First existing and planned wind farms on U.S. Indian lands, located on the Campo Reservation in San Diego County.
http://apps1.eere.energy.gov/tribalenergy/pdfs/
　course_biz0904_connolly.pdf
High Pass Energy, 8880 Rio San Diego Dr., Suite 800,
　San Diego, CA 92108

(619) 307-0160

tipaay@aol.com or m.connolly@highpassenergy.com

Energy for Māori

This organization provides governance and resource valuation studies and reports on the progress of pilot energy studies. The project has made particularly good progress with developing plans for a wind farm with the Kaimanawa Trust.

http://www.niwa.co.nz/publications/wa/vol13-no4-december-2005/energy-in-rural-m%C4%81ori-communities

Te Kūwaha o Taihoro Nukurangi/National Institute of Water and Atmospheric (NIWA) Research/ Māori Research and Development Unit, P.B. 14901, Kilbirnie, Wellington, NZ

64 (4) 386 0511 or 64 (27) 246 6177

c.severne@niwa.co.nz

Lummi Nation Renewable Energy Projects

Lummi Nation projects include conducting a wind energy development feasibility assessment, lighting a walking trail with solar LEDs, installing a geothermal heat pump system for a new administrative building, and developing a strategic energy plan.

http://lnnr.lummi-nsn.gov/LummiWebsite

http://tribalclimate.uoregon.edu/the-lummi-nation-pursuing-clean-renewable-energy/

2616 Kwina Road, Bldg I, Bellingham WA 98226

(360) 384-2267

merlej@lummi-nsn.gov or jeremyf@lummi-nsn.gov or richardj@lummi-nsn.gov

Shishmaref Erosion & Relocation Coalition

The community of Shishmaref, Alaska, has determined that the threat to life and property from reoccurring beachfront erosion requires immediate action. The community has taken the first step by establishing an erosion and relocation coalition made up of the governing members of the City, Indian Reorganization Act Council and Shishmaref Native Corporation Board of Directors.

(907) 649-2289 or 649-3821

http://www.shishmarefrelocation.com

P.O. Box 72100, Shishmaref AK 99772

http://www.shishmarefrelocation.com/contact_us.html7.

7. HABITAT PROTECTION

Consider strategies to unite tribes around habitat protection, looking ahead to the inevitable effects of climate change.

National Tribal Environmental Council (NTEC)

Formed in 1991 as a membership organization (now including 184 federally recognized tribes) dedicated to working with and assisting tribes in the protection and preservation of tribal environments.

http://www.ntec.org

4520 Montgomery Blvd. NE, Suite 3, Albuquerque, NM 87109

(505) 242-2175 office • (505) 242-2654 fax

BGruenig@ntec.org

Northwest Indian Fisheries Commission (NWIFC)

A support service organization for twenty treaty Indian tribes in western Washington. The role of the NWIFC is to assist member tribes in their role as natural resources co-managers, and it provides a forum for tribes to address shared natural resources issues.

http://www.nwifc.org

6730 Martin Way E., Olympia WA 98516

http://www.nwifc.org/contact/

Columbia River Inter-Tribal Fish Commission (CRITFC)

The Warm Springs, Yakama, Umatilla, and Nez Perce tribes joined in 1977 to renew their authority in fisheries management and to organize intertribal representation in regional planning, policy, and decision making. The organization's website includes information and workshop materials on climate change.

http://www.critfc.org

729 NE Oregon St., Ste. 200, Portland, OR 97232

(503) 238-0667

croj@critfc.org

Great Lakes Indian Fish and Wildlife Commission (GLIFWC)

Intertribal natural resources agency of the Ojibwe reservations around Lake Superior committed to the implementation of off-reservation treaty rights.

http://www.glifwc.org

PO Box 9, Odanah, WI 54861

(715) 682-6619

jzorn@glifwc.org or serikson@glifwc.org

Southwest Tribal Fisheries Commission

Nonprofit organization that provides technical skills and support needed to move individual tribal fisheries programs and projects from concept to reality.

http://www.swtfc.org

PO Box 190, Mescalero, NM 88340

(505) 464-8768

mmontoya@matisp.net

Intertribal Timber Council

Explores issues and identifies practical strategies and initiatives to promote social, economic, and ecological values in protecting and utilizing forests, soil, water, and wildlife.

http://www.itcnet.org/index.html

1112 NE 21st Avenue, Suite 4, Portland, OR 97232

(503) 282-4296

itc1@teleport.com

Indigenous Peoples' Restoration Network
Ecocultural restoration, traditional ecological knowledge, part of the Society for Ecological Restoration (SER) International.

http://www.ser.org/iprn/default.asp

iprn@ser.org

Iwi Estuarine Monitoring Toolkit
Ngā Waihotanga Iho [what is left behind, lift up], an estuary monitoring toolkit, has been developed to provide Māori Iwi [tribes] with tools to measure environmental changes in their estuaries.

http://www.niwa.co.nz/our-science/te-kuwaha/
research-projects/all/ngA-waihotanga-iho-a-iwi-
estuarine-monitoring-toolkit

Ngati Whanaunga, PO Box 178, Coromandel, NZ

64 (7) 866 7229

d.rickard@niwa.co.nz

Native American Fish & Wildlife Society
Incorporated in 1983 to develop a national communications network for the exchange of information and management techniques related to self-determined tribal fish and wildlife management.

http://www.nafws.org/news/climate-change.html

8333 Greenwood Blvd., Ste. 260, Denver, CO 80221

(866)-890-7258 or (303) 466-1725

fmatt@nafws.org

8. INTERNATIONAL PARTICIPATION
Get actively involved as sovereign governments in UN climate change negotiations and pressure national governments to reduce emissions.

United Nations Permanent Forum on Indigenous Issues (UNPFII)
Meets annually for ten-day sessions to develop recommendations to other UN bodies. Its seventh session special theme in 2008 was "climate change, bio-cultural diversity and livelihoods: the stewardship role of Indigenous peoples and new challenges."

http://social.un.org/index/IndigenousPeoples/
UNPFIISessions/Seventh.aspx

United Nations, 2 UN Plaza, Room DC2-1454, New
York, NY 10017

(917) 367 5100 office • (917) 367 5102 fax

indigenous_un@un.org

Guide on Climate Change and Indigenous Peoples
2008 UNPFII publication to enhance Indigenous peoples' knowledge on climate change so that they will be better equipped to participate more effectively in shaping relevant policies and actions taken to address this issue.

http://www.tebtebba.org/index.php/content/160-
2nd-edition-of-guide-on-climate-change-and-
indigenous-peoples-now-released

Tebtebba
Indigenous Peoples' International Centre for Policy Research and Education, covering Indigenous peoples' participation in global climate change discussions and negotiations.

http://www.tebtebba.org

Indigenous Climate Portal

http://www.indigenousclimate.org

No.1 Roman Ayson Road, Baguio City 2600,
Philippines

+63 (74) 444-7703

vicky@tebtebba.org

Indigenous Peoples' Global Summit on Climate Change
April 2009 summit in Anchorage, Alaska, whose purpose was to enable Indigenous peoples from all regions of the globe to exchange their knowledge and experience in adapting to the impacts of climate change and to develop key messages and recommendations.

http://www.cakex.org/case-studies/2823

The Northern Forum, 716 W 4th Avenue, Suite 100,
Anchorage, AK 99501

(907) 561 6645 or (907) 561 3280

info@indigenoussummit.com

United Nations Framework Convention on Climate Change (UNFCCC)
United Nations global treaty regulating carbon emissions; the signatory member states meet annually in Conferences of the Parties (COP).

http://unfccc.int/2860.php

UN Intergovernmental Panel on Climate Change (IPCC)
Scientific panel honored with 2007 Nobel Peace Prize for its role in documenting and predicting climate change.

http://www.ipcc.ch

c/o World Meteorological Organization, 7bis
Avenue de la Paix, C.P. 2300, CH-1211 Geneva 2,
Switzerland

+41-22-730-8208/54/84 office • +41-22-730-8025/13
fax

IPCC-sec@wmo.int

Indigenous Peoples Climate Change Assessment (United Nations University)
Studies on the impacts of climate change and climate change responses on Indigenous people by the United Nations University's Institute of Advanced Studies' Traditional Knowledge Initiative.

http://www.unutki.org

Building 1, Level 3, Red Precinct, Charles Darwin
University, Casuarina Campus, Ellengowan Drive,
Darwin, NT 0909, Australia

+61-8-8946-6792/7652 office • +61-8-8946-7720 fax

tki@ias.unu.edu

World Peoples' Conference on Climate Change and the Rights of Mother Earth
Global conference held in Cochabamba, Bolivia (as a response to the failed UNFCCC session in Copenhagen), to emphasize the economic and ecological debt of the Global North to the Global South, April 2010.
http://pwccc.wordpress.com

9. YOUTH INVOLVEMENT

Get youth involved in cultural education and defend the future of the tribe from harmful climate change.

Native Movement
A national nonprofit organization that supports cultur-ally based leadership development and sustainability programs in Alaska (through the Indigenous Leadership Institute) and the Southwest, inspiring and supporting young Indigenous leaders across North America.
http://www.nativemovement.org
Indigenous Leadership Institute, 3535 College Road, Suite 204, Fairbanks, AK 99709
(907) 374-6899
Native Movement Southwest, PO Box 896, Flagstaff, AZ 86002
(928) 213-9063
nativemovement@gmail.com

Tribal Campus Climate Challenge
Indigenous Environmental Network youth organizing project in partnership with the Energy Action Coalition Campus Climate Challenge, aiming to involve more than forty tribal and community colleges across the United States and Canada.
http://www.ienearth.org/tccc.html
(701) 214-1389 or (701) 214-1389 or (928) 213-5909
iencampusclimate@igc.org or chelsea.rc@gmail.com

Campus Climate Challenge (Environmental Justice and Climate Change Initiative)
A diverse coalition of U.S. environmental justice, religious, climate justice, policy, and advocacy networks working for climate justice.
http://ejcc.org/campus/ccc/
Message to the leaders of the world: http://www.youtube.com/watch?v=NG9T9B5q6o8
1904 Franklin Street, Suite 600, Oakland, CA 94618
(510) 444-3041 x305 office • (510) 444-3191 fax
http://ejcc.org/forward/email/

Northwest Indian College
Supported a student video on climate change by tribal students and sponsors the Red Alert climate change conferences.

Video project: *Where Words Touch the Earth*
http://www.teachersdomain.org
(360) 676-2772
skinley@nwic.edu
Red Alert conferences on fisheries and timber
http://www.climate.org/publications/Climate%20 Alerts/Summer2009/TribalFisheries.html
(360) 392-4307
spavlik@nwic.edu

Tribal Canoe Journeys
Annual summer cultural event of Native American and First Nations Indigenous people of the Pacific Northwest, honoring the ancestral modes of communications, travel, and commerce. They are family-oriented activities focused on building healthy communities.
http://tribaljourneys.wordpress.com
http://tribaljourneys.wordpress.com/contact-info/
http://www.PaddleToSquaxin2012.org

American Indian Higher Education Consortium (AIHEC)
The collective spirit and unifying voice of tribal colleges and universities. It provides leadership and influences public policy on American Indian higher education issues through advocacy, research, and program initiatives, including an Indian Education Renewable Energy Chal-lenge and the NASA Research Experience.
http://www.aihec.org
121 Oronoco Street, Alexandria, VA 22314
(703) 838.0400 office • (703) 838-0388 fax
info@aihec.org

Thunder Valley Community Development Corp. (South Dakota)
Empowering Lakota youth and families to improve the health, culture, and environment of their community through the healing and strengthening of cultural identity and construction of a new community based on sustain-able energy, food, and housing.
http://www.thundervalley.org
PO Box 290, Porcupine, SD 57772
(605) 455-2700 office • (605) 441-7485 cell
nick@thundervalley.org

Strategic Watershed Analysis Teams (SWAT) (British Columbia)
Gitxsan youth using GPS and Geographic Information Systems (GIS) to map their First Nations territory, to guard it against timber companies and non-Native claims. Possible model for tribal youth climate adaptation teams.
http://www.conservationgis.org/native/native2.html
Gitxsan Treaty Office, PO Box 229. Hazelton, B.C., Canada
swat@mail.bulkley.net

Alaska Youth for Environmental Action

Inspires and trains diverse youth leaders to affect environmental issues through community action projects and campaigns; offers skills training in leadership, environmental education, civic engagement, and community organizing.

http://ayea.org/

750 W 2nd Ave, Suite 200, Anchorage, AK 99501,

(907) 339-3907 office • (907) 339-3980 fax

ayea@nwf.org

NASA Cryospheric Sciences Branch (Tribal College and University Project)

Education and research opportunities for tribal college and university students, including climate change–related topics.

http://neptune.gsfc.nasa.gov/csb/personnel/index.php?id=35

(301) 614-6572

Nancy.G.Maynard@nasa.gov

10. INDIGENOUS COOPERATION

Work with other Indigenous nations in a treaty relationship transcending colonial international boundaries.

United League of Indigenous Nations

Indigenous Nations representatives from the United States, Canada, Aotearoa (New Zealand), and Australia, meeting at the Lummi nation, signed a Treaty of Indigenous Nations on August 1, 2007, establishing a political and economic alliance to advance their common interests regarding the impacts of climate change, to promote trade and commerce among the more than eighty-four signatory nations, to bring their cultural properties under the protection of the laws of Indigenous nations, and to assert rights to cross international borders.

http://www.ulin-web.org

(360) 867-6614 or (360) 384-2337

NIARI, SEM 3122, TESC, 2700 Evergreen Pkwy. NW, Olympia, WA 98505

parkeral@evergreen.edu or jewell@lummi-nsn.gov

Coast Salish Gathering

This organization declares: "We, the Indigenous peoples of the Salish Sea, with our autonomous status as sovereign Tribes and First Nations and our inherent responsibility as protectors of our Mother Earth, will continue to speak with One Voice for the preservation, restoration, and protection of the Salish Sea Eco Region for the sustainability of our sacred inherent family rights and values that have been passed on to us by our ancestors."

http://www.coastsalishgathering.com

c/o SITC, PO Box 11, LaConner, WA 98257

(360) 391-5296

dlekanof@swinomish.nsn.us

National Congress of American Indians (NCAI)

Working since 1944 to inform the public and Congress on the governmental rights of American Indians and Alaska Natives, including more recently on issues of climate change, environment, and natural resources policy.

http://www.ncai.org/Land-Natural-Resources.24.0.html

1516 P Street NW, Washington, DC 20005

(202) 466-7767

Jose_Aguto@NCAI.org

Assembly of First Nations (AFN)

Canadian First Nations governments' assembly, including the Environmental Stewardship Unit, which deals with climate change issues.

http://www.afn.ca/index.php/en/policy-areas/environmental-stewardship

Trebla Building, 473 Albert Street, Suite 810, Ottawa, ON K1R 5B4 Canada

(613) 241-6789 office • (613) 241-5808 fax

swuttke@afn.ca

International Indian Treaty Council

Founded in 1974 to support grassroots Indigenous struggles through information dissemination, networking, coalition building, technical assistance, organizing, and facilitating the effective participation of traditional peoples in local, regional, national and international forums, events, and gatherings.

http://www.treatycouncil.org

456 N. Alaska Street, Palmer, AK 99645

(907) 745-4482 office • (907) 745-4484 fax

andrea@treatycouncil.org and alberto@treatycouncil.org

Affiliated Tribes of Northwest Indians (ATNI)

Working since 1953 to promote tribal self-determination and the sovereignty of Pacific Northwest tribal nations. In 2007 member tribes signed the Intertribal Economic and Trade Treaty.

http://www.atnitribes.org

NE 44th Ave., Suite 130, Portland, OR 97213-1443

(503) 249-5770 office • (503) 249-5773 fax

atni@spiritone.com

United Indian Nations of the Great Lakes

Formed in 2005 at a meeting of 140 tribal and First Nation representatives, this organization emerged from the same process and was led by the same leadership that formulated the Great Lakes Water Accord and the St. Mary's River Treaty. In 2004, 120 Tribes and First Nations signed the Great Lakes Water Accord, rejecting the Great Lakes Charter Annex. The St. Mary's River Treaty was signed by two Michigan tribes and two Ontario First Nations in 2006.

http://www.turtleisland.org/discussion/viewtopic.php?f=2&t=3575

Little Traverse Bay Bands of Odawa Indians, 101

Greenwood St., Petoskey, MI 49770

(231) 242-1406 or (231) 838-9735

Chairman Frank Ettawageshick, fettawa@freeway.net

First Nations Summit (British Columbia)
Comprised of a majority of First Nations and tribal councils in B.C., this organization provides a forum for First Nations in British Columbia to address issues related to treaty negotiations as well as other issues of common concern.

http://www.fns.bc.ca

Suite 1200 - 100 Park Royal South, West Vancouver, BC V7T 1A2 Canada

(604) 926-9903 office • (604) 926-9923 fax

info@fns.bc.ca

Union of B.C. Indian Chiefs (British Columbia)
Working since 1969 to support the work of Indigenous people, whether at the community, national, or international level, in their common fight for the recognition of their aboriginal rights and respect for their cultures and societies.

http://www.ubcic.bc.ca

500 - 342 Water Street, Vancouver, BC, V6B-1B6 Canada

(604) 684-0231 office • (604) 684-5726 fax

http://www.ubcic.bc.ca/contact/

Inuit Circumpolar Council
Major international non-government organization representing approximately 150,000 Inuit of Alaska, Canada, Greenland, and Chukotka (Russia). The organization holds Consultative Status II at the United Nations.

http://www.inuitcircumpolar.com/index. php?Lang=En&ID=1

3000 C Street, Suite N201, Anchorage, AK 99503

(907) 565-4052 or (907) 274-9058 office • (907) 274-3861 fax

pcochran@aknsc.org

Arctic Athabaskan Council
International treaty organization established to represent the interests of U.S. and Canadian Athabaskan member governments. Its Arctic Biodiversity Assessment combines and evaluates data on biodiversity, and will be used to inform governments, policy makers, scientists, and residents in the Arctic and sub-Arctic regions.

http://www.arcticathabaskancouncil.com

2166 2nd Avenue, Whitehorse, YK Y1A 4P1 Canada

(867) 393-9214 office • (867) 668-6577

Cindy.Dickson@cyfn.net

Yukon River Inter-Tribal Watershed Council
Indigenous grassroots organization, consisting of sixty-six First Nations and tribes in Yukon and Alaska, dedicated to the protection and the preservation of the Yukon River Watershed.

http://www.yritwc.org

815 2nd Avenue, Suite 201, Fairbanks, AK 99701

(907) 451-2530 office • (907) 451-2534 fax

jwaterhouse@yritwc.org

Bering Sea Sub-Network
An international community-based observation alliance for the Arctic-observing network in Siberia and Alaska to increase our understanding and knowledge of pan-Arctic processes, thereby enhancing the ability of scientists, Arctic communities, and governments to predict, plan, and respond to environmental changes.

http://www.bssn.net/NewsArticles/BSSNTheBeringSea.html

Aleut International Association, 333 W. 4th Avenue, Suite 301, Anchorage, AK 99501

(907) 332-5388 or (907) 332-5380

victoriag@alaska.net

Pacific Islands Forum
Political grouping of sixteen independent and self-governing states in the South Pacific whose aim is to develop a collective response on a wide range of regional issues including trade, economic development, civil aviation, maritime shipping, telecommunications, energy, and political and security matters.

http://www.forumsec.org.fj

Secretary General, Pacific Islands Forum Secretariat, Private Mail Bag, Suva, Fiji

+679 3312 600 office

info@forumsec.org.fj

Asia-Pacific Network on Climate Change (APNet)
A knowledge-based online clearing house for the Asia-Pacific region on climate change issues; provides platform for policy dialogues and consultation within the region; provides access to latest information and data on climate change issues and developments focusing on the Asia-Pacific; provides capacity building for developing countries.

http://www.climateanddevelopment.org/ap-net

Overseas Environmental Cooperation Center, Shibakoen Annex 7th Floor, 3-1-8, Shibakoen, Minato-ku, Tokyo 105-0011, Japan

+81-3 -5472-0144 office • +81-3 -5472-0145 fax

ap-net@ap-net.org

Te Whare Wananga o Awanuiarangi (Māori University)
Since 1992, recognizing the role of education in providing positive pathways for Māori development. The institution offers a range of qualifications, from community education programmes to certificates and diplomas, bachelors, masters, and doctoral degrees.

http://www.wananga.ac.nz

Alliance of Small Island States (AOSIS)
Coalition of small island and low-lying coastal countries that share similar development challenges and concerns about the environment, especially their vulnerability to

the adverse effects of global climate change.
http://aosis.info

Mission of Grenada to the UN, 800 2nd Ave., Suite 400k, New York, NY 10017

(212) 599-0301 office • (212)-599-1540 fax

grenada@un.int

Climate Change and the Pacific Islands
Ministerial Conference on Environmental and Development in Asia and the Pacific, 2000.

http://www.unescap.org/mced2000/pacific/
background/climate.htm

Environment and Natural Resources Development Division, UN ESCAP,

Rajadamnern Avenue, Bangkok 10200, Thailand

(66-2) 288-1614 office • (66-2) 288-1059 fax

karim.unescap@un.org

Pacific People's Partnership
Canadian NGO that since 1975 has worked with Indigenous and civil society partners across the South Pacific, and First Nations across Canada, on rights-based sustainable development initiatives.

http://www.pacificpeoplespartnership.org

620 View St., Ste. 407, Victoria BC V8W1J6 Canada

(250) 381-4131

info@pacificpeoplespartnership.org

Seventh Generation Fund for Indian Development
Indigenous non-profit organization dedicated to promoting and maintaining the uniqueness of Native peoples throughout the Americas.

http://www.7genfund.org

P.O. Box 4569, Arcata, CA 95518

(707) 825-7640

info@7genfund.org

I just found out about this group, and there seems to be space on 208

at the end of Resources:

Indigenous Peoples Global Network on Climate Change & Self-Determination
The IPCCSD brings together indigenous organizations from different countries that have on-the-ground work related to climate change and climate-sensitive development.

No. 1 Roman Ayson Road, 2600 Baguio City, Philippines

(63) (74) 4447703

vicky@tebtebba.org and raymond@tebtebba.org

Pacific Climate Change Blog
http://climatepasifika.blogspot.com

COMMUNITY ORGANIZING BOOKLET ON CLIMATE CHANGE

Debra McNutt

The Evergreen State College
Master of Public Administration—Tribal Concentration

The 16-page booklet that follows, entitled *Northwest Tribes: Meeting the Challenge of Climate Change*, is a tool for members of Indigenous communities in the Pacific Northwest, to educate each other about the challenge that climate change poses to tribal cultures, economies and treaty rights, and the tribal responses to that challenge.

In 2006, the Northwest Indian Applied Research Institute (NIARI) at The Evergreen State College in Olympia, Washington published the 79-page report *Climate Change and Pacific Rim Indigenous Nations: A Report to Tribal Leadership*, which later inspired the publication of this book *Asserting Native Resilience*.

Debra McNutt edited the community organizing booklet in order to summarize the detailed information in the report. Her hope was that tribal members could share the booklet around their community and within their families, beginning conversations about what climate change means to the future, and what can be done to respond or adapt.

The first edition of the booklet was published in 2007, and this second edition was originally published in 2010. The booklet was given away at no cost, for purely educational purposes, and hard copies are now out of print. Some Washington tribes have distributed it at their General Council meetings. The information has also been shared at intertribal meetings and conferences, Indigenous youth trainings, high school and college classes, and elsewhere.

By including this community organizing booklet in *Asserting Native Resilience*, the editors hope that readers will copy these 16 pages and use them to educate their own communities—from middle-school and high school students to elders. The booklet cannot cover all aspects of climate change, but covers some main points, so community members can be more informed and get involved.

This reduced version of the booklet can be enlarged to 110% when printing on letter-sized paper. An Adobe Acrobat PDF version in color can be downloaded from http://osupress.oregonstate.edu/book/
asserting-native-resilience
or http://academic.evergreen.edu/g/grossmaz/
climatechangebooklet.pdf

Northwest Tribes:
Meeting the Challenge
of Climate Change

Home on Hoh Reservation falling into the ocean.
(Photo: Hoh Tribal Archives).

Edited by Debra McNutt for the
Northwest Indian Applied Research Institute
The Evergreen State College, Olympia, Washington
http://nwindian.evergreen.edu

CLIMATE CHANGE:

IT'S NOT JUST HYPE…
IT'S NOW A REALITY

1

Native peoples have become an early-warning system for the rest of humanity—that climate change is already altering our environment, our economies, and our cultures. *Indian Country Today* **states that "indigenous people are quietly reminding the rest of the world that they are the ones living with the consequences, in the here and now." But Native people can also help lead the way in showing how to take a stand and respond to the climate crisis.**

Alaska and B.C. forests have been damaged by the spruce bark beetle, increasing the risk of wildfire.

(Photo: KPB Spruce Bark Beetle Task Force).

The latest global scientific evidence shows us that climate change is speeding up at a much faster rate than scientists originally thought. We can see climate change coming, whether in Hurricane Katrina (itself worsened by warmer Gulf of Mexico temperatures), or winter megastorms here in

the Pacific Northwest. It is no longer a question whether the climate crisis is coming. It is already here, and we have to go out and meet the challenge by preparing *today.*

Global warming is caused by the emission (release) of carbon gases into our atmosphere that trap the sun's heat from rising back into space, so the Earth is becoming a hotter "greenhouse" (see diagram). The Earth has gone through natural warming cycles in its long history, but nothing like today. Our use of fossil fuels (mainly oil and coal) has released so many "greenhouse gases" that the Earth's climate is becoming unstable and extreme weather more common. Our summers in the past few years have been the hottest on record. As carbon dioxide increases in the atmosphere, global temperatures also increase (see graph on page 4).

The United Nations Intergovernmental Panel on Climate Change (IPCC) released a definitive report by the world's leading climate scientists in 2007, which laid to rest any remaining doubts about the human and industrial origins of global warming. It stated that "Global atmospheric concentrations of carbon dioxide, methane and nitrous oxide have increased markedly as a result of human activities since 1750 and now far exceed pre-industrial values determined from ice cores spanning many thousands of years. The global increases in carbon dioxide concentration are due primarily to fossil fuel use and land-use change…".

We are now experiencing climate instability caused by greenhouse gases released *decades ago*, so the crisis will only grow worse in coming years. Scientists now have strong evidence of abrupt climate change, with sudden and dangerous shifts that can cause catastrophic loss of lives and property. Communities that begin to prepare will suffer fewer consequences than those that ignore or trivialize this "greatest challenge in human history." While some climate change is now inevitable, we can do something to help our communities survive it.

Temperatures have risen about 1.5° Celsius in the past century. Scientists expect temperature increases at the end of this century to increase 2 to 11.5° C, according to a 2009 U.S. Global Change Research Program report. While these average changes may not seem like much, in the delicately balanced global climate system, they can cause massive

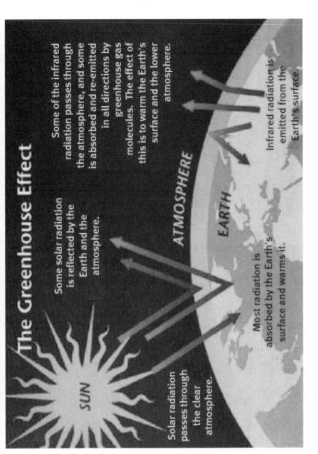

The Greenhouse Effect

SUN

Some solar radiation is reflected by the Earth and the atmosphere.

Solar radiation passes through the clear atmosphere.

Some of the infrared radiation passes through the atmosphere, and some is absorbed and re-emitted in all directions by greenhouse gas molecules. The effect of this is to warm the Earth's surface and the lower atmosphere.

ATMOSPHERE

EARTH

Most radiation is absorbed by the Earth's surface and warms it.

Infrared radiation is emitted from the Earth's surface.

---

instability. Sudden temperature shifts can cause heat waves and drought in some regions, and blizzards or floods in other regions. *The effects include sea-level rises (submerging coastal areas), melting of glaciers, shifts and extinction of species, and crop and human health issues.*

Sea level rises. As the ice sheets melt near the North Pole (Arctic) and South Pole (Antarctic), massive amounts of freshwater are being added to the oceans. At the same time, warmer ocean temperatures cause seawater to expand in volume. Both changes contribute to rises in ocean levels which will make storm waves and coastal flooding far more damaging. Oceans have risen about 8 inches in the past century. Scientists now predict at least a 3- to 5-foot sea-level rise by 2100.

South Pacific indigenous peoples are already finding their low islands inundated by rising seas, erosion from intense storms, and saltwater in freshwater supplies.

A sudden melting of the Greenland ice sheet in the Arctic could cause the evacuation of coastal cities, and possibly shift warm North Atlantic ocean currents away from Europe, causing a "little ice age" on the continent.

Melting glaciers. Even at current warming levels, half of the glaciers in the Pacific Northwest are already gone, and the rest would disappear in this century. Glaciers in Alaska are melting twice as fast as previously predicted. This means less water in streams and rivers, drastically affecting fish and other aquatic life.

The drying up of streams can affect vegetation, water temperatures, and freshwater supplies, further damaging the resources and communities that depend on them.

The melting of permafrost (frozen ground) in the Far North can release huge amounts of methane, adding more of this harmful greenhouse gas to the atmosphere.

Shifts in species, pests and disease. As warmer temperatures creep northward (or up mountain slopes) every year, they drive some species out, and create habitat for new species to come into our area. These species include trees, plants, fish, wildlife, insects, and microbes. Some species can move with the temperature shifts, but others cannot move fast enough, and may face extinction. Climate shifts allow invasive species to displace traditional species, insects and pests to run rampant in new areas, and disease to flourish in warmer temperatures.

Many species have reproductive cycles tied to the seasons, or are more vulnerable to pests or predators at particular times in a season. Climate instability can create havoc with these balanced natural systems.

"Emerging diseases" can become threats to the health of people, livestock, wildlife, and crops, and cause epidemics in water and soil where they have not existed before.

IT'S NOT JUST IN THE FUTURE...
IT'S HAPPENING NOW

2

In North America, climate changes have already drastically affected indigenous peoples' hunting and fishing, economic infrastructure, water and housing availability, forest and agricultural resources, and even their health. Using traditional ecological knowledge, Native harvesters are describing *today* the same drastic shifts in the environment that Western scientists had predicted would occur in the future. For the past decade, Native peoples have been meeting to document these changes. This scale of change will present severe challenges to all tribal cultures, resources and well-being.

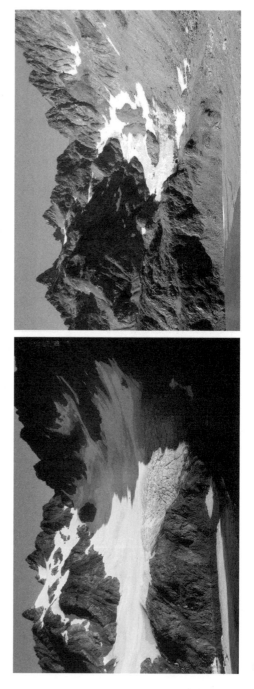

The Anderson Glacier in Washington, in 1978 (left) and 2006 (right). Photos: Larry Workman

Far North. Native nations of the Arctic and Subarctic are already feeling catastrophic effects of warmer temperatures, in the melting of the sea ice, glaciers, and permafrost (frozen ground), and increase in fires, insects, flooding and drought patterns.

The polar bear is becoming an endangered species. Ice floes are the polar bears' home, where they mate, give birth, and raise their cubs. Now many polar bears are leaving their young to search for food, and the cubs are drowning because the ice floes are too thin. If the polar bear become extinct, a key part of the Inuit (Eskimo) culture will be lost. Northern peoples also hunt seals (who can only give birth on the ice) and caribou (which are getting harder to find). Hunters crossing ice are more often falling into open water and drowning.

The Alaskan village of Shishmaref (inhabited for 4,000 years) is facing evacuation. Due to a reduction of sea ice and permafrost, the village is no longer protected from erosion by violent storms. Homes in many other northern villages are sinking into the melting permafrost.

Many new species are migrating northward in the Arctic, such as the robin (for which the Inuit language has no word). These invasive bird species can carry diseases such as the West Nile Virus. Other species such as orcas are eating new species that they had never before been seen eating.

Southwest. Drought has affected the water table levels and limited water sources that depend on the little rain the region get to replenish them, killing plants and livestock.

Droughts have caused beetles to suck all the sap of trees (such as the piñon) for water, and caused the death of medicinal plants.

The 1993 Hanta Virus outbreak was a mystery to scientists until Navajo elders noticed that increased rainfall had caused an explosion in the population of mice (which fed on piñon nuts). The rainfall had been caused by intensified "El Niño" fluctuation in ocean temperatures. Hanta has also spread to rodents and humans in other regions.

Great Plains. Increased extreme weather such as flooding, blizzards and drought are threatening tribal economies where livestock and grain are the primary sources of income.

Summer heat and severe weather has increased health risks of children and elders.

Water resources on the surface and in underground aquifers are becoming depleted before they can be recharged by rain.

Eastern Woodlands. Climate Change proves an ecological global risk that disputes traditional food gathering and forestry. Severe weather events include ice storms and flowing from rapid snowmelt.

Culturally significant trees such as sugar maples and birch are shifting northward out of the Northeast and Great Lakes regions.

Greater numbers have been seen of stinging black flies, disease-spreading mosquitoes, and predatory fish. Fewer numbers have been seen of hummingbirds and frogs.

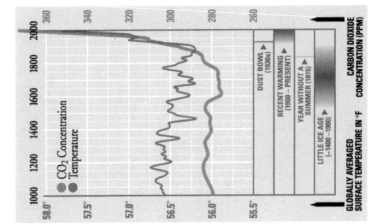

Graph of temperature and carbon dioxide rise

4

CLIMATE CHANGE:

IT'S NOT JUST HAPPENING ELSEWHERE...... BUT IN OUR OWN REGION

3

We usually hear about the effects of climate change devastation in far-off places, such as the Arctic and South Pacific. But here in the Pacific Northwest, we have already begun to see the climate become unstable, and begin to affect tribal communities. The U.S. Global Change Research Program stated in a 2009 report, *"Impacts related to changes in snowpack, streamflows, sea level, forests, and other important aspects of life in the Northwest are already underway, with more severe impacts expected over coming decades in response to continued and more rapid warming."*

Shifts in weather patterns and glaciers. An unstable climate affects different regions in different ways. Our region, where the ocean meets the coast and the mountains, is especially vulnerable to extreme variations in weather. The concern is not that there is more wind and rain, but that it comes with increased intensity in shorter bursts.

Scientists debate whether specific recent windstorms, mudslides, floods and blizzards can be tied to climate change. But it is not debatable that weather pattern shifts are occurring in the Pacific Northwest.

Normally, weather systems come in from the west, bringing in ocean moisture. When the moisture hits the mountains, it rises and condenses into rain or snow. But increasingly, the winds are coming instead from the north or south, and are following the ridgelines, resulting in less rain and snow in the mountains.

Melting glaciers and snowpack. Warmer temperatures are reducing both glaciers and snowpack in the mountains, and changing the timing of their seasonal melts. With shorter winters, there is less time for snow to pile up in the mountains. Smaller glaciers mean less freshwater in the streams for fish and other aquatic life. More than 90 percent of North Cascades glaciers could disappear in 40 years if annual temperatures increase by 2° C. Between 20 and 40 percent of the glaciers' volume has been lost in 25 years. At least one Olympic glacier has melted entirely.

The snow melts more quickly in the Spring, not only causing floods, but damaging salmon habitat—scouring and stirring up sediments, when the salmon are still smolts. Less of the runoff has time to seep into the groundwater.

In the summer, there is less runoff and it flows more slowly. The blueback salmon are threatened in the Quinault River because of reduced runoff from the Olympics. Other fishers report seeing salmon with lesions associated with warm-water diseases.

Flooding at Skokomish (Photo provided by The Sounder)

Changes in fishing.

Northwest tribes have spent decades fighting for the right to fish in "usual and accustomed places." But these places are already being affected by the reduction of rainfall and snowmelt in the mountains, the melting glaciers, and warmer temperatures and shifts in ocean currents.

A "dead zone" has been growing off the Washington and Oregon coasts, where fish and crab are being starved of oxygen by wild "upwellings" of micro-organisms that feed on oxygen. This crisis was caused by ocean current shifts tied to climate change. The Quinault Tribe is now using a submarine to study hypoxia (oxygen starvation) in the ocean.

Increasing levels of domoic acid in the warming ocean affects the gathering of crab and shellfish due to human health concerns, such as short-term memory loss and other neurological problems. It also affects birds and marine mammals who eat shellfish. The Quileute Tribe is now testing shellfish for domoic acid.

Shifts in species.

Warmer ocean temperatures have caused marine species to shift northward. Some plant species are also shifting northward, or up mountain slopes. Once the species reach the top of a land mass or a slope and can go no farther, they may go extinct. Tribes may no longer have access to some natural resources guaranteed to them in treaties, as the species shift outside of treaty-ceded territories.

Quinault and Quileute fishers report catching anchovies and sunfish for the first time off their coast, and a major reduction in the smelt catch and kelp beds. Some fish, eggs and plants are no longer available for the First Foods Feast.

The brown pelican and giant Humboldt squid have migrated up from central California in recent years; both are voracious eaters of fish. Large turtles and Great White Sharks are also newcomers to the area.

Some grasses are no longer available to coastal basketmakers, because the species have shifted northward. Invasive plants arriving from the south, such as Japanese Knotweed, have also displaced native plants.

Changes in forestry.

Northwest forests are already being affected by inland drought conditions, and the northward shifts of diseases and pests. It is not known whether specific fires have been caused by climate change, but certain patterns of fires have been caused by climate instability.

Warmer winters mean greater survival for many pests and disease. Pest infestations make the trees useless for harvest, and dry logs are vulnerable to very hot fires.

The spruce bark beetle has infested more than 12 million acres in B.C., Yukon, and southern Alaska. Huge swaths of forest have been killed by the infestations and the resulting catastrophic wildfires.

In another "fingerprint of climate change," mountain pine beetles have killed more than 40 million acres of forest from Colorado to Alaska—twice the area of Ireland.

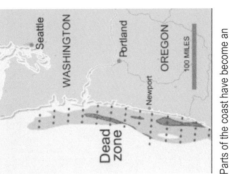

Parts of the coast have become an oxygen-starved "Dead Zone" (Map: U.S. Global Change gram).

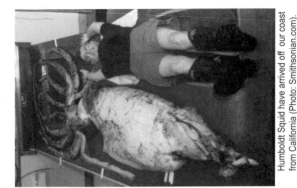

Humboldt Squid have arrived off our coast from California (Photo: Smithsonian.com).

IT DOESN'T AFFECT EVERYONE THE SAME... NATIVE PEOPLE HAVE MORE AT STAKE

Climate change is a potential Culture Killer. Native rights are primarily placed-based rights, dependent on a longtime attachment to local tribal territories. Climate change shifts and disrupts plant and animal habitats, and in doing so forces tribal cultures to move, adapt to new conditions, or die. Politicians and the media treat climate change as a large-scale global or national crisis, with all communities evenly affected in the same ways. But Native peoples have more at stake than other North Americans, particularly here in the Northwest.

A landslide on the Skokomish Reservation closes a major highway for more than four days. (Photo by Mark Warren, *Sounder*, 2006)

7

Photo by Sheila McCloud,
Northwest Indian Fisheries Commission

Species and treaty boundaries. According to the Tulalip Natural Resources Department, "For the tribes, range shifts in native species will threaten their cultural existence. The treaty-protected rights of tribes to hunt, fish, and gather traditional resources are based on reservation locations and usual and accustomed areas on public lands. These locations are chosen to ensure access to culturally significant resources, whose locations were thought to be fixed. If the traditionally significant plants, animals, and aquatic species shift out of these areas, tribes will no longer have the same legal rights to them."

The Tulalip add, "Even if rights to these species could be secured, without access, to use of these species will be virtually impossible…. Few tribes can afford to purchase large territories of new land, and federal laws prohibit the transfer or expansion of tribal jurisdiction."

Loss of traditional knowledge. The loss or mitigation of culturally important species on which traditional knowledge depends will make it more difficult for elders to practice and pass their knowledge to the next generation. Some climate stresses will fall to the elders who are more vulnerable to heat waves, and food and water stress.

Tribal youth are also in danger of losing touch with traditional hunting, fishing, and gathering, as warmer weather keeps them indoors (up to 4 degrees by 2040).

Western scientific studies often take many years to fund, implement, review, and publish. Traditional knowledge comes directly from the harvesters themselves on the land and waters. This knowledge provides an early warning system for policymakers, saving precious time in an era of rapid climate changes.

Hoh tribal headquarters are permanently sandbagged due to chronic flooding.
(Photo: Elaine Thompson /Associated Press).

Coastal sea level rises. In the Pacific Northwest, many reservations are located on the coast or near the fishing grounds at the mouths of rivers. This makes them vulnerable to rises in sea level and coastal erosion. Higher seas will intensify storm damage, coastal flooding, and the risk of saltwater contaminating freshwater.

The Quileute Reservation at LaPush, Washington, has asked to be relocated to higher ground because of the risk of tsunamis and storm surges, which will only be made worse by rising seas. Waves from one severe storm pushed huge driftwood up next to the tribal school.

The Hoh Tribe is receiving federal assistance to move to higher ground. Their tribal headquarters are now permanently sandbagged due to severe flooding, and homes have fallen into the water. A tribal council member says, "How are we going to save what land base we have left, because you can't create more land."

A moderate 14-inch rise in ocean levels would inundate 40 percent of Puget Sound mudflats, wiping out a significant habitat for shellfish and waterfowl.

IT'S NOT JUST DEPRESSING...
NATIVE COMMUNITIES ARE RESPONDING.

5

Native communities have a number of unique advantages in dealing with climate change that non-Native communities rarely possess:

* Traditional Ecological Knowledge: Indigenous cultures have centuries of experience with local natural resources, so they recognize environmental changes before Western scientists detect them, and can develop our own ways to respond to these changes.

*A Sense of Community: In contrast to the non-Native population, *we still have community*. We still have extended families that care for each other, assume responsibility for each other, and extend hospitality in times of need.

* Political Sovereignty: Because tribes have a unique status as nations, we can develop our own models of dealing with climate change, and managing nature in a sustainable way.

Gather and share information. As tribal people who have survived against all odds in the past, we will survive the changes associated with global warming--if we prepare now. Climate change is too large a concern to leave to the sole concern of a tribal department. The tribal government and members together need to gather and share the information.

In Alaska, Inuit and Aleut villages have held community workshops of elders, hunters, harvesters, and youth to document changes in the resources, to collect samples, and educate their communities.

Native Peoples/Native Homelands conferences have brought together tribal members from around the country to document changes in their regions, based on both Western and traditional Native science, and plan for responses.

Northwest tribal agencies have met to discuss not only the existing and predicted effects of climate change, but how to mitigate (lessen) the effects or adapt to them in ways that can help protect cultural lifeways.

Involve the youth. The young people of today are going to be the ones most affected by climate change. If youth become aware of these issues and get active, youth can educate their community. Middle school and high school students (and other youth) can form climate action groups.

Alaska Youth for Environment Action collected thousands of signatures on their climate change petition, which was presented to Congress. Youth from the Indigenous Environmental Network have been involved in the Powershift youth movement on climate change.

First Nations youth in British Columbia have gathered data and mapped their traditional resources to protect them from timber companies and non-Native claims. Similar youth teams could also look at how to "harden" their communities against climate change.

Tribal youth trainings have included discussion and videos on climate change. Native youth could also set up their own groups to educate their community using booklets, films, art, theater, music, etc. to educate families and communities about the climate crisis.

Use treaties to protect habitat. Because dumping carbon into the atmosphere destroys habitat for tribal resources, climate change can be seen as a violation of treaty rights. By using the treaties to protect the habitat of fish, shellfish, wild game and plants, tribes can strengthen their case against these violations.

Tribes can pressure the federal government to fulfill its trust duty to protect tribal homelands, by reducing U.S. carbon emissions.

Intertribal fish commissions are already protecting or co-managing fish habitats (under federal court decisions such as the Boldt and Culvert cases) but their work also protects the resources from climate change, for future generations.

Tribes on the Olympic Coast have used their treaty standing to negotiate agreements with the State to prolong fishing seasons after stormy weather, and to jointly manage a National Marine Sanctuary.

Develop renewable energies. Tribes are in a unique position to develop renewable energies, to convert from fossil fuels to cleaner and more locally controlled power sources. Tribes can select the most appropriate energy sources for the natural region: wind, solar, biomass, wave, tidal, and others. These energy sources can provide a source of tribal income, through selling the power to non-Native communities, and also provide an example to their neighbors.

At Native Renewable Energy Summits, city governments committed to reducing their carbon emissions began to discuss purchasing renewable energy (such as wind power) from tribes.

The Tulalip Tribes in Washington state have developed a biogas energy project with local dairy farms, to generate electricity by burning methane (thereby keeping cattle waste out of the river).

Yakama Power is monitoring wind on the reservation as part of a wind energy assessment. The Makah Tribe has studied potential electrical power from wave energy.

Get involved in the global process. For the past decade, indigenous organizations from around the world have attended the annual conferences of the United Nations Framework Convention on Climate Change (UNFCCC).

Native non-governmental organizations have asked for special status for indigenous nations as the people most affected by climate change. Their case can be strengthened by the direct participation of tribal governments, who have greater powers to pressure federal government.

An Inuit lawsuit in an interamerican human rights court charges the U.S. with violating their human rights to their culture and hunting. This is the first example of an indigenous people using international law to protect their homeland from climate change. The Alaskan community of Kivalina has sued the oil companies for damages in U.S. court.

Dairy cattle waste that previously would have contaminated salmon habitat now is used at Tulalip's Qualco Energy partnership to produce methane energy.

IT'S NOT JUST HOPELESS...
THERE'S A LOT THAT TRIBES CAN DO!

6

We can see that climate change is going to be devastate us if we are not prepared, so we have to go out and meet it. The people of the world and especially our Native communities, no longer have 5 to 10 years to begin planning. We must begin *today!*

Tribes can respond to climate change by securing freshwater supplies and food supplies, protecting our region's fish, wildlife and plant life, planning with local governments, and building intertribal relationships. There are many responses that tribes can take, but it is urgent that *we begin now.*

Secure freshwater supplies. Tribes need to secure access to fresh water supplies. Tribes need to assert their water rights not just for present development needs, but can start thinking ahead to future shortages.

Some mountain streams can be redirected so the water runoff recharges the aquifers (underground water) so the area does not dry up.

Tribes can work with local governments to plan water conservation measures, water treatment, and protection of local supplies. Concerns about sea-level rise contributed to a partnership between the Nisqually Tribe and Olympia to relocate their freshwater source from McAllister Springs to wells on higher ground. The move will also help restore water flows to Medicine Creek.

Secure food supplies. Native communities need to be thinking ahead to a situation of food shortages, and not be completely dependent on supermarkets for basic needs. Tribes can plan both perishable and non-perishable food storage facilities. Tribal nations can become part of the growing trend toward emphasizing Native traditional foods, to protect the health of tribal members and to keep local control.

Tribes that have agricultural crops should research and adopt drought-resistant seed varieties.

Tribes that do not have an agricultural base can create agreements with tribes that have food crops and animal herds, so members can have greater food security in times of need.

Prepare for impacts on species. We need to prepare for some culturally important plant and animals species to move out of our region, and determine whether adaptation will provide alternative sources of plant and animal species. As climate change changes our landscape, inevitably we will see the disappearances of fish runs, and entire fisheries may be replaced by other runs of species that are shifting north.

Droughts will also influence plant and animal species that have sustained Indigenous peoples as subsistence food sources or as essential to their ceremonial life. Tribes may have to adapt as fish and other marine life, and land-based animals and plants shift northward.

Cooperate with other tribes. In order to survive, tribes have to work with other Indigenous nations across imposed colonial boundaries, on the basis of being part of the same natural region. First Nations in the U.S. and Canada need to cooperate to decolonize ancestral territories and protect our common property (the air and water) for future generations.

First Nations can share skills and knowledge about shifting animal and plant species, and exchange ideas and training about community adaptation, climate change mitigation, and renewable energies.

Indigenous nations around the Pacific Rim region can use the United League of Indigenous Nations Treaty process as a way to build political alliances and practical cooperation to jointly respond to climate change.

Traditional varieties of white flint corn, grown by the Oneida Nation at its Tsyunhehkwa Farm in Wisconsin. Photo: Michele Shaw.

Tribal communities can begin to teach each other how to develop and harvest the new resources coming into their area. We can begin to draw on relationships with other tribal communities to their south to anticipate the species that are moving into their area. Likewise, tribes can also draw on relations to their north to teach them about the useful or harmful species that will be migrating in that direction.

Plan locally with neighbors. Tribes and local communities have the ability to work together as neighbors around common interests, such as land-use planning to prevent climate change problems, and emergency planning for the more disastrous impacts of climate change. Tribes can search for common ground with local, municipal and county governments, and provide models for them to learn from. In unstable climate conditions, local relationships are the most important. *We cannot rely on state or federal assistance: look what happened after Hurricane Katrina.* The smaller size of tribal and local governments can make them more flexible.

Tribal and local governments can cooperate in joint land-use planning to prevent climate change problems, such as moving or building homes above floodplains, conserving and treating water, protecting shorelines and beaches from erosion, building and retaining floodplain walls, and controlling pests and diseases through local education.

The Swinomish Tribe has faced severe seasonal flooding, but is now working with neighboring Skagit River Delta governments on an adaptation and mitigation plan (Swinomish Climate Change Initiative).

There is also a need for cooperation with local governments in emergency planning, such as identifying community shelters, identifying evacuation procedures and routes, sharing emergency food, water, and heat, and cooperating on medical and fire services.

RECOMMENDATIONS FOR TRIBAL LEADERS

From the *Climate Change and Pacific Rim Indigenous Nations report*, by the Northwest Indian Applied Research Institute (NIARI), 2006

1. Educate tribal membership on the present and future effects of climate change on tribal homelands.

2. Secure sources of fresh water now to meet future needs of tribal communities located in drought- and flood-impacted areas.

3. Secure a future source of food stocks, long-term storage capacity, and production capabilities (including intertribal trade) for crops that can adapt to climate change.

4. Prepare for impacts on culturally significant wild plant and animal species, and teach each other about both harvestable and invasive species that are shifting northward.

5. Develop relationships with neighboring governments and communities regarding land use planning to prevent worsening storm effects, and emergency plans for the more disastrous impacts of climate change.

6. Consider alliances with local governments to build and market renewable energy capacity, such as wind and biogas power.

7. Consider strategies to unite tribes around habitat protection, looking ahead to the inevitable effects of climate change.

8. Get actively involved as sovereign governments in U.N. climate change negotiations, and pressuring national governments to reduce emissions.

9. Get youth involved in cultural education, and defending the future of the tribe from harmful climate change.

10. Work with other Indigenous nations in a treaty relationship transcending colonial international boundaries.

Northwest Indian Applied Research Institute (NIARI) http://nwindian.evergreen.edu
SEM 3122, 2700 Evergreen Parkway NW, Olympia, WA 98505
Phone: (360) 867-6614 Fax: (360) 867-6553 E-mail: nwindian@evergreen.edu

NIARI Climate Change and Pacific Rim Indigenous Nations Project
Copies of booklet: http://academic.evergreen.edu/g/grossmaz/climatechangebooklet.pdf
Report: http://academic.evergreen.edu/g/grossmaz/IndigClimate2.pdf
Powerpoint: http://academic.evergreen.edu/g/grossmaz/IndigClimate2009.ppt
Article: http://academic.evergreen.edu/g/grossmaz/Indigenous Nations' Responses.pdf
Other Links: http://academic.evergreen.edu/g/grossmaz/climate.html

RESOURCES AND LINKS

Northwest Indian Fisheries Commission (NWIFC) www.nwifc.org

NWIFC video Shadow of the Salmon and Resource Guide www.salmondefense.org/projects/shadow-of-the-salmon http://education.wsu.edu/nativeclearinghouse/achievementgap/ShadowoftheSalmonCurriculumGuide(4).pdf

Columbia River Inter-Tribal Fish Commission (CRITFC) www.critfc.org/tech/climate/cc_workshop.html

United League of Indigenous Nations www.ulin-web.org

UN Permanent Forum on Indigenous Issues (UNPFII) www.un.org/esa/socdev/unpfii/en/session_seventh.html

Guide on Climate Change and Indigenous Peoples (UNPFII) www.tebtebba.org/index.php?option=com_docman&task=cat_view&gid=18&Itemid=27

Anchorage Declaration on Climate Change (2009) www.indigenoussummit.com/servlet/content/declaration.html

UN Framework Convention on Climate Change (UNFCCC) http://unfccc.int/2860.php

UN Intergovernmental Panel on Climate Change (IPCC) www.ipcc.ch

Native Climate Commons (UN & Tulalip Tribes) http://climate.nativecommons.net

National Tribal Environmental Council (NTEC) www.ntec.org

Native Peoples - Native Homelands climate change workshops www.usgcrp.gov/usgcrp/Library/nationalassessment/native.pdf (1998) www.nativepeoplesnativehomelands.org (2009)

Native Communities and Climate Change (2007) www.colorado.edu/Law/centers/nrlc/publications/ClimateChangeReport-FINAL%20_9.16.07_.pdf

Planning for Seven Generations (2008) www.cbp.ucar.edu/tribalconfhome.html

Global Climate Change Impacts in the United States (2009) www.globalchange.gov/publications/reports/scientific-assessments/us-impacts/download-the-report

Tribal Climate Change Forum (2009) www.sustainablenorthwest.org/programs/policy/tribal-climate-change-policy-training-meeting-materials/?searchterm=tribal%20climate

Indigenous Environmental Network (IEN) Climate Justice Campaign www.ienearth.org/climatejustice.html

Native Energy / Intertribal Council on Utility Policy (COUP) www.intertribalcoup.org Native Wind www.nativewind.org

National Wildlife Federation www.tribalclimate.org

Coast Salish Gathering www.coastsalishgathering.com

Climate Change at Quileute and Hoh http://academic.evergreen.edu/g/grossmaz/papiez.html

Tribes and Climate Change (N. Arizona Univ.) http://www4.nau.edu/tribalclimatechange/tribes/northwest.asp

Swinomish Climate Change Initiative www.swinomish-nsn.gov/departments/planning/climate_change/climate_main.html

Energy Planning: A Guide for Northwest Indian Tribes www.nwseed.org/documents/NWSEED_Tribal%20GB_Final.pdf

Climate Impacts Group (Pacific Northwest) www.cses.washington.edu/cig

Sea-level Rise on Pacific Northwest Coast www.nwf.org/sealevelrise/Maps_of_the_Pacific_Northwest_Coast.cfm

State of the Salmon www.stateofthesalmon.org

Environmental Justice and Climate Change (EJCC) Initiative www.ejcc.org

Energy Justice Network www.energyjustice.net

Qualco Energy (Tulalip biogas energy plant) www.qualcoenergy.com

THANK YOU to the San Manuel Band of Serrano Mission Indians for funding this project www.sanmanuel-nsn.gov

"The message that I would send to young people in terms of the impacts of future problems associated with global warming would be to keep passing on your knowledge, our values associated with nature, the love for Mother Earth, and protection of the environment. The other day I talked to kids about some of the things I feel were occurring and it seems to go in one ear and out the other. But then I think back about my youth, and my father talked to me and taught me. I think my father felt the same way, that it all went in one ear and out the other and that I didn't learn anything. But when I think back, I value those things that I was told. Whether we think they are listening or not, we need to protect the environment. They'll pick it up. There is no doubt in my mind that the younger generation will continue with the things that we believe in . . ."

"We will be the last ones to have the last green areas remaining on the Earth in the future because of all the development in other areas. And we should continue to hold our values. We treasure nature and it shows in our Native lands."

--Caleb Pungowiyi, Alaskan Yupi'k leader

"Our elders used to tell us that they knew it was coming. They would tell us: 'One day this water, these mountains, you will see a time of change where it is going to hurt our people.' That almighty dollar...is what is hurting our climate today, and hurting our people, our animal life, and our water supply, and all our mountains where our berries are. We never used to get the kind of heat we see today in Washington and Oregon. We never did see our waters, our rivers, or our lakes get so warm. Who is creating it? Not the Indian people."

--Chief Johnny Jackson,
Klickitat Band of Yakama
1998

CONTRIBUTORS' BIOGRAPHIES

Jeannette Armstrong, a member of the Okanagan Syilx Nation, is an author and artist and executive director of the En'owkin Centre (the culture, language, and arts education institution of the Okanagan Nation), as well as an assistant professor in Indigenous Studies at the University of British Columbia–Okanagan. She is also a member of the National Aboriginal Traditional Knowledge subcommittee of the Committee On the Status of Endangered Wildlife in Canada and has received a number of honorary doctorates and awards, including EcoTrust USA's Buffett Award for Indigenous Leadership. She has an interdisciplinary Ph.D. in Environmental Ethics and Syilx Indigenous Literatures from the University of Griefswald, Germany.

Laural Ballew is a member of the Swinomish Tribe in La Conner, Washington, and has lived on the Lummi Reservation near Bellingham, Washington, for more than thirty-five years. Her professional experience has been in finance and management. She is currently employed by Northwest Indian College in the position of comptroller. She received her associate of arts and science degree in business administration from Northwest Indian College in 2000. She graduated from Western Washington University in 2002 with a bachelor of arts degree in American cultural studies. In 2006 she graduated from The Evergreen State College in Olympia, Washington, with a master's in public administration–tribal governance degree. She is active on several committees with the Lummi Nation, Northwest Indian College, and the Ferndale school district.

Jamie Bown earned her bachelor of arts and bachelor of science degrees from The Evergreen State College (with emphases in Native American and World Indigenous Peoples' Studies and ecological informatics), after earning her associate of technical arts in database management degree at South Puget Sound Community College. Her research interests include treaty and Indigenous rights, government and tribal relations, adaptive natural resource management and environmental policies, climate change, and the restoration of the Elwha River. Originally from Southern California, she lives in the Olympia area with her three sons: Joshua, Isaiah, and Josiah.

Bradford Burnham earned a bachelor's degree at Connecticut College, a master's degree in biology at City College of New York, and a master's in public administration from The Evergreen State College in Olympia, Washington. He has taught at a community college and worked at museums, nature centers, and aquariums. He has also written science books for children.

William Charlie, Jr., "Chaquawet" has been vice president of the Union of British Columbia Indian Chiefs since 2009. He is the elected chief and appointed CEO of the Chehalis (Sts'ailes) Indian band on the banks of the Harrison River, about 100 km east of Vancouver. Growing up in the close-knit Coast Salish village of Sts'ailes, William was able to grasp and utilize the cultural teachings of his people. Many communities and organizations throughout the Fraser Valley call on William to be a voice for special ceremonies, gatherings, or meetings, and he is a much sought-after lecturer and consultant. He has been a band staff member and an elected council member for many years. He was instrumental in restructuring the band leadership as a director/management team structure. Chaquawet is presently the chairperson for a group of independent leaders who oversee the operation of the Kwikwexwelhp Healing Village. He also sits on two B.C. First Nations Chiefs working groups: the First Nations Interim Health Governance Committee and the Interim First Nations Child and Family Wellness Council. He is the co-owner of an award-winning, family-owned Aboriginal tourism business called Sasquatch Tours, which provides authentic First Nations cultural experiences.

Billy Frank, Jr., is a member of the Nisqually Indian Tribe and has been chairman of the Northwest Indian Fisheries Commission (NWIFC) since 1981. In this capacity, he "speaks for the salmon" on behalf of twenty Treaty Indian Tribes in western Washington. In the 1960s and early 1970s, Frank was a grassroots political activist who was frequently jailed for civil disobedience because he took part in numerous "fish-ins" in opposition to state authority over the tribes. Years of resistance finally paid off when federal court ruled in favor of the tribes in the 1974 Boldt decision. NWIFC was formed in 1975 to support tribal fisheries management activities and to enable the tribes to speak with a united voice. With Frank's leadership, the NWIFC and the tribes it serves are working to protect and restore the salmon resource for Indians and non-Indians alike. Celebrated regionally, nationally, and internationally as an outstanding Native American leader, Frank has been the recipient of numerous recognition awards, including the 1991 Albert Schweitzer Prize for Humanitarianism and the 2004 Indian Country Today Inaugural American Visionary Award.

Zoltán Grossman is a member of the faculty in geography and Native American and World Indigenous peoples studies at The Evergreen State College, in Olympia, Washington. He has been a senior research associate at the Northwest Indian Applied Research Institute (NIARI) and its Climate Change and Pacific Rim Indigenous Nations Project. He was co-chair of the Indigenous Peoples Specialty Group of the American Association of Geographers (AAG) in 2008–2010 and an International Geographical Union observer at the 2008 climate change session of the UN Permanent Forum on Indigenous Issues. He earned a Ph.D. in geography with a minor in American Indian studies in 2002 as a Udall Fellow at the University of Wisconsin–

Madison and taught at the University of Wisconsin–Eau Claire in 2002–05. His doctoral dissertation studied "Unlikely Alliances: Treaty Conflicts and Environmental Cooperation between Native American and Rural White Communities." He was a co-founder of the Midwest Treaty Network during the Wisconsin Ojibwe spearfishing conflict and later helped bring together Native nations with their former adversaries in sport fishing groups to protect the fish from metallic mining projects.

Preston Hardison is a watershed policy analyst for the Tulalip Tribes in Marysville, Washington, and has coordinated the Chinook Salmon Recovery Plan, programmed databases for the Cultural Stories Project, developed data policy, and supported natural resources negotiations. Since 1996, he has helped negotiate terms in the Convention on Biological Diversity (CBD) relating to Indigenous knowledge and has provided expert testimony to the World Intellectual Property Organization. He serves on several international biodiversity information network committees and promotes the development of information exchange standards and protocols.

Greg Mahle is an enrolled member of the Upper Skagit Indian Community, located near Sedro-Woolley, Washington. In 2005 he graduated from Northwest Indian College (NWIC) with a two-year degree in Native American studies. He then transferred to Western Washington University, where he is currently pursuing a history degree with an emphasis on Coast Salish studies. As a culture curriculum developer for NWIC, he has been given the unique opportunity of working side-by-side with some of the most traditional elders and respected language speakers throughout Coast Salish territory. He has learned many traditional ways of the ancestors while learning to speak three Salish languages: Halkomelem, Lushootseed, and Lummi. He says, "I would like to thank the many elders who have taken the time to help me learn the ways of our people. I would like to give special thanks to Sharon Kinley, director of the Coast Salish Institute, and Northwest Indian College for allowing me to participate in the many culturally rich activities. Hyshqe."

Dennis Martinez is of O'odham, Chicano, and Swedish heritage. He has worked in ecocultural restoration for nearly 39 years in temperate terrestrial, and tropical terrestrial and aquatic-marine ecosystems as a restoration and ethno-ecologist. He is Founder and Co-Chair of the Indigenous Peoples' Restoration Network of the Society for Ecological Restoration International and is Co-Director of the Takelma Intertribal Project. He works internationally with community-based Indigenous Peoples on cultural rights, resource access and protection, climate change, forest restoration, and bridging Western Science with Traditional Ecological Knowledge. He is a well-known speaker and writer, has received awards in restoration and social justice, and was an awardee in the Ecotrust-Buffet Award for Indigenous Conservation Leadership in northwestern North America. He currently is focusing on climate change, is on the steering committee of the Indigenous Peoples Climate Change Assessment, and consults with the National Congress of American Indians and the American Indian and Alaska Native Climate Change Network on Indigenous adaptation and mitigation of climate change.

Debra McNutt is a graduate student at The Evergreen State College (Olympia, Washington), where she is enrolled in the Master of Public Administration / Tribal Concentration program. She has coordinated NIARI's Climate Change and Pacific Rim Indigenous Nations Project and edited the institute's community organizing booklet on climate change for Northwest tribal members. She was an observer at the 2008 climate change session of the UN Permanent Forum on Indigenous Issues and researched the health effects of climate change on Native women and children for the Community Alliance and Peacemaking Project. As a co-founder of the Midwest Treaty Network, Debra coordinated witnesses for treaty rights during the Wisconsin Ojibwe spearfishing crisis in the late 1980s and early 1990s to deter and document anti-Indian harassment. She also spent ten years working to build environmental anti-mining alliances between Native American and white communities in Wisconsin.

Larry Merculieff has over thirty-five years of experience serving his people, the Aleuts of the Pribilof Islands, and other Indigenous peoples locally, nationally, and internationally in a number of leadership capacities. He was the first Alaska Native commissioner of the State Department of Commerce and Economic Development. Larry also served as the chair of the Indigenous knowledge sessions of the Global Summit of Indigenous Peoples on Climate Change. He is the co-founder of the Indigenous Peoples' Council for Marine Mammals, the Alaska Forum on the Environment, the International Bering Sea Forum, and the Alaska Oceans Network. He has received a number of awards, including the Buffet Finalist Award for Indigenous Leadership, the Environmental Excellence Award for lifetime achievement from the Alaska Forum on the Environment, the Rasmuson Foundation Award for Creative Nonfiction, and the Alaska Native Writers on the Environment Award. He is featured as one of ten Native Americans in the Second Story Press book *Native American Men of Courage*. He co-authored the 2009 book *Aleut Wisdom: Stories of an Aleut Messenger*.

Chelsie Papiez earned her bachelor's degree in botany from the University of Washington and her master's of environmental studies from The Evergreen State College. Highlighted in this book is the part of her thesis that focuses on climate change impacts on coastal Indigenous people of Washington state. She blended traditional knowledge with natural science to formulate a unique multidisciplinary project. After receiving her master's degree, she was awarded a National Oceanic and Atmospheric Administration (NOAA) Coastal Management Fellowship. During the two-year fellowship, she assisted the state of Maryland by evaluating how and where sea level rise inundation will affect coastal areas. As part of the project she identified coastal land characteristics for conservation in response to sea level rise in the Chesapeake Bay. This research will enhance Maryland's ability to prioritize land conservation activities based on the identification of adaptive land benefits.

Alan Parker is the director of the Northwest Indian Applied Research Institute (NIARI) at The Evergreen State College, in Olympia, Washington, where he has been a faculty member since 1997 and co-founded the nation's first graduate program in tribal public administration. He is a citizen of the Chippewa Cree Nation of the Rocky Boy's Reservation in northern Montana. He graduated from the UCLA School of Law in 1972 and practiced law for over twenty years in Washington, D.C. There he directed research on tribal governments for the American Indian Policy Review Commission and was the first Native American to serve as chief counsel to the U.S. Senate Committee on Indian Affairs, in 1977–81 and 1987–91. He was instrumental in securing passage of the Indian Child Welfare Act, the American Indian Religious Freedom Act, the Native American Graves Protection and Repatriation Act, the Tribal Self-governance Act, American Indian Development Finance Corporation Act, and numerous tribal land and water claims settlements. He served as president of the American Indian National Bank in 1982–87. He later organized the National Indian Policy Center at George Washington University and was appointed the first Native American attorney on the Washington State Gambling Commission.

Renée Miller Klosterman Power. See the tribute in front of this book.

Rudolph C. Rÿser is Taidnapum and the chair of the Center for World Indigenous Studies, in Olympia, Washington. He has for more than forty years worked in the field of Indian affairs as a writer/researcher and Indian rights advocate. He has served as a senior advisor and speechwriter for major Indigenous leaders, including Quinault President Joe DeLaCruz and his successor, President Fawn Sharp; Grand Chief George Manuel, Lummi Chairman Henry Cagey, and Yakama nation leader Russell Jim. Since 1975, Dr. Rÿser extended his work in Indian affairs to encompass Indigenous peoples throughout the world. He has contributed to policy development activities of the Affiliated Tribes of Northwest Indians, the American Indian Policy Review Commission, the Conference of Tribal Governments, the National Congress of American Indians, and the World Council of Indigenous Peoples, and he worked from 1984 onward on language for the UN Declaration on the Rights of Indigenous Peoples. A well-known essayist among Indigenous peoples throughout the world, he is author of *NationCraft*, a new book focused on fourth world political development, and he is actively engaged in international negotiations over Climate Change Treaty language for Indigenous nations.

Ata Brett Stephenson enjoys a strong *whanau* (extended family) lifestyle underpinned by genealogical attachment to Te Kapotai, Waikare, a sub-tribe of Ngapuhi, in Aotearoa (New Zealand). He received traditional *wananga*-based training from early childhood in boatmanship, fishing, and woodcarving practices, which were to provide an essential grounding for his professional participation in the marine field while holding an appointment as curator in marine biology with the Auckland Museum for thirty-two years. At the invitation of tribal trustees of Te Runanga o Ngati Awa, Brett was appointed in 1998 to plan and prepare a course for tertiary level study leading to a bachelor of environmental studies in which Maori knowledge was used to inform the teaching of Western environmental science. As senior lecturer Brett taught in a multidisciplinary capacity that embraced *matauranga putaiao* (Maori concepts of environment) and acted as the Pou Turuki (Support Vine) at the tribal university Te Whare Wananga o Awanuiarangi in Whakatane. Recently retired, he has continued to advise former students in wetland and water management projects and also established new interests in environmental ethics and advocacy through his *hapu* (tribal) links and as a member of the New Zealand Bioethics Council.

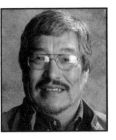

Mike Williams is a Yupiaq Eskimo and a member of the Akiak Native Community, located on the Kuskokwim River in Alaska. He is the vice chairman of the Akiak Native Community, a federally recognized tribe, and serves as vice chairman of the Alaska Inter-Tribal Council, which represents 229 tribes in Alaska. Mr. Williams is also on the executive committee of the National Congress of American Indians.

Lexie Tom is a member of the Lummi Nation who graduated in 2007 from Northwest Indian College (NWIC) with a degree in Native studies. She is a cultural curriculum developer for the NWIC Coast Salish Institute and is working toward her degree in anthropology and linguistics at Western Washington University. She says, "I really enjoy my work. I learned so much here about the history of my own people that I wouldn't have learned anywhere else. I started working here right out of high school, which was a confusing time for me. Trying to retain your identity as a Native American student in a public high school was difficult. But when I came to Northwest Indian College and started working with elders and my mother, I became very proud of who I am. And I owe everything to them. Hyshqe."

Terry Williams has served since 1982 as a Fisheries and Natural Resources Commissioner for the Tulalip Tribes, in Marysville, Washington. Since 1985, he has also served on the Northwest Indian Fisheries Commission (NWIFC) and the Pacific Fisheries Management Council, and since 1997 he has been a member of the Pacific Salmon Commission. He was the director of the U.S. Environmental Protection Agency's American Indian Environmental Office in 1995–96, and as chair of the Tribal Committee of the National Environmental Justice Advisory Committee in 2003–04. In 1997, the Secretary for Policy and International Affairs Office of the U.S. Department of the Interior appointed Williams to represent Indigenous peoples on the U.S. delegation to the United Nations Conference on Biodiversity. He served in 1985–95 on the Puget Sound Water Quality Authority, Williams has received the Washington State Environmental Award and the Seventh Generation Legacy Award for his work, and he was a finalist for the Buffett Award for Indigenous Leadership in 2004.

Michele (Shelly) Vendiola (Swinomish/Lummi/Filipina) is a mediator, educator, and community activist who works as a consultant for environmental, economic, and social justice organizations. As part of the Swinomish Climate Change Initiative, she cofounded the Climate Change Education and Awareness Group. Ms. Vendiola was formerly the campaign director for the Indigenous Environmental Network and continues to advocate environmental justice initiatives for tribes in the Pacific Northwest. She serves on the board of the Progressive Technologies Project, a national nonprofit raising the level of technical resources available to grassroots organizations. She provides training and technical assistance to the Lummi CEDAR Project, a community-based nonprofit that provides youth leadership programs. Ms. Vendiola provides conflict resolution training and facilitation with her mother and a cadre of trainers. She has a M.Ed. in adult and higher education and practices popular education methodology in all aspects of her work as an educator, activist, and community organizer.

INDEX

Page numbers written in italics denote photographs or illustrations.